# BORON HYDRIDES

William N. Lipscomb

DOVER PUBLICATIONS, INC.
MINEOLA, NEW YORK

*Bibliographical Note*

This Dover edition, first published in 2012, is an unabridged republication of the work originally published by W. A. Benjamin, Inc., New York, in 1963.

*International Standard Book Number*
*ISBN-13: 978-0-486-48822-6*
*ISBN-10: 0-486-48822-5*

Manufactured in the United States by Courier Corporation
48822501
www.doverpublications.com

# Preface

No one should begin a short monograph on boron hydrides and related compounds without tribute to the research that Alfred Stock and his collaborators did during the period from 1912 to 1936. The characterization of $B_2H_6$, $B_4H_{10}$, $B_5H_9$, $B_5H_{11}$, $B_6H_{10}$, $B_{10}H_{14}$, and many of their chemical reactions, at a time when he and his students had to develop new experimental methods, is a remarkable achievement. No guides for these studies were available from valence, mechanistic, or theoretical chemistry. Indeed, it has only been within the last ten years that the geometric structures have been elucidated, that a valence theory has developed, and that the chemistry, both known and predictive, has been formulated into a widely applicable framework.

In this monograph I have covered only the general principles of structure and reactions of the boron hydrides and related compounds. The function of theory has been to formulate a consistent description of known compounds and reactions in such a way that useful predictions can be made, both of new compounds and new reactions. The development of techniques for the growth and study of single crystals by X-ray diffraction methods at low temperatures, and the extensions of valence theory to a variety of three-center bonds in the less compactly arranged compounds and to molecular orbitals in the more compact compounds, have provided the experimental and theoretical foundations for the present formulation. Parallel to these developments there has been a huge expansion in the number of compounds and derivatives prepared and in our knowledge of the physical and chemical properties of known compounds. Unfortunately, there are probably some interesting and relevant studies in the classified literature with which I am not conversant. Even the published literature is expanding so rapidly that soon it will not be feasible to compile a monograph on the general subject of boron compounds but only on more specific areas of research. Therefore, no apology seems necessary for some restriction of the range of interest as developed here. Finally, for some time now, I have wanted to cover in a single work the many separate parts of the structural theory and its relation to chemistry in order to make these ideas more widely accessible.

For the information of the reader, the last communications included in the Concluding Remarks (pages 196 through 200) were received in late July 1963, which was also the approximate time of the last changes in the proof.

It is impossible to acknowledge properly those members of my research group who have contributed so much to the X-ray diffraction, theoretical, and chemical studies so briefly summarized here. I can only hope, therefore, that the reader will note the references carefully. For permission to use nuclear magnetic resonance spectra, I wish to thank J. D. Baldeschwieler, M. F. Hawthorne, T. Heying, and R. E. Williams. Also, special acknowledgment is due to R. Hoffmann and P. G. Simpson, the most recent of my principal collaborators; their studies form much of the basis of this monograph. I am particularly grateful to the Office of Naval Research, whose interest and support have made much of this research possible.

WILLIAM N. LIPSCOMB

*Cambridge, Massachusetts*
*August 1963*

# Contents

# 1

# Boron Hydrides and Related Structures

The boron hydrides, first described and characterized by Stock,[299] form an unusual set of compounds. A systematic description of their chemistry is now possible, owing to the elucidation of their molecular structures, a valence-theory description, and many new studies of their chemical reactions. Just on the horizon is a detailed mechanistic description of their chemical reactions, and a predictive chemistry approaching that of carbon chemistry. In this monograph an attempt is made to systematize the molecular structure properties, the valence structures in terms of three-center bonds and molecular orbitals, the nuclear magnetic resonance (NMR) spectra, and, finally, the known chemical reactions.

## 1–1 The Hydrides

Tentative evidence, mostly of a physical type, places the number of presently identifiable distinct hydrides somewhere near thirty, but only ten have been isolated and well characterized. These known hydrides are $B_2H_6$, $B_4H_{10}$, $B_5H_9$, $B_5H_{11}$, $B_6H_{10}$, $B_9H_{15}$, $B_{10}H_{14}$, $B_{10}H_{16}$, $B_{18}H_{22}$, and iso-$B_{18}H_{22}$. The structural evidence for them is outlined below, and is followed by similar discussions concerning the boranates (boron hydride ions), the amine derivatives, the halides of boron, and the carboranes.

### $B_2H_6$

That the bridge type of structure is the correct one for diborane is generally recognized to have been established by Price[242,243] on the basis of the infrared spectrum. Prior to this study, however, there did exist a general realization[17,186,203,237,297,298] that the physical evidence was overwhelmingly in favor of this structure. The failure of general acceptance of, for example, the high barrier to internal rotation and the earlier infrared evidence[297] as indicative of the bridge structure, must be attributed to the insistence[8,10] that the early electron diffraction evidence favored the ethane type of structure. A later electron-diffraction study,[105] however, confirmed the bridge structure and was in disagreement with the ethane type of structure. The structure is shown in Fig. 1–1.

The correct molecular geometry was first proposed by Dilthey.[52] However, interpretations in terms of valence bonds or molecular orbitals were made only comparatively recently.[203,237,317] Whether one thinks in terms of a "protonated double bond," i.e., a four-atom bridge bond, in an ethylenic type of structure, or in terms of the interaction of two $BH_3$ groups, is, in any complete analysis, merely a matter of taste, but analogy of bonding geometry and electronic structure with these properties of ethylene is very striking.

### $B_4H_{10}$

Both hydrogen and boron positions (Fig. 1–2) were established unambiguously by the X-ray diffraction study[207,208] in which there were 616 observed reflections to determine 14 parameters counting the boron, scale, and temperature parameters. Hydrogen atoms, not included in the counting of parameters, appeared both in the presence of the boron atoms and when the boron atoms were subtracted out in the electron-density maps. This model was independently suggested in the recent electron-diffraction study[139] on the basis of its plausible relation to the other boron hydrides and of its consistency with the electron-diffraction data, but these results are not unambiguous, for a slight modification of the model decided upon in the earlier diffraction study[9] is equally consistent with the electron-diffraction data. Tests of other possible

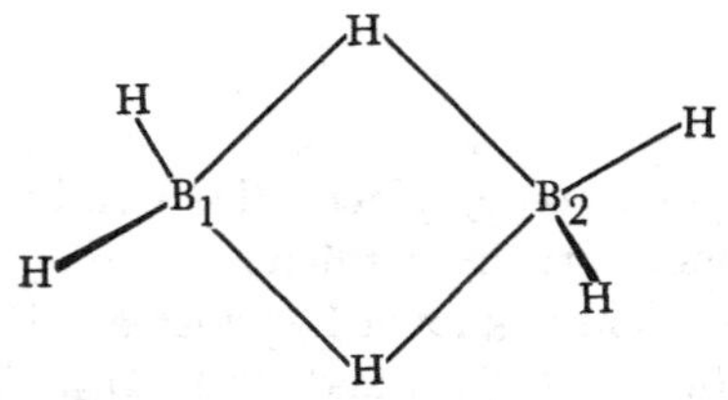


**Figure 1–1** *The structure of $B_2H_6$. The $B_1$—$B_2$ distance is 1.77 A, the B—H distance is 1.19 A, the B—$H_\mu$ (bridge) distance is 1.33 A, and the H—B—H angle is 121.5°.*

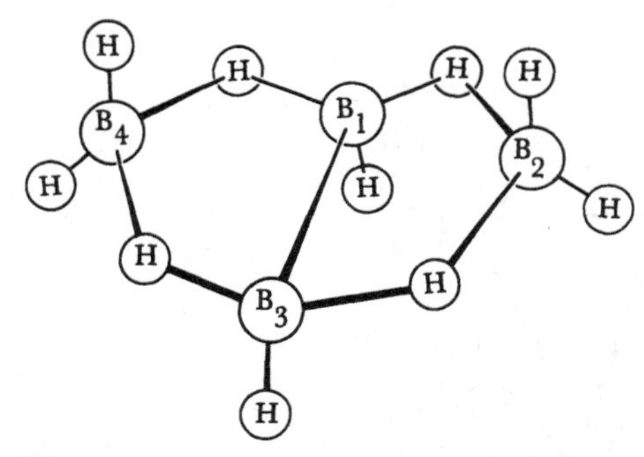


**Figure 1–2** *The structure of $B_4H_{10}$. The $B_1$—$B_3$ distance is 1.71 A, and the other four close B · · · B distances are 1.84 A. The B—H distances average to 1.19 A, while the B—$H_\mu$ (bridge) distances are 1.33 A toward $B_1$ and $B_3$ and 1.43 A toward $B_2$ and $B_4$.*

models were not made in this most recent electron-diffraction study, but the results of the earlier study were shown to be incorrect.

Comparison of the results of the X-ray and electron-diffraction study indicates the molecular parameters, $B_1$—$B_2$ = 1.84 A, $B_1$—$B_3$ = 1.71 A, $\angle B_2B_1B_4$ = 98°, B—H = 1.19 A, $B_1$—$H_\mu$ = 1.33 A, and $B_2$—$H_\mu$ = 1.43 A. This is a model obtained from the X-ray values of boron-boron distances and bond angles, including hydrogen angles, and from the electron-diffraction values of the B—H distances, but with the asymmetry of the hydrogen bridges reversed to agree with the X-ray results. This combination avoids the systematic errors introduced by the lack of complete convergence of the Fourier series in the hydrogen distances of the X-ray study and the very large uncertainties in the bond angles in the electron-diffraction study.

The boron arrangement can be considered as a fragment of either the icosahedron or the octahedron because the $B_2$—$B_1$—$B_4$ bond angle is between 105° and 90°.

### $B_5H_9$

The certainty with which this structure (Fig. 1–3) is established is based upon the crystal-structure study,[58,59] in which, at the outset, molecules of $C_{4v}$ symmetry were required by the space group of the crystal. Hydrogen atoms were located by subtraction of boron contributions and, in addition, are also uniquely placed by the packing in the crystal. This structure was arrived at independently, but not uniquely, in an electron-diffraction study.[107,108] Because these electron-diffraction data were also consistent with at least three other structures,[109] suggestions of a molecule of high symmetry from the infrared

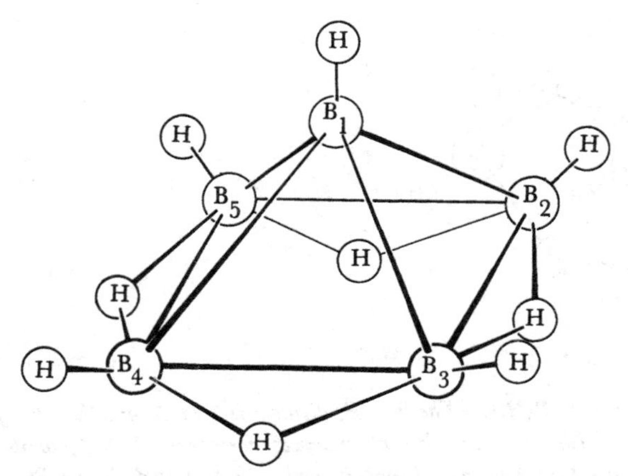


**Figure 1-3** *The structure of $B_5H_9$. The close B—B distances to $B_1$ are 1.69 A, and those among atoms 2, 3, 4, and 5 are 1.80 A.*

spectrum and entropy were included in the argument which led to electron-diffraction results. The molecular parameters of these studies are in reasonably good agreement, and in view of the confirmation of this structure by the microwave study,[131] the most probable molecular parameters are B—B (base) = 1.80 A and B—B (slant) = 1.69 A.

The boron arrangement is closely related to the octahedron found in the cubic borides.[229] If one removes one boron atom from the octahedron, stitches up the open face with bridge hydrogens, and attaches one hydrogen to each boron atom, the $B_5H_9$ structure can be expected. This boron arrangement was advocated by Pauling during both the early electron-diffraction study,[12] which discarded it in favor of an incorrect model, and the recent electron-diffraction study,[107,108] which initially erroneously eliminated this model.[275]

### $B_5H_{11}$

Although the X-ray diffraction data[164] were less complete than for $B_4H_{10}$, the use of 299 observed reflections to determine the 17 parameters ($x,y,z$ for each boron atom plus scale and temperature factors) yielded the boron positions with certainty. Subtraction of these boron atoms from the electron-density map yielded the hydrogen positions as the eleven highest remaining peaks. Unlike the $B_4H_{10}$ study, where the remaining background was low, there were false peaks nearly as high as these hydrogen peaks, but they were too far from the boron atoms to correspond to bonded atoms. Hence the probability is very high indeed that the structure shown in Fig. 1-4 is correct, independent of any chemical assumptions. A recent electron-diffraction study[140] is consistent with the boron arrangement but yields no direct information concerning the

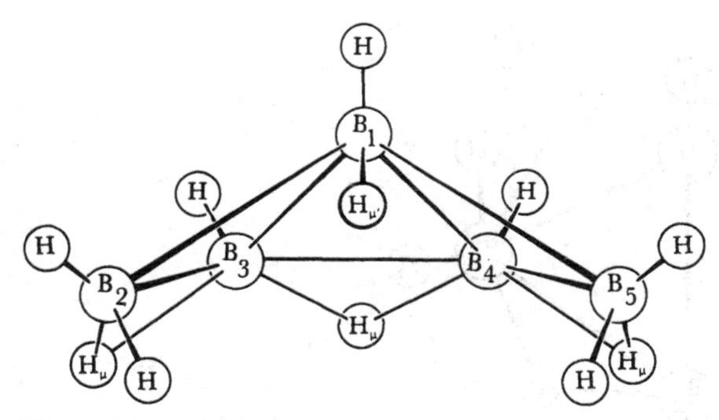


**Figure 1–4** *The structure of $B_5H_{11}$. The B—B distances in A are 1.87 for 1—5 and 1—2, 1.77 for 4—5 and 1.75 for 2—3, 1.70 for 1—4 and 1.73 for 1—3, and 1.77 for 3—4.*

hydrogen positions. Certainly the previous electron-diffraction results[9] are incorrect.

Except for the two B—B distances of $1.86_5$ A from the apex to the outer two boron atoms, the B—B distances have an average value of 1.74 A. The B—H distances are probably all in the normal range, even though they appear to be systematically about 0.1 A short because of the lack of convergence of the Fourier series. The case of $B_5H_{11}$ was unlike the cases of $B_4H_{10}$, $B_5H_9$, and $B_{10}H_{14}$, where the X-ray studies proved that the H bridges were, respectively, unsymmetrical, symmetrical, and unsymmetrical, in that no definite conclusion about the symmetry of the bridge hydrogens in $B_5H_{11}$ could be drawn from the experimental data.

The position of the unique H atom attached to the apical B atom of $B_5H_{11}$ but extending toward the plane of the other four B atoms has been reinvestigated twice since the original structure determination. In least-squares refinements of the X-ray diffraction data it has remained approximately 1.09 A from the apex B, and 1.77 and 1.68 A, both $\pm 0.19$ A, from the two outer B atoms. Moreover, the apparent amplitude of thermal motion of this H atom has remained, when varied separately, at about the average of the apparent amplitudes of motion of all the H atoms. Hence, a statistical, or actual, displacement toward the two outer B atoms seems to be ruled out. No doubt some bonding of this H atom occurs to the two outer B atoms, as is suggested by the coupling anomaly in the $B^{11}$ NMR spectrum. However, as a first approximation we shall regard this atom as completely bonded to the apex B atom, which then is part of a $BH_2$ group.

### $B_6H_{10}$

The structure,[50,71,114] shown in Fig. 1–5, has six B atoms from the icosahedral arrangment to which are bonded six terminal H atoms and four bridge

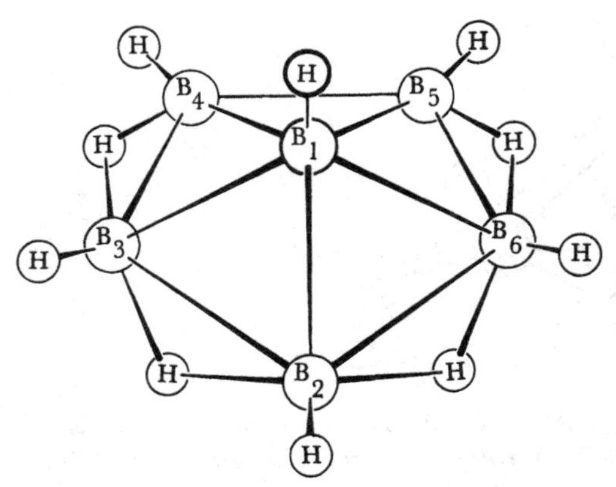


**Figure 1-5** *The structure of $B_6H_{10}$. The B—B distances in A are 1.74 for 1—2, 1.79 for 2—6 and 2—3, 1.75 for 1—6 and 1—3, 1.80 for 1—5 and 1—4, 1.74 for 5—6 and 3—4, and interestingly 1.60 for 4—5, the shortest observed distances among the boron hydrides.*

H atoms. Another structure, in which there are three bridge H atoms and one $BH_2$ group, is consistent with the theory of valency to be described below, and will be invoked as an intermediate in the H rearrangement suggested by the $B^{11}$ NMR spectrum; but this latter H arrangement has been eliminated conclusively for the molecule in the crystal on the basis of the behavior of the apparent amplitudes of H atoms during least-squares refinement.[250]

### $B_9H_{15}$

Although no chemical analysis has yet been made on $B_9H_{15}$, the complete structure has been established[50,51,290] from the X-ray data alone (Fig. 1-6). Structurally the molecule resembles $B_4H_{10}$ in the region of the $BH_2$ group, and $B_5H_{11}$ in the region of the three neighboring bridge H atoms. A recent reinvestigation[290] of the X-ray diffraction data has indicated that the single crystals were pure $B_9H_{15}$. A recent study of the synthesis and additional chemical properties has appeared.[31]

### $B_{10}H_{14}$

In an X-ray diffraction study[142] of the molecular structure of decaborane, the molecular geometry, including both boron and hydrogen positions, has been uniquely established (Fig. 1-7). The probable, but not unique, structure assigned in an electron-diffraction study[287] is incorrect, but these electron-diffraction data have since been shown[189] to be consistent with the correct structure.

The boron arrangement is closely related to the icosahedron found

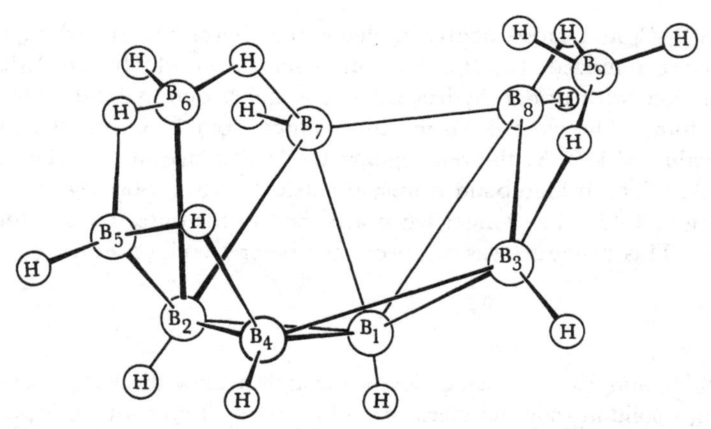


**Figure 1-6** *The structure of $B_9H_{15}$. The B—B bond distances in A averaged over the mirror plane of the molecule are 1.76 for 1—3 and 1—8, 1.80 for 3—8, 1.86 for 3—9 and 8—9, 1.95 for 3—4 and 7—8, 1.75 for 1—4 and 1—7, 1.77 for 1—2, 1.82 for 2—4 and 2—7, 1.76 for 2—5 and 2—6, 1.78 for 5—6, and 1.84 for 4—5 and 6—7.*

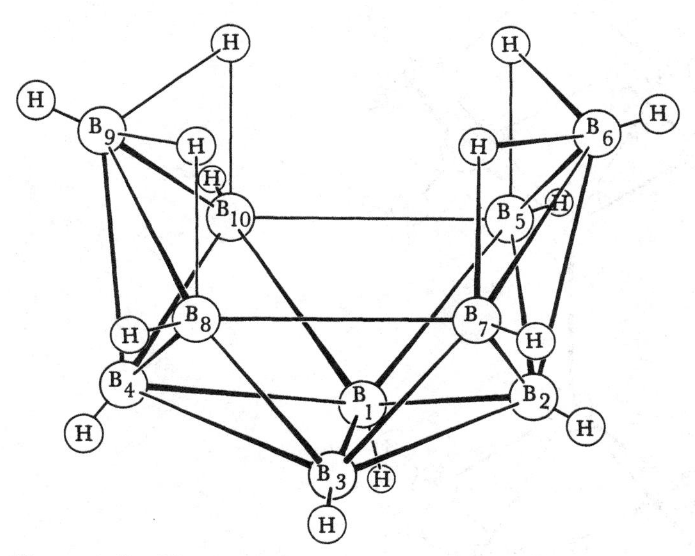


**Figure 1-7** *The structure of $B_{10}H_{14}$. The B—B bond distances in A are 1.71 for 1—3; 1.80 for 1—2 and 3—4; 1.78 for 2—3 and 1—4; 1.76 for 2—5 and 4—8; 1.80 for 2—7 and 4—10; 1.77 for 5—6, 6—7, 8—9, and 9—10; 1.72 for 2—6 and 4—9; 1.78 for 1—5 and 3—8; 1.77 for 3—7 and 1—10; and interestingly 2.01 for 5—10 and 7—8.*

earlier[42,239] in $B_{12}C_3$ and subsequently in elementary boron.[116] If two adjacent boron atoms are removed, the $B_{10}$ skeleton remains, on which one "stitches up" the open face with bridge hydrogens and attaches one hydrogen atom to each boron atom. The close B—B distances range from 1.71 to 2.01 A, with an average value of 1.76 A; the ten regular B—H distances are in the range 1.20 to 1.30 A. Each bridge bond is unsymmetrical with a short leg of 1.34 A and a long leg of 1.42. The longer leg is attached to the outer boron atom of the structure. This structure was not predicted before its discovery.

### $B_{10}H_{16}$

When $B_5H_9$ and $H_2$ are passed slowly through a glow discharge between Cu electrodes, a solid at room temperature is formed.[87] This compound, which is easily separable from $B_{10}H_{14}$ because of its greater volatility, has been shown

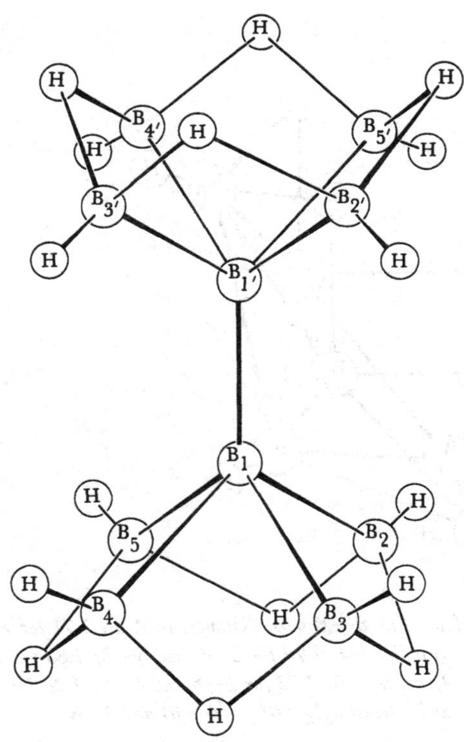


**Figure 1-8** *The structure of $B_{10}H_{16}$. The B—B bond distances (±0.06 A) are 1.74 for 1—1', 1.76 A for 1—3, and 1.71 A for 2—3, averaged, where possible, over all symmetry equivalent positions. The molecule has $D_{4h}$ symmetry.*

to consist of two $B_5H_8$ groups joined by a single B—B bond. Each $B_5H_8$ group is like that obtained by removal of an apex H atom from the tetragonal pyramidal $B_5H_9$. The resulting $B_{10}H_{16}$ structure (Fig. 1–8) has been shown to have the centrosymmetric arrangement about the B—B bond in the solid, but a large barrier to internal rotation is not expected. There exists a possibility that this structural principle is a general one, and that other stable hydrides may be formed by these methods.

### $B_{18}H_{22}$

The structure of $B_{18}H_{22}$ has recently been established[291] by X-ray diffraction methods. The structure, shown in Fig. 1–9, is like two $B_{10}$ units from decaborane sharing a pair of B atoms. In the outer parts of the molecule the bridge H arrangement is quite similar to that in $B_{10}H_{14}$. However, the bonding region near the molecular center of symmetry has features substantially different from those of the other hydrides. Here, for the first time, are two unusual B atoms each of which is coordinated to six other B atoms. In addition, a bridge H atom, a seventh neighbor, also connects each of these unusual B atoms to one of the six neighboring B atoms. It is possible to believe that there are many new hydrides having structural formulas based upon this central arrangement, all in the $B_6$—$B_{20}$ range. In the valence theory, developed below, the two unusual B atoms are the most positive, and also sterically inaccessible. Hence, a basis will be found for the unusual stability of this molecule to electron pair donors. It does form a $B_{18}H_{21}^-$ anion readily, and may therefore be similar in this respect to $B_{10}H_{14}$, which is also a strong acid, forming $B_{10}H_{13}^-$.

### Iso-$B_{18}H_{22}$

A second molecular structure and chemical formula have just been established by X-ray diffraction methods by P. G. Simpson, K. Folting, and W. N. Lipscomb. This is the first example of stable isomers among the boron hydrides. The relation of the structures of iso-$B_{18}H_{22}$ of symmetry $C_2$ and $B_{18}H_{22}$ of symmetry $C_i$ is shown in Fig. 1–10. These structures may be described as decaborane—14 cages joined by atoms in common at the 5—6 positions (the circled B atoms) in such a manner that the cages open up in opposite directions. The manner of joining is 5 in common with 5′ (i.e., 5—5′) and 6—6′ in iso-$B_{18}H_{22}$, and 5—6′ and 6—5′ in $B_{18}H_{22}$. No chemical studies have yet been made of this molecule. The relatively exposed 6—6′ B atom attached to two bridge H atoms should be the site of nucleophilic attack, while the electrophilic sites of reactivity should be similar to those in $B_{18}H_{22}$.

### Evidence for Other Hydrides

The two methods that provide some evidence for the possible existence of a fairly large number of other hydrides are X-ray diffraction and mass spec-

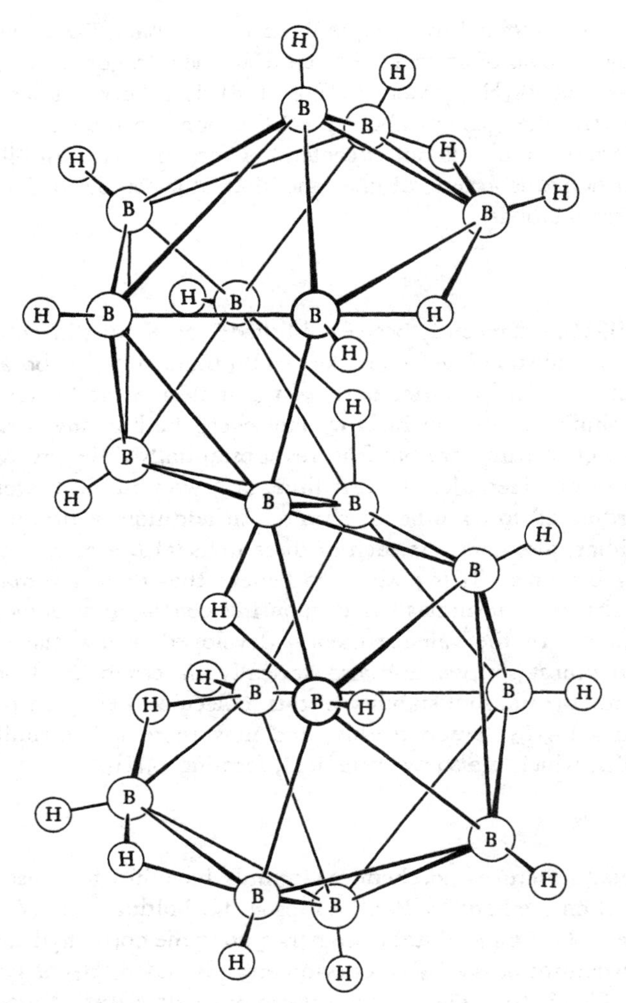


**Figure 1–9** *The structure of* $B_{18}H_{22}$. *The geometrical structure is centrosymmetric, and resembles two decaborane-14 molecules which share a common pair of B atoms.*

trography. Each different powdered crystalline substance normally gives rise to a large number of very narrow X-ray diffraction lines. Moreover, the relative intensities for each crystalline substance are characteristic in the usual case that the orientation of crystals is approximately random, and the probability of overlap of a large fraction of lines from two different crystalline substances is small. The possibility that the same molecule can crystallize in two or more different crystalline systems having different X-ray diffraction

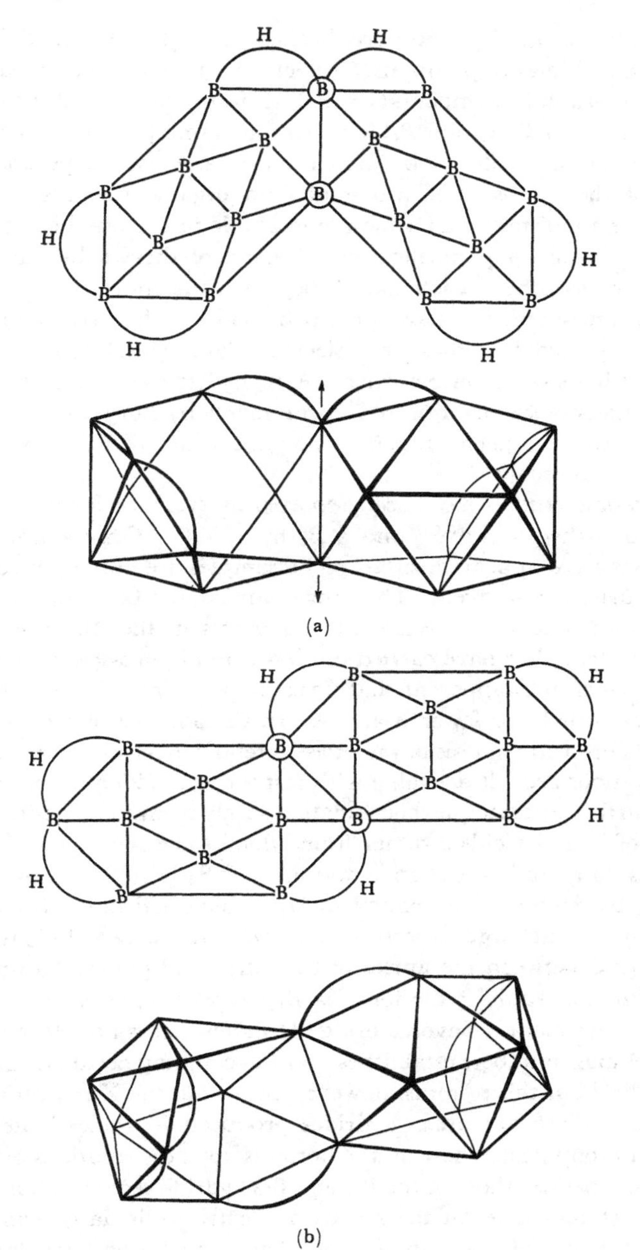


**Figure 1–10** *Comparison of the molecular structures of iso-*$B_{18}H_{22}(a)$ and $B_{18}H_{22}(b)$. *The twofold axis of iso-*$B_{18}H_{22}$ *is indicated by arrows. One terminal H atom has been omitted from each B atom except for the encircled B atoms, which do not have terminal H atoms. The two bridge H atoms, above and below the plane of the drawing (a), are attached to the most positively charged B atoms (encircled) in iso-*$B_{18}H_{22}$.

patterns may occur, but is usually quite rare, because these different modifications usually demand different conditions. Experiments[86] have indicated at least eight distinct crystalline compounds resulting from the glow-discharge polymerization reactions of $B_2H_6$ and $B_5H_9$ in various atmospheres. Usually only two to four different crystalline substances occur in any given experiment. Moreover, many of these substances are soluble in organic solvents, some reactive toward air and moisture, and some are insoluble in the usual solvents. We feel that the separation and purification of many of these substances is feasible, and that studies of their reactions will be worth the effort.

Vapor-pressure measurements have provided evidence for a second $B_6$ hydride, originally reported by Stock and Siecke,[301] later withdrawn,[299] but resuggested[299] on the basis of newer evidence. A $B_8$ hydride was suggested as a possibility on the basis of vapor-pressure measurements by Burg and Schlesinger,[31,33] and may also have been present in preparations[259] of $B_9H_{15}$, which showed vapor-pressure anomalies.

Mass-spectrographic studies have been reported as evidence for a second $B_6$ hydride,[79,143] two $B_7$ hydrides,[160,257] and a $B_8$ hydride.[280] Only extremely brief communications have appeared so far. The samples are either stated to be impure, or the purity is not given. The range of mass numbers outside the range of those attributed to the new hydride fragments in the spectrometer has remained unreported. We have carried out less careful mass-spectroscopic studies,[50,51,71] from which it is apparent that peaks occur at all mass numbers for possible hydrides with 10 or fewer B atoms. In the absence of a detailed experimental and theoretical analysis of the more careful[78] mass-spectrographic studies, some reservations are felt desirable with respect to the identity of these new hydrides, but further attempts at their isolation surely seem to be desirable.

Pyrolysis[220,245] of $B_{10}H_{14}$ yields a dimer from which some hydrogen of the original material is lost, and a tentative formula of $B_{20}H_{24}$(?) is assigned. Irradiation of $B_5H_9$ by deuterons produces[93] $B_{10}H_{16}$, which had been obtained previously from a spark-discharge process and shown[87] to be two $B_5H_8$ units joined by a B—B bond between the apices of the tetragonal pyramidal units. In the earlier preliminary study[274] the formula $B_{10}H_{18}$ was reported, and the dimerization was then thought to involve bridge H atoms. It is not clear how the determination of mass numbers establishes that the coupling occurs through the apical B atoms,[93,274] but this result is, however, proved in the X-ray diffraction study.[87] Clearly there are other hydrides produced in these deuteron bombardments, and group peaks showing the boron isotope distributions occur at other mass numbers besides those in the $B_{20}$ and $B_{10}$ hydride regions. Similar mass-spectrographic studies here on the results of electrical-discharge experiments also indicate the presence of a number of other stable boron hydrides.

A new boron hydride, considerably less volatile than $B_{10}H_{14}$ but volatile enough to be sublimed into beautiful single crystals, has been isolated from experiments by L. Friedman and W. N. Lipscomb, in which $B_{10}H_{14}$ is passed through an electric discharge. X-ray data (R. D. Dobrott and W. N. Lipscomb) are as follows: space group I $4_1/a\ cd$; dimensions, $a = 9.65$ and $c = 29.64$ A; molecular weight 235 ± 2 for 8 molecules of a $B_{20}H_x$ hydride.

## 1–2 Negative Ions and Equivalent Coordination Derivatives

Here we introduce one of the fairly general principles, that replacement of a terminal $H^-$ by an electron pair donor such as $:NCCH_3$, $:N(CH_3)_3$, or $:P(C_6H_5)_3$ is a formal procedure which allows classification of these derivatives as boranates, or boron hydride ions. In the same sense we could, for the purposes of the elementary theory of bonding, regard the hydrides themselves as also representing the corresponding alkyl or aryl derivatives, where, for example, $\cdot CH_3$ replaces $H\cdot$ in the position of the terminal H atom. Immediately we add, however, that some of the more interesting detailed chemistry is thereby obscured. Thus, we first derive the gross details of bonding, and only later elucidate more detailed variations of charge distribution in these molecules and ions. Perhaps the situation can be best appreciated by comparison with the aromatic hydrocarbons and their derivatives, in which one searches for charge distribution only after a fairly detailed structural, topological, and valence theory is available.

A second apology will become more vivid when derivatives of $B_2H_6$ are considered, for not only is the number of $CH_3$ groups which can replace H limited to the four terminal positions, but a different type of amine derivative from those considered here is present in[32,106] $B_2H_5NH_2$ and $B_2H_5N(CH_3)_2$. Here the N occupies a bridge position, but there are two electron pairs in the bonds from N to B, rather than the one electron pair in the BHB bridge bond. Finally the halogen derivatives may also be considered as included in the hydrides themselves, by formal replacement of H by a halogen, but at this initial stage one loses the remarkable detail in which the only $B_2H_5Br$ does not have Br in the bridge, as it is[252] in $BeCl_2$ or $Al_2Cl_6$, but rather in the terminal position.[45]

### $BH_4^-$

The tetrahedral nature of $BH_4^-$, isoelectronic with $CH_4$, has been established by the infrared work of Price,[244] and supported by the X-ray powder-diffraction pattern.[292] This structure has been confirmed by an NMR study.[212]

### $B_3H_8^-$ and $B_3H_7N(CH_3)_3$

The $B_3H_8^-$ ion was first recognized and isolated by Hough, Edwards, and McElroy.[126] Its structure has essentially three $BH_2$ groups arranged at the vertices of a triangle and joined by two bridge H atoms and by one single B—B bond.[230] This structure was predicted on the basis of the valence theory[174,175]

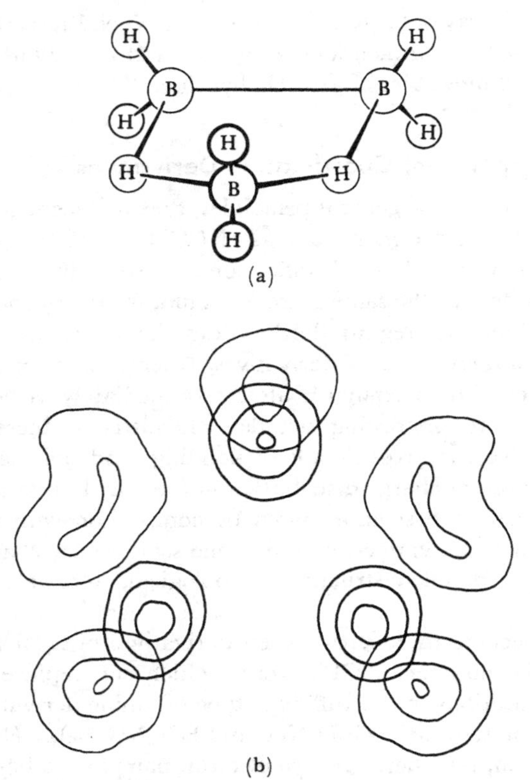


**Figure 1-11** *(a) The structure of $B_3H_8^-$ ion. (b) The electron density map, which omits one terminal H on each B atom. Note the possible indication of a statistical arrangement of H atoms.*

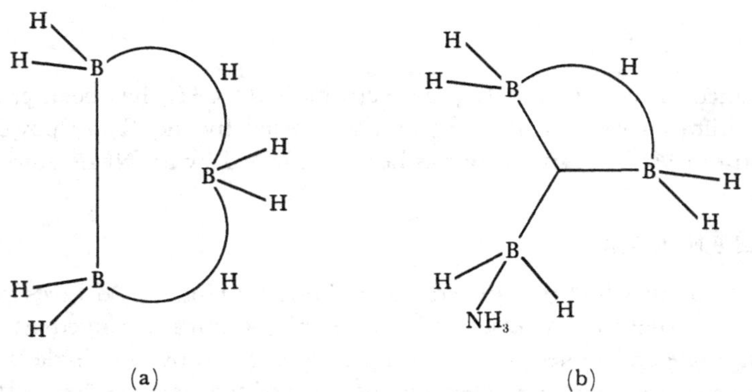


**Figure 1-12** *Extreme structures like $B_3H_8^-$ (a) or like $BH_3^-$ joined to $B_2H_5$ in place of a bridge H atom (b). The correct structure is intermediate between these two.*

and independently[152,154] on the basis of the reaction of $B_4H_{10}$ with $2NH_3$ to give $H_2B(NH_3)_2^+$ and $B_3H_8^-$. The X-ray diffraction study definitely proved the essential correctness of the structure (Fig. 1-11a), and thus eliminated two incorrect suggestions.[227,231] The appearance of the bridge H atoms in the electron-density map (Fig. 1-11b) is especially interesting, because of the possibility that the bridge H atoms are unsymmetrical, although statistically arranged in the crystal. Studies of the NMR spectra are described below.

The structure of $B_3H_7NH_3$ has been completed very carefully by X-ray methods,[209,210] which indicate a H arrangement intermediate between the two formulas of (Fig. 1-12). This intermediate structure for one of the bridge H atoms has been confirmed in an independent study[136] of the X-ray data.[210]

### $C_2H_5NH_2B_8H_{11}NHC_2H_5$ (Hypothetical $B_8H_{13}^-$)

The salt first thought[83] to be $EtNH_3^+B_9H_{12}NH_2Et^-$ has been shown to be the molecular compound $EtNH_2B_8H_{11}NHEt$ by a three-dimensional X-ray diffraction study[167] (Fig. 1-13). All H atoms of the boron framework were located unambiguously, but one H of the $NH_2$ unit and three H's of one $CH_3$ (possibly rotating) were not completely resolved. This compound is the first $B_8$ derivative, and the first higher hydride in which $NR_2$ (R = H or alkyl) formally replaces a bridge H, as it does[106] in $B_2H_6$ to form $B_2H_5NR_2$. Very closely related structural formulas for $B_8H_{12}NMe_3$ and for $B_8H_{13}^-$ had been suggested[176] previously on the basis of the valence theory.

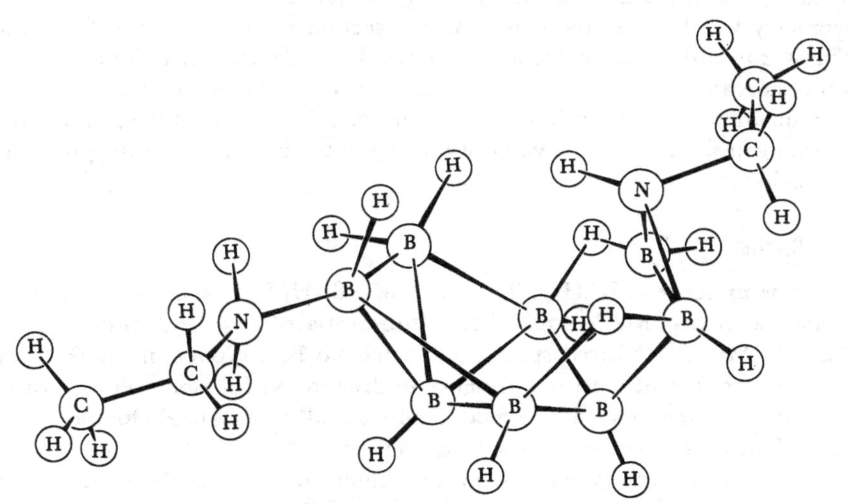


**Figure 1-13** *Geometrical structure of $C_2H_5NH_2B_8H_{11}NHC_2H_5$. The $B_8$ unit is an icosahedral fragment.*

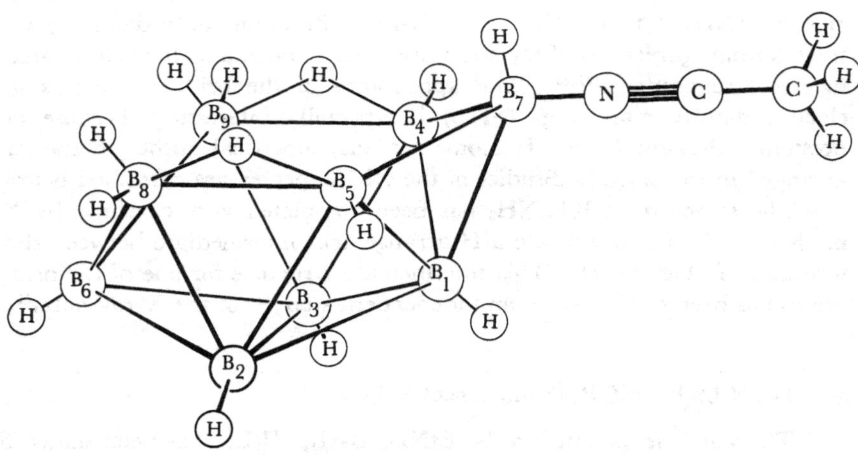


**Figure 1–14** *The structure of $B_9H_{13}NCCH_3$.*

## $B_9H_{13}NCCH_3$ (i.e., Hypothetical $B_9H_{14}^-$)

The $B_9H_{14}^-$ ion has not been described,* but the existence[83,102] of $B_9H_{13}NCCH_3$ and other electron pair donors to the $B_9H_{13}$ residue make its existence quite likely. The geometrical structure[319,320] of the $B_9H_{13}L$ type of compound, where L is an electron pair donor, is shown in Fig. 1–14. This structure was derived from the simple relations among various hypothetical bonding arrangements from the rules of valency to be described later, but the geometry will be introduced here in the section of well-established structures. The X-ray diffraction evidence[320] for the H arrangement, uniquely expected from the valence theory, has been obtained from the X-ray diffraction study with no chemical assumptions, and is different from the H arrangement (which is also inconsistent with the valence theory) in the first report of the preparation of this ion.[102]

## $B_{10}H_{10}^=$

The presence of $B_{10}H_{10}^=$ ion[182] in "ionic $B_{10}H_{12}L_2$," where L is an electron donor such as $Me_3N$, has been substantiated by the preparation of $B_{10}H_{10}(NMe_4)_2$. Infrared spectra indicated no $BH_2$ groups, no BHB bridges, and were consistent with the presence of discrete $NMe_4^+$ ions. The NMR $B^{11}$ spectrum showed only a low-field doublet and a high-field doublet of area about four times that of the low-field doublet.

No satisfactory valence structure based on semilocalized three-center bonds in the low-symmetry $B_{10}$ framework of $B_{10}H_{14}$ has been found.[248] On

* However, see Concluding Remarks, p. 198.

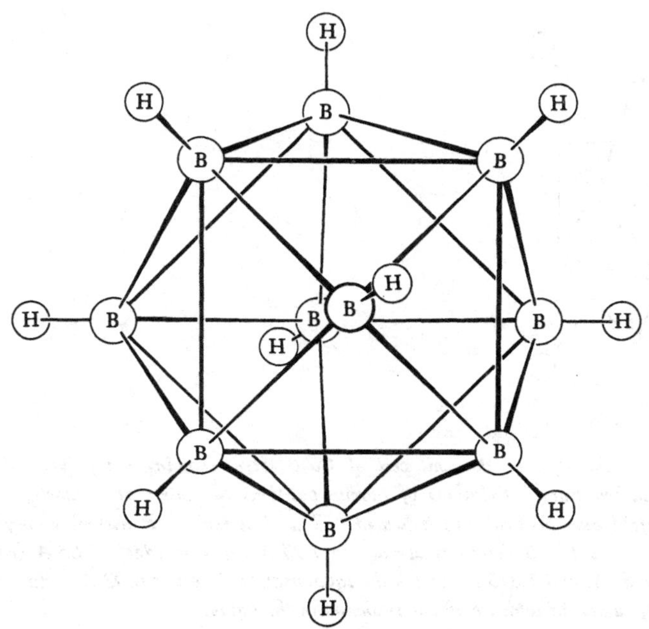


**Figure 1–15** *Structure of the $B_{10}H_{10}^{=}$ ion viewed approximately along the $\bar{8}$ axis.*

the other hand, a molecule of high symmetry is suggested by the NMR results. A simple $B_{10}H_{10}^{=}$ polyhedron, closely related in structure to $B_5H_9$ and based on satisfactory molecular orbitals, was discovered.[182]

The molecular structure (Fig. 1–15) has only two kinds of boron atoms in the ratio of 2 (apex) to 8 (approximately equatorial) in a polyhedron of symmetry $D_{4d}$. The valence orbitals of the apical boron atoms are slightly less symmetrical than those of the equatorial boron atoms, and hence the temperature-independent paramagnetism should be slightly greater for the apical boron atoms. This explanation, based upon a choice of origin at the boron nucleus, forms a more satisfactory basis for understanding $B^{11}$ resonance shifts than that based strictly on formal charge and diamagnetism. Thus the NMR results are well satisfied.

An X-ray diffraction investigation[55,141] of $Cu_2B_{10}H_{10}$ has recently confirmed the structure and has proved the existence of covalent bonds between Cu and the B atoms forming edges to the apices of the $B_{10}H_{10}^{=}$ ion (Fig. 1–16).

### $B_{12}H_{12}^{=}$

The $B^{11}$ resonance,[233] which indicates a doublet only, suggests that all B atoms are equivalent and that each B is bonded only to a single H atom. This spectrum is consistent with at least two static structures, the icosahedron and

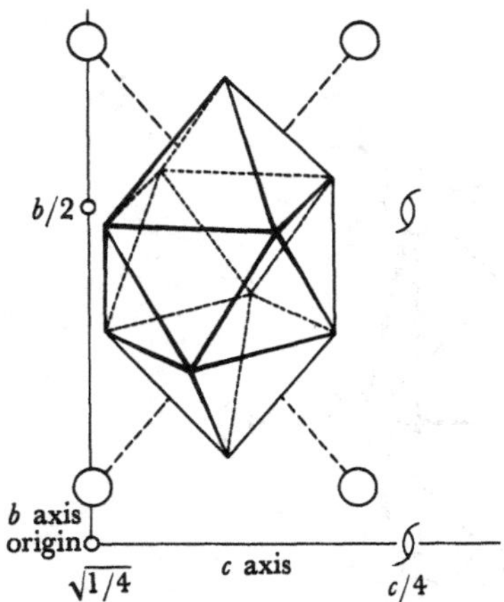


**Figure 1–16** *One-eighth of the unit cell of $Cu_2B_{10}H_{10}$, showing Cu–polyhedral edge interactions as dashed lines. Relations of atomic positions to centers of symmetry (small circles) and twofold axes of Pcab are indicated. The close Cu · · · B distances vary from 2.14 to 2.33 A. The B—B distances average to 1.73 A (apex to edge), 1.86 A (edge to edge within pyramids), and 1.815 A (across the equatorial belt), all ±0.02 A. Each Cu is bonded to two $B_{10}$ units, of which only one is shown in the figure.*

the cube-octahedron, both of which appear in the borides and both of which are expected to have filled molecular orbitals if the charge is −2, as is shown in Chapter 3. These data also might be consistent with a molecule in which there might be time-averaging effects, or protons showing anomalous coupling phenomena (e.g., $H_\mu$ in $B_5H_{11}$), or in which rapid icosahedral transformation back and forth to cube-octahedral structures might occur.

These ambiguities were resolved by the X-ray diffraction study,[336] which showed that in $K_2B_{12}H_{12}$ there are $B_{12}H_{12}^=$ ions of very nearly icosahedral symmetry (Fig. 1–17). The space group of the cubic crystals, having $a$ = 10.61 A, require at least $T_h - m3$ symmetry for the $B_{12}H_{12}^=$, and the atomic positions eliminate the cube-octahedral arrangement. There are 6 B—B distances of 1.755 A and 24 B—B distances of 1.780 A in the polyhedron. This slight inequality is just over three times the standard deviation of ±0.007 A, and hence has a relatively high level of significance. There thus appears to be a steric interaction of B—H with the $K^+$ ions which causes this distortion. It is therefore reasonable to expect that the isolated ion is of icosahedral symmetry, with a B—B distance of 1.77 A.

### $B_{10}H_{12}(NCCH_3)_2$ (i.e., $B_{10}H_{14}^=$)

Removal of the electron pair donor, and substitution of $H^-$ with its two electrons at each of the two positions in $B_{10}H_{12}(NCCH_3)_2$ yields $B_{10}H_{14}^=$. The structure of the known[306] ion, $B_{10}H_{14}^=$, is not known, but there are theoretical

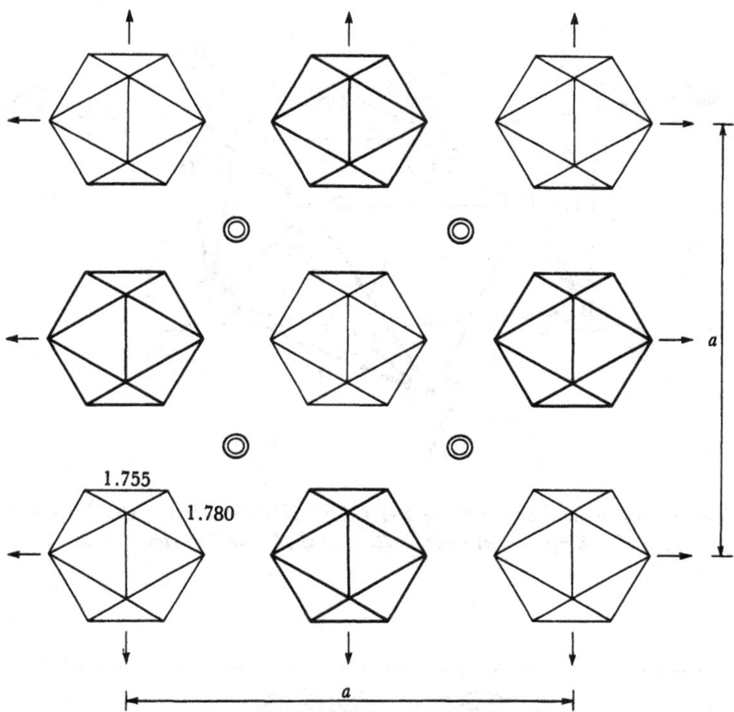


**Figure 1–17** *The unit cell of $K_2B_{12}H_{12}$, with the center of symmetry at the origin. The icosahedra have been drawn at two-thirds of their proper size in order to avoid overlap. Icosahedra at level zero are indicated by lighter lines, and those at level $\frac{1}{2}$ are indicated by heavier lines. Double circles indicate $K^+$ at levels $\frac{1}{4}$ and $\frac{3}{4}$.*

reasons for believing that it is related to the structure described here for $B_{10}H_{12}(NCCH_3)_2$ by the conceptual replacement described above.

The structure,[246–248,256] shown in Fig. 1–18, has a boron arrangement similar to that in $B_{10}H_{14}$, but with more nearly uniform B · · · B bond distances in the range 1.74 to 1.88 A, as compared with the range[197] 1.71 to 2.01 A in $B_{10}H_{14}$. The H arrangement (Fig. 1–19), however, is different from that in $B_{10}H_{14}$. Although there is a terminal H atom attached to each B atom in each of these compounds, the four bridge H atoms of $B_{10}H_{14}$ are missing in $B_{10}H_{12}(NCCH_3)_2$ in which two H bridges are present in quite a different place in the open face of the molecule. If two $H^-$ are substituted for the two $NCCH_3$ groups, then two $BH_2$ groups would be present in $B_{10}H_{14}^=$, and in either $B_{10}H_{14}^=$ or $B_{10}H_{12}(NCCH_3)_2$ the boron framework would have to absorb an extra pair of electrons. Interestingly, this process receives an adequate description in terms of the valence theory, and, indeed, was predicted[174] at about the same time as the discovery[307] of $B_{10}H_{14}^=$ itself.

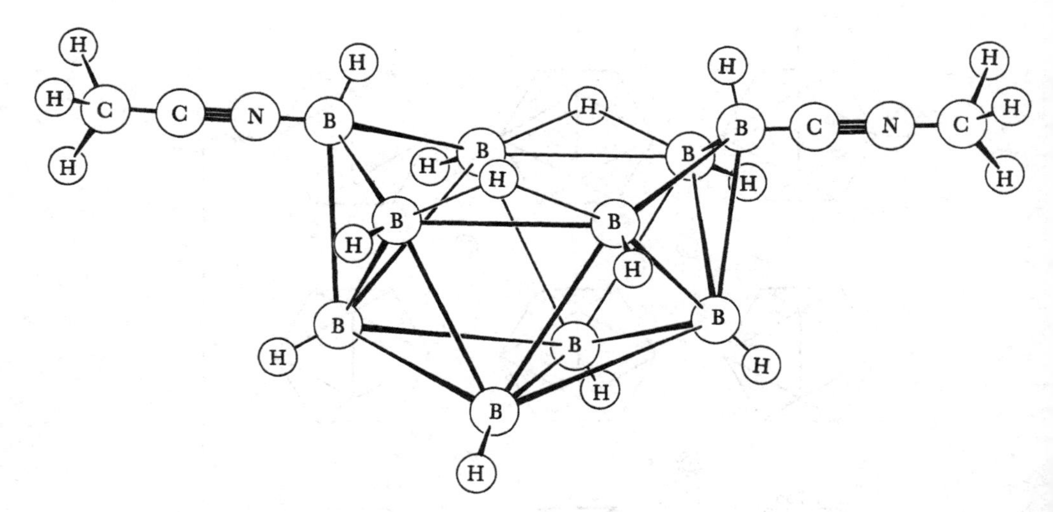


**Figure 1-18** *Isometric drawing of the $B_{10}H_{12}(NCCH_3)_2$ molecule. $B_{10}H_{12}(SMe_2)_2$ has a similar structure.*[256] *Large circles indicate B, C, and N; small circles indicate H.*

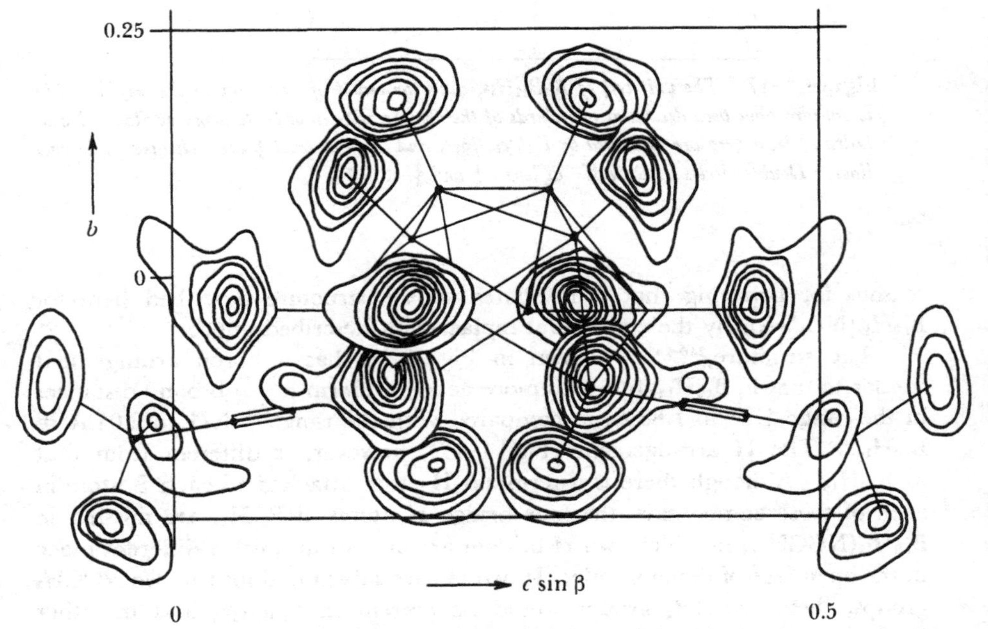


**Figure 1-19** *Final three-dimensional-difference electron-density map showing hydrogen atoms. The molecular skeleton is indicated for reference. Contours start at 0.17 $eA^{-3}$ and are given at intervals of 0.08 $eA^{-3}$.*

### Other Negative Ions and Amine Derivatives

**$B_2H_7^-$.** As we shall see, the orbital description of bonding in $B_2H_7^-$ ion, the preparation of which was published by Brown, Mead, and Tierney,[27] leads to only one reasonable structure involving two $BH_3$ groups linked by a B—H—B bridge bond.[28] However, no direct proof of this structure is as yet available, and, in particular, one can only suspect that the B—H—B bond is unsymmetrical. The $B^{11}$ resonance is of little help, because all protons are equivalent,[328] an effect which is undoubtedly an intramolecular rearrangement. The solvent appears to play an important role in the stabilization of this ion, and it is therefore possible that a more complex description of the $B_2H_7^-$–solvent interaction is required.

**$B_5H_9X^-$.** The preparation of $NaB_5H_9CN$ from $B_5H_9$ and NaCN has been reported.[1] One possibility is that there is a $B_5H_9CN^-$ ion, which may be a substituted $B_5H_{10}^-$ ion, structures for which are proposed in Appendix A.

**$B_5H_9 \cdot 2NR_3$.** This compound is formed by reaction[29,130] of $B_5H_9$ with $N(CH_3)_3$ as a white sublimable solid. Its structure is not known, but a possibility is $(NMe_3)_2BH_2^+ \cdot B_4H_7^-$.

**$B_9H_{12}^-$.** Proton abstraction[83,102] from $B_9H_{13}SMe_2$, with the use of the reagent $(C_6H_5)_3PCH_2$ yields the unstable $B_9H_{12}^-$ ion. The $B^{11}$ resonance is inconclusive, because of the instability of the ion, but is suggestive of more than two kinds of B atoms. The X-ray diffraction study[318] is also inconclusive because of a rapid falloff of X-ray intensities with increasing angle of scattering, thus suggesting some disorder, and because of a transition which makes it difficult to obtain a crystal stable at lower temperatures. The crystal symmetry, and positions of the $N(CH_3)_4^+$ ions in the cubic crystal of $N(CH_3)_4^+B_9H_{12}^-$, require the ion to have, statistically or actually, a threefold axis. At least one conclusive result has been obtained. The packing in the crystal which satisfies the gross intensity relations indicates that the $B_9H_{12}^-$ ion is not dimerized or further polymerized, and the disorder suggests that this ion is nearly spherical in Van der Waals contacts.

**Other Ions.** The available experimental evidence for ionic boron hydride fragments suggests that much careful experimental work is desirable. Stock[299] mentions $Na_2B_2H_6$ and $Na_2B_4H_{10}$, from which the original hydrides $B_2H_6$ and $B_4H_{10}$ are at least partly recoverable. He also describes[300,302] $Na_2B_5H_9$, and discusses $Na_2B_4H_8$ as a decomposition product of $Na_2B_4H_{10}$. Some recent unpublished work suggests that $LiB_7H_{13}$(?) and $NaB_8H_{13}$(?) exist.[151]

## 1–3 Boron Halides

### $BCl_3$

Of course, this molecule is planar, with the B atom at the center of an equilateral triangle of Cl atoms.[294] The B—Cl distance of 1.75 A is a bit

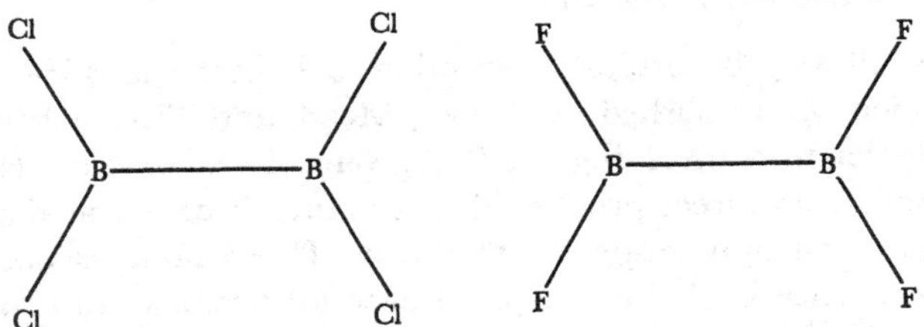


**Figure 1–20** *Planar structure from the X-ray crystallographic studies of $B_2Cl_4$ (B—B = 1.75, B—Cl = 1.72, ∠Cl—B—Cl = 120.5°) and $B_2F_4$ (B—B = 1.66, B—F = 1.32 ∠F—B—F = 120°).*

larger than the B—Cl distance in the other halides to be described below. The Cl · · · Cl repulsions may be the important factor, but resonance interaction with the vacant orbital on B is perhaps also a bit less in $BCl_3$ per B—Cl bond because three Cl interact with only one vacant B orbital.

### $B_2Cl_4$ and $B_2F_4$

These molecules have a direct B—B bond between two $BX_2$ groups.[5,6,104,309] In the solid phase both of these molecules are centrosymmetric and therefore planar (Fig. 1–20). On the other hand, planarity of $B_2Cl_4$ is not strongly constrained in the gas phase, and even the equilibrium configuration is not certain.[104]

Reactions of $B_2Cl_4$ with $C_2H_4$ results in addition of $BCl_2$ units to the double bond of ethylene.[313] The resulting compound is centrosymmetric[196] and has distances as shown in Fig. 1–21.

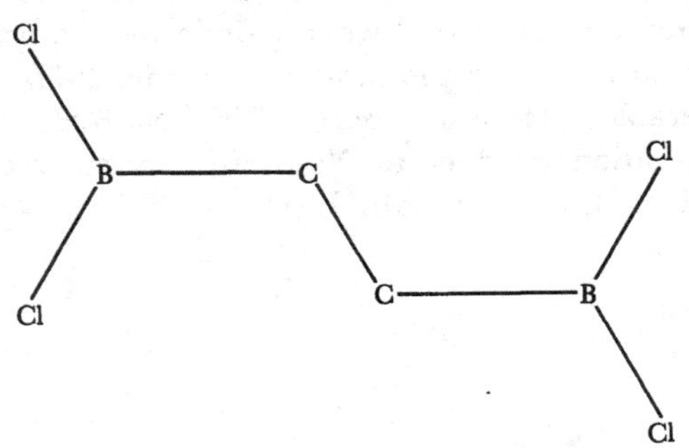


**Figure 1–21** *The $B_2Cl_4 \cdot C_2H_4$ adduct. Bond distances are C—C = 1.46, B—Cl = 1.76, B—C = 1.58 A, ∠Cl—B—Cl = 117°, and ∠C—C—B = 116°.*

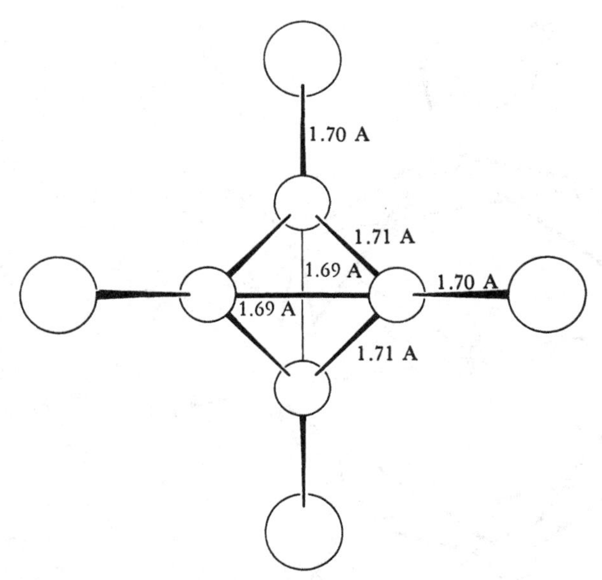


**Figure 1–22** *The tetrahedral $B_4Cl_4$ structure.*

### $B_4Cl_4$

The geometry[3,4] of $B_4Cl_4$ reflects both the electron deficiency of the boron framework, which is a tetrahedron, and the ability of the Cl to back-coordinate to the vacant orbitals on B, as we shall discuss in Chapter 3. The molecule has a tetrahedral arrangement of four B—Cl groups, with B—B = 1.71 and B—Cl = 1.70 A (Fig. 1–22). Actually there is a small distortion of the tetrahedral shape in the solid such that there are four Cl · · · Cl at 4.50 ± 0.01 A and two Cl · · · Cl at 4.46 ± 0.01 A, but the packing of molecules in the crystal suggests that this small distortion is due to nearest intermolecular interactions.

### $B_8Cl_8$ and $B_{12}Cl_{11}$

The complete structure[134,135] of $B_8Cl_8$ was determined at a time when the compound was thought to be $(BCl_{0.9})_x$,[314] where $x$ is unknown. However, it has been clearly established that these two compounds are distinct, and $(BCl_{0.9})_x$, which may be $B_{12}Cl_{11}$, has been shown to be paramagnetic.[310,311] The $B_8Cl_8$ structure is shown in Fig. 1–23. The boron polyhedron upon which this structure is based is a distorted Archimedian antiprism. As we shall show in Chapter 3, the molecular orbitals are filled for this geometrical arrangement in spite of its high symmetry, which is $D_{2d}$.

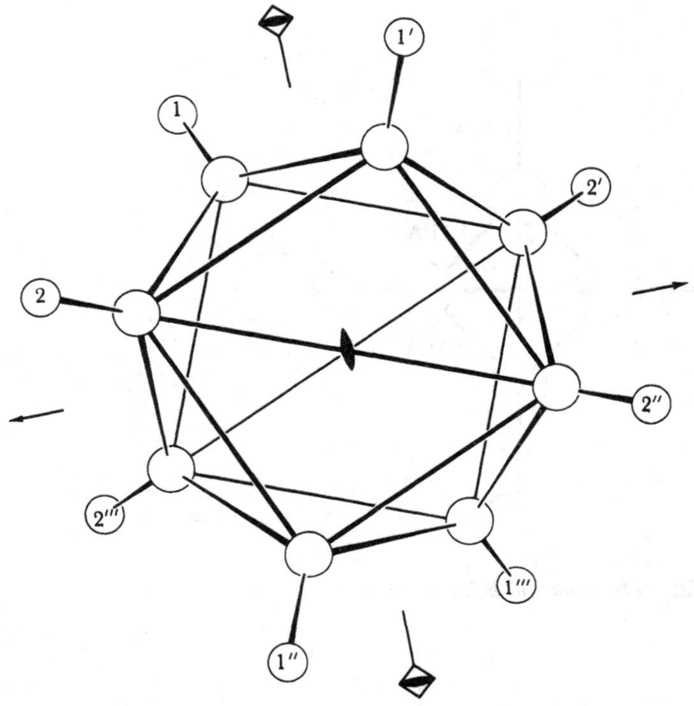


**Figure 1–23** *The $B_8Cl_8$ structure. Small circles represent B and large circles represent Cl. Average Cl · · · Cl distances (±0.04 A) are 4.05 A for the eight symmetry-equivalent distances of type 1—2, 3.51 A for the other four of type 1—2, 4.45 A for the four of type 2—2, and 3.94 A for the two of type 1—1. Corresponding average B—B distances (±0.05A) are, respectively, 1.79, 1.85, 2.07, and 1.78 A. The average B—Cl distance is 1.70 ± 0.04 A.*

### Other Halides

A sublimable $(BCl_2)_n$ and $(BBr_2)_m$, where *n* and *m* are not yet determined, and possibly not constant, have been prepared.[166] In addition, the action of $I_2$ on $B_{10}H_{14}$ produces[321] $(B_{10}I_8)_x$, which is insoluble in the usual solvents.

## 1–4 Structures of the Boranates (Borohydrides)

Since the original discovery[34,270,271,273] of the boranates of Al, Be, and Li, the boranates of over 35 metals have been made, ranging from highly ionic $BH_4^-$ salts to covalent molecules bonded from the metal through two bridge H atoms for each boranate group. Excluding the H atoms, which have some unusual properties in the NMR spectra, the heavier atoms are linear in $Be(BH_4)_2$, and planar in $Al(BH_4)_3$, and the H bridges are unsymmetrically displaced toward the B atoms in each case that has been studied.

## 1-5 Carboranes (Heteroatom Hydrides)

The topological identity of $B_{12}H_{12}^{=}$ with $C_2B_{10}H_{12}$, or of hypothetical $B_6H_6^{=}$ with $C_2B_4H_6$, or of hypothetical $B_5H_5^{=}$ with $C_2B_3H_5$ has been a factor in the search for and identification of one kind of heteroatom hydride, called carboranes, in which C substitutes for $B^-$. Of course, the chemical details are completely modified by such a formal substitution. Nevertheless, the chemical properties are probably reasonably well treated by methods used for aromatic molecules, especially heterocyclics, as we shall develop in Chapter 3. The idea of the existence of such compounds is almost as old as the first papers on these polyhedral ions.[69,187,188] It is not clear when the first such compound was prepared,[179,ftnt 9] but NMR evidence for one isomer of $C_2B_3H_5$ and two isomers of $C_2B_4H_6$ have appeared,[282,335] and some evidence that $C_2B_{10}H_{12}$ exists is now clear.[41,277,315,316] Further studies of the charge distributions[121] and probable rearrangements[124] will be discussed below. It seems clear that this area of chemistry will produce a very large class of new compounds and reactions.

Compounds related to the carboranes, but not quite so stable, include $B_4H_8C_2H_4$, which has $CH_2CH_2$ linked[96,97] to B at each end of a $B_4H_8$ residue of $B_4H_{10}$, and the dihydro carboranes[218] of the formula $B_4C_nH_{2n+4}$ (more generally, $B_mC_nH_{m+n}$), structural formulas of which are not yet uniquely proved by the $B^{11}$ NMR evidence.

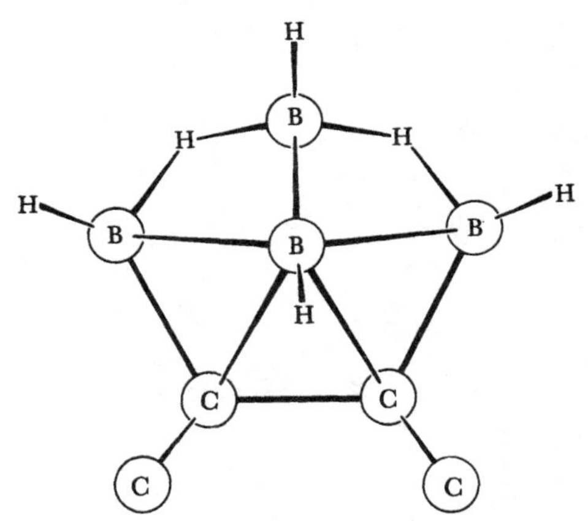


**Figure 1-24** *The molecular of $B_4H_6C_2(CH_3)_2$. Open circles are B atoms and solid circles are C atoms. The C—C distance in the ring is only 1.435 A.*

### $B_4H_6C_2(CH_3)_2$

The only carborane for which a detailed structure is known is $B_4H_6C_2(CH_3)_2$, recently elucidated by X-ray diffraction methods by W. Streib, F. P. Boer, and W. N. Lipscomb. No rearrangement[124] has occurred here, and hence the structure proposed[218] from the NMR spectrum is correct, although it was not the only possibility.[124] A most interesting feature of this molecule (Fig. 1–24) is the short C—C distance of 1.435 A in the five-membered ring. The molecular geometry of this molecule is similar to that of $B_6H_{10}$, where there is a correspondingly short distance of 1.60 A for a B—B bond. Both of these short distances suggest that a small amount of multiple bonding occurs.

### $C_2B_{10}Cl_{12}$

Preliminary crystallographic studies (J. Potenza and W. N. Lipscomb) of $C_2B_{10}Cl_{12}$ support the highly symmetrical, essentially icosahedral structure. There are four molecules in a cubic unit cell ($a$ = 12.5 A) of symmetry Pa3. The molecular site symmetry is $\bar{3}$m, even though the two C atoms are adjacent to one another in the molecule. Thus the crystal does not distinguish among the orientations of the molecule which make B and C essentially equivalent, but leaves the Cl in apparently ordered but also statistically equivalent positions. A determination of the C—C distance in a derivative which is not statistically disordered would be of great interest.

# 2

# Three-Center Bonds and Their Applications

As noted in the introduction to Chapter 1, three-center bonds will be used to describe the more open hydrides, such as those based upon icosahedral fragments of $B_n$ units, whereas molecular orbitals of a more general sort will be used to describe bonds in the more condensed types of $B_n$ arrangements in Chapter 3. First, the three types of three-center bonds are described. Second, after a brief discussion of bonding in $B_2H_6$, we develop the electron, orbital, and atom accounting procedures. Third, a set of rules which are abstractions of structural principles present in the known hydrides are introduced. The predictions of this theory are then given, and finally compared with the experimental results so far known.

## 2–1 Three-Center Bonds

The only unfamiliar concept is the "three-center bond" (not to be confused with the three-electron bond). The normal covalent bond may be called a two-center bond; two atoms supply two orbitals, one centered on each atom, and these interact to form one bonding orbital and one antibonding orbital; if two electrons are available, they will just fill the bonding orbital and constitute a covalent bond. Similarly, three atoms may supply three orbitals, one on each atom, and these will interact as discussed below to form one bonding and two antibonding orbitals; two electrons may just fill the bonding orbital to constitute a three-center bond. Both of these bond types are indeed special cases of the more general multicenter molecular orbital represented as a linear combination of atomic orbitals. The special value that attends their use lies in the intuitive feeling for the chemical bonding which results from localizing the bonding electrons in particular regions of the molecule and which frequently makes possible the extension of understanding from simple structures to more complicated ones. The multicenter molecular orbital can of course be described in terms of resonance involving single bonds, but we find the explicit recognition of the three-center bond concept to be most perspicuous in the treatment of the boron hydrides.

It is at once clear that the three-center bond will be most useful in "electron-deficient" molecules, i.e., where there are more usable atomic orbitals than valence electrons. The normal two-center bond provides a comfortable home for as many electrons as it uses atomic orbitals: Two atomic orbitals are used, one bonding orbital is formed, and two electrons can be housed. If $n$ atomic orbitals are used to form two-center bonds, they can form $n/2$ bonding orbitals and accommodate $n$ electrons; any more would have to go into the antibonding orbitals which are energetically unfavorable. If these $n$ atomic orbitals are combined into three-center bonds, they form only $n/3$ bonding orbitals, which can accommodate only $2n/3$ electrons; any more electrons would have to go into undesirable antibonding orbitals. Of course, if $n$ atomic orbitals remain unshared they can be filled with $2n$ electrons as "lone pairs," which are neither bonding nor antibonding to a first approximation. We see that if the atoms forming a molecule provide altogether $n$ orbitals and $m$ electrons then when $m = n$, as in hydrocarbons, normal bonds will be favored; three-center bonds would be unfavorable because some electrons would be forced into antibonding orbitals. When $m > n$, as in $NH_3$ and $H_2O$, "lone pairs" will be used as far as possible to take off the excess. When $m < n$, as in the boron hydrides, three-center bonds will be used in so far as they are energetically favorable.

Moreover, if we specify that all bonding orbitals be filled and no antibonding orbitals be used, then the valence pattern for a given set of atoms is restricted to a few well-defined possibilities. This principle, with some empirical additions, provides the basis of our approach.

There are clearly other cases, such as $Al_2Me_6$, where the same approach may be useful. The three-center bond is very useful in considering bent triatomic structures, such as nitrosyl halides or the carboxylate ion. Likewise there are many boron compounds other than hydrides whose valence structure is clarified by this approach; thus our picture of $B_2H_6$ is readily modified[179] to account for[106] $B_2H_5N(CH_3)_2$, where the extra electrons from the N permit "normal" B—N bonds to be formed. But in this chapter we shall consider only the hydrides.

## 2-2 Basic Assumptions[69]

The fundamental assumptions underlying the approach suggested here may be enumerated as follows:

1. Only the 1*s* orbital of hydrogen and the four $sp^3$ orbitals of boron are

used (Fig. 2–1). The boron orbitals are hybridized as suggested by the environment of each boron.

2. Each external B—H bond is regarded as a "normal," localized, single bond requiring the hydrogen orbital, one hybridized boron orbital, and one electron each from the hydrogen and the boron. Because of the very small difference in electronegativity of hydrogen and boron, these bonds are assumed to be nonpolar. It is a structural fact that every boron forms zero or one but never more than two such external bonds in the compounds to be considered here.

3. Each B—H—B bridge bond (Fig. 2–2) is regarded as a filled, three-center, localized bonding orbital requiring the hydrogen orbital and one hybrid orbital from each boron. For purposes of accounting, the hydrogen is assumed to contribute one electron, and each boron one-half electron to this orbital. These bonds are also assumed to be essentially nonpolar.

4. The orbitals and electrons of any particular boron are allotted so as to satisfy first the requirements of the external B—H single bonds and the bridge B—H—B bonds. The remaining orbitals and electrons are contributed to framework molecular orbitals.

5. The framework molecular orbitals are constructed from hybridized boron orbitals directed as suggested by the immediate environment of each atom or, where alternatives exist, in directions convenient for forming linear combinations transforming properly according to the irreducible representations of the symmetry group associated with the molecular geometry. The relative energies of these molecular orbitals may be estimated through the usual LCAO secular equations in terms of resonance integrals related qualitatively to the degree of overlap between orbitals. In fitting electrons into the resultant energy scheme only the bonding orbitals are used for any given stable molecule, and these are completely filled with the available framework electrons.

6. The framework dipole moment is calculated to a first approximation by assigning electrons to an atom as dictated by the squares of the coefficients

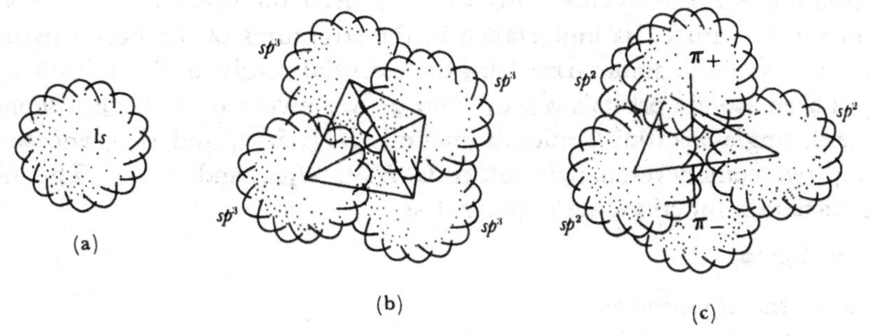


**Figure 2–1** *The spherically symmetrical 1s orbital for H; (b) tetrahedral hybrid orbitals for B; (c) trigonal hybrid orbitals for B, showing the π orbital extending above and below the plane of the $sp^2$ hybrids.*

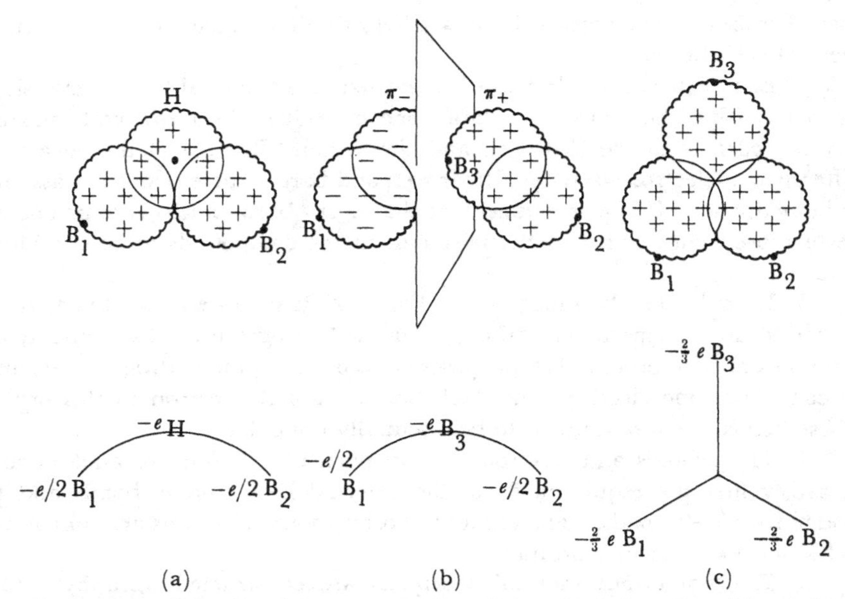


**Figure 2-2** *Examples of three-center bonds. Boron orbitals from $B_1$ and $B_2$, and from $B_3$ of (c) may be $sp^2$ or $sp^3$. The simplest LCAO calculations distribute the bonding electron pair equally among the three atoms of the central three-center bond in (c), and place $-e$ on H or on $B_3$ of the open three-center bonds, (a) or (b). The effect of electron correlation*[48] *is neglected.*

of the corresponding atomic orbital in the simple LCAO wave functions for the filled molecular orbitals. As a somewhat better approximation the Coulomb integrals[202] appearing in this simple approach are modified as suggested by the self-consistent LCAO treatment of Moffitt.[194]

## 2-3 Three-Center Orbitals

Although the three-center problem has been discussed in detail several times in the literature, its importance in the structures of the boron hydrides makes it advisable to summarize here the essential results of the LCAO treatment. Three atomic orbitals $\phi_A$, $\phi_B$, and $\phi_C$ are involved, each on a separate atom, with single-electron Coulomb integrals $H_{AA}$, $H_{BB}$, and $H_{CC}$, and single-electron, two-center resonance integrals $H_{AB}$, $H_{BC}$, and $H_{AC}$. The usual LCAO variation functions are expressed as

$$\psi_i = \Sigma_j a_j^i \phi_j$$

and lead to the equations[69]

$$\Sigma_j a_j^i (H_{jk}^i - E^i S_{jk}) = 0 \qquad j,k = \text{A,B,C}$$

For the applications envisaged here, nonorthogonality effects may be ignored, i.e., $S_{jk} = \delta_{jk}$, and the resonance integrals may be evaluated symmetrically such that $H_{AB} = H_{BC} = \beta$ and $H_{AC} = \gamma$. To a first approximation, all the Coulomb integrals may be evaluated as suggested by the self-consistent treatment of Moffitt in terms of the coefficients $a_j^i$, to be considered different for each molecular orbital. The resonance integrals are considered to be the same for all the molecular orbitals.

In the approximation of constant $H_0$, the variation equations lead immediately to values of the energy given by the expressions

$$E_+ = H_0 + \frac{\gamma}{2} + \left[\left(\frac{\gamma}{2}\right)^2 + 2\beta^2\right]^{1/2}$$

$$E_0 = H_0 - \gamma$$

$$E_- = H_0 + \frac{\gamma}{2} - \left[\left(\frac{\gamma}{2}\right)^2 + 2\beta^2\right]^{1/2}$$

These three levels are sketched in Fig. 2–3 as a function of $\gamma/\beta$. It will be noticed that for all values of this parameter only one bonding orbital is available; $E_+$ and the other orbitals are either nonbonding or antibonding. For the special case $\gamma/\beta = 1$, the equilateral triangular configuration, the two excited levels are degenerate.

The forms of the wave functions corresponding to these levels may be written down readily. In the limit $\gamma/\beta = 0$, the bonding three-center orbital is given explicitly by the expression $\psi_+ = \frac{1}{2}(\phi_A + \sqrt{2}\phi_B + \phi_C)$ and lies entirely along the (bent) line joining the three atoms. In the limit $\gamma/\beta = 1$, the bonding orbital takes the form $\psi_+ = (1/\sqrt{3})(\phi_A + \phi_B + \phi_C)$ and is distributed evenly among the three atoms. In the limit $\gamma/\beta = \infty$, the bonding orbital represents a single bond between atoms A and C, and B is not bonded to the system. In the applications to follow we shall find three general types to be of special interest: The B—H—B bridge bonds are all considered as examples of the "open" type, with $\gamma/\beta = 0$ and the electron density concentrated along the line joining the atoms. The coefficients of this type of orbital place one-half an electron on each of the two boron atoms and one electron on the hydrogen; consequently the hydrogens are electrically neutral. The boron framework orbitals are of two types: an "open" bond with $\gamma/\beta = 0$ using a $\pi$ orbital of the central boron and two nearly $\sigma$ AO's from the end borons, and a "central" type with $\gamma/\beta = 1$ using $\sigma$ AO's from all three borons. In the central type the electron density is concentrated near the center of the triangle joining the atoms.*

* A simplified description of the open three-center bond can be given in such a way that the secular determinant is not required. The molecular orbital is formulated as a linear combination, $a_1\psi_A + a_2\psi_B + a_3\psi_C$, from the atomic orbitals $\psi_A$, $\psi_B$, and $\psi_C$ of atoms A, B, and C, respectively. The possible linear combinations yield only one lowest-energy state for reasonable assumptions of the geometries and interaction parameters. Thus for the ground state only a single electron pair is required. The detailed discussion[69] will not be repeated here but the results are summarized in Fig. 2–3. As an example,

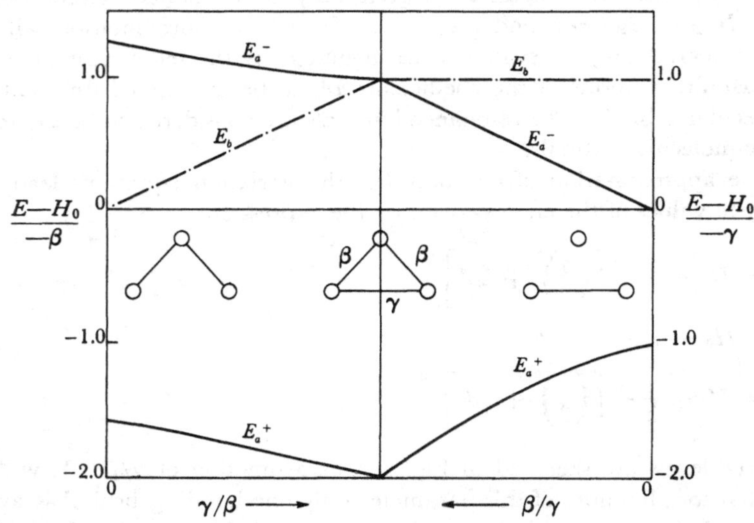


**Figure 2–3** *Results of simplified LCAO discussion of the three-center orbital. Orthogonality of the AO's and constancy of the Coulomb integrals have been assumed. The ratio of resonance integrals $\gamma/\beta = 0$ corresponds to an open three-center bond, that for $\gamma/\beta = 1$ to a central three-center bond, and that for $\beta/\gamma = 0$ to an ordinary electron pair bond. The lowest-energy state is $E_a^+$.*

Deviations of the Coulomb integrals from equality will, of course, change the energy levels and coefficients. In particular, an increase in the absolute magnitude of a Coulomb integral will result in an increase in the coefficient of the corresponding atomic orbital in the LCAO form at the expense of the

---

however, consider the bonding level of the open three-center bond where overlap is neglected. Symmetry requires that $\psi_A$ and $\psi_C$ enter equally, and normalization requires that the sum of squares of coefficients shall be unity. Hence the form of this orbital must be

$$\psi = a(\psi_A + \psi_C) + \sqrt{1 - 2a^2}\,\psi_B$$

Now neglect the interaction between the nonadjacent pair $\psi_A$ and $\psi_C$, and call $\int\psi_A H\psi_B\,d\tau = \int\psi_B H\psi_A\,d\tau = \int\psi_C H\psi_B\,d\tau = \int\psi_B H\psi_C\,d\tau = \beta$, and $\int\psi_A H\psi_A\,d\tau = \int\psi_B H\psi_B\,d\tau = \int\psi_C H\psi_C\,d\tau = H_0$. Then the energy $E = \int\psi H\psi\,d\tau$ becomes

$$E = H_0 + 4a\sqrt{1 - 2a^2}\,\beta$$

The energy minimum can be found by setting $dE/da = 0$, from which we find $a = \frac{1}{2}$, so that $E = H_0 + \sqrt{2}\,\beta$, and therefore

$$\psi = \frac{1}{2}\psi_A + \frac{1}{\sqrt{2}}\psi_B + \frac{1}{2}\psi_C$$

other atomic orbitals. The effect of such distortions is to introduce an asymmetry into an otherwise symmetric three-center bond.

## 2–4 Three-Center Orbitals in the Known Structures

### $B_2H_6$

The structure of diborane has been discussed previously in terms of a protonated double bond,[237] half-bonds,[251] and also from the standpoint of two three-center orbitals for the bridge H bonds.[69,184] In this molecule each boron forms two normal B—H bonds at an angle[105] of 121.5 ± 7°, as shown in Fig. 2–4. Assuming the angle between these bonds to be identical with that between the maxima of the two $sp^3$ hybrid orbitals used for their formation, it is possible to construct two other *equivalent* boron orbitals as indicated by Coulson[46] and by Torkington.[308] These two orbitals lie in a plane normal to the plane of the external B—H bonds and make an angle of 103° with each other. As indicated in the Fig. 2–4, the two bridge hydrogens are found by experiment[105] to lie in this plane and make an H—B—H angle of 97 ± 3°. Thus, it is evident that the hybrid boron orbitals can be regarded as being pointed almost directly toward the bridge hydrogens and the conditions for the formation of two localized, three-center bridge bonds are very favorable. On the basis of our assumptions, each boron contributes one-half an electron to each of these two orbitals, and together with the hydrogen electrons, these electrons just suffice to fill the bonding orbitals. Since the orbitals formed correspond closely to the "open" limit $\gamma/\beta = 0$, the coefficient of the boron atomic orbital in each three-center orbital is $\frac{1}{2}$, and each filled three-center orbital then puts $2 \times (\frac{1}{2})^2 = \frac{1}{2}$ electron on each boron. The resultant charge distribution is then such that each atom is electrically neutral in so far as the assumption of equal Coulomb integrals is acceptable. A more refined treatment taking into account small changes in the Coulomb integrals and also possibly finite values for $\gamma$ would lead to some polarity in these bonds, but such a treatment is not within the scope of the present work.

The atoms are neutral in this simplified valence theory. However, a thorough theoretical study of the molecular orbitals, in a complete LCAO SCF discussion of the four-electron problem involving the bridge H atoms and the boron orbitals directed toward them, indicates[95] that the bridge protons have about $-0.2e$ excess charge, a result confirmed by a more extensive study.[337]

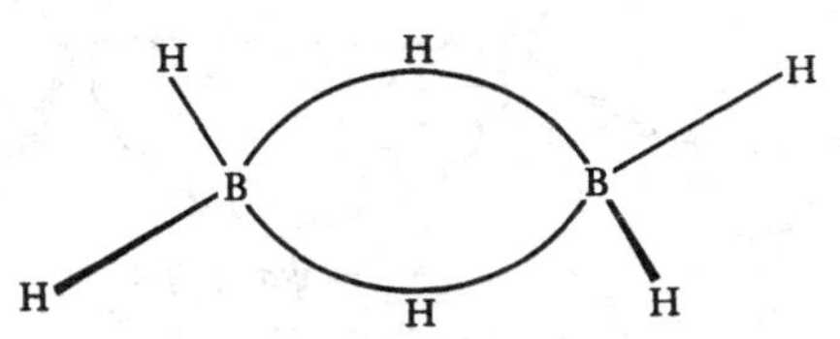


**Figure 2–4** *Bonding in $B_2H_6$.*

A recent study of the infrared vibrational intensities indicates a large change in electric moment as the bridge hydrogens move toward the same boron and a small change when a terminal hydrogen moves.[268] In our discussion of the more complex boron hydrides, we shall make some use of these indications of charge distribution about bridge and terminal hydrogen atoms in discussions of their dipole moments.

Comparison of the bonding in diborane and ethylene is illuminating. In the usual description, each C atom of ethylene has planar $sp^2$ orbitals, two directed toward H and the third directed toward the other C atom. Overlap of the two $sp^2$ orbitals directed toward each other, and of the remaining $\pi$ orbital normal to the plane of $sp^2$ hybridization with the corresponding orbitals of the other C atom, give the two molecular orbitals of the double bond. If we call these molecular orbitals $\psi_\sigma$ and $\psi_\pi$, we may then form equivalent orbitals[165] $\chi_1 = \psi_\sigma + \psi_\pi$, and $\chi_2 = \psi_\sigma - \psi_\pi$, without altering the description of the ground state or of the molecular wave function. But if we remember that $\psi_\sigma$ has the same sign above and below the plane of the molecule, while $\psi_\pi$ changes sign in this molecular plane, we then observe that $\chi_1$ is almost completely above the molecular plane, and $\chi_2$ is almost completely below this plane (Fig. 2–5). The equivalent molecular orbital, $\chi_1$, for example, can be thought of as being formed from two atomic approximately $sp^3$ hybrids, one from each C atom. This two-center orbital can be thought of as an electron-pair bond. In order to obtain the three-center electron-pair bond in $B_2H_6$, one has only to superimpose the $1s$ orbital of hydrogen and replace C by B. Thus the general appearance of the orbitals is not greatly changed, and, in

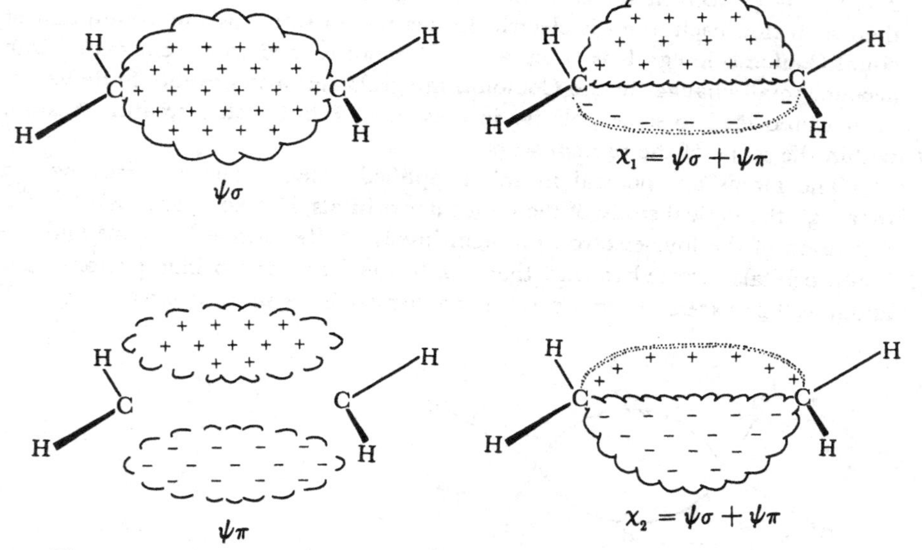


**Figure 2–5** *Equivalent orbitals in the ground state of ethylene.*

particular, the symmetry properties and number of nodes remains exactly the same in $C_2H_4$ and $B_2H_6$.

### $B_4H_{10}$

The structural data for this molecule are presented in Chapter 1. The $B_2$ and $B_4$ atoms in this structure each form two external B—H bonds and two B—H—B three-center bridge center bonds (Fig. 2–6). These four bonds account completely for the four $sp^3$ hybrids of the boron atom and the three boron electrons. It is to be noted, however, that the angle between the bridge H bonds is considerably greater in this structure than in $B_2H_6$ because of the triangular configuration of the borons. Hence the overlapping of the boron hybrids and the hydrogen orbital is not as favorable and the bonds may well be somewhat weaker than in $B_2H_6$. The $B_1$ and $B_3$ atoms each form only one external B—H bond and two three-center bonds. These bonds account for three of the boron orbitals and two electrons and hence leave a single orbital and one electron which may be used to form essentially a normal single bond between $B_1$ and $B_3$. It is also of interest that the configuration of the three hydrogens surrounding borons $B_1$ and $B_3$ is approximately trigonal and suggests approximately tetrahedral hybridization for these borons. Such hybridization would predict a 120° angle between the two boron-triangle planes (experimentally determined as 118.1°) and would lend further support to the assignment of the bond between $B_1$ and $B_3$ as a "normal" single bond. Although the observed length of this bond (1.71 A) is considerably less than the bridge B lengths (1.84 A), it does not appear anomalously short and suggests a deviation from strictly tetrahedral hybridization in a direction which may be anticipated as weakening the B—B single bond but strengthening the B—H—B bonds. Such an effect would result in asymmetry of the B—H—B bridge so as to reduce the $B_1$—H and $B_3$—H distances at the expense of the $B_2$—H and $B_4$—H distances. This is indeed the direction of asymmetry observed (1.16 and 1.37 A,

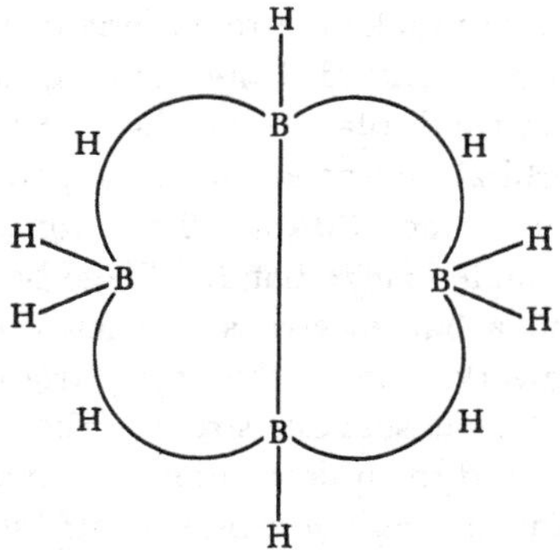


**Figure 2–6** *Three-center bonds in $B_4H_{10}$.*

respectively). Such asymmetry may also be influenced by other factors, notably the unfavorable H—$B_2$—H angle as compared with the corresponding angle in $B_2H_6$ and the change in Coulomb integrals consequent to the over-all charge distribution. In the case of $B_4H_{10}$ all borons are electrically neutral to a first approximation, and the change in Coulomb integrals is expected to be of secondary importance. The reduced overlap resulting from the unfavorable H—$B_2$—H angle should also lead to asymmetry in the direction observed.

Many of the more detailed features of the structure may be ascribed to H—H repulsion. Thus, the angle $B_1$—$B_3$—H is observed to be 118°, and the corresponding H—H distance is 2.80 A. Similarly, repulsion between the two hydrogens attached to $B_1$ and $B_3$ and separated by a distance of 2.81 A, probably accounts for the slight twist of the boron tetrahedra which makes the plane of $H_\mu$—$B_2$—$H_\mu$ incline at a slight angle to the plane of $B_1$—$B_3$—$B_2$. No attempt is made to include such fine details within the general approach suggested here.

As pointed out above, all the borons are electrically neutral and the dipole moment due to framework electrons has been predicted to be very small. An unpublished, more detailed molecular orbital study by C. W. Kern and W. N. Lipscomb indicates a value $<2$ D(debyes). The observed value[323] is 0.6 D.

### $B_5H_{11}$

The structural data for this compound are given in Chapter 1. In this molecule (Fig. 2-7), the unique boron, $B_1$, forms two external B—H bonds which use up two of the boron orbitals and two electrons. This atom then contributes two orbitals and a single electron to the framework. There is no unambiguous way of choosing these orbitals on the basis of the molecular geometry, but, for convenience, one may be chosen as the third member of an $sp^2$ set lying in the plane of symmetry and designated $1d$, and the other may be chosen as essentially a pure $p$ orbital normal to this plane and designated $1p$. The $sp^2$ type of orbital of this pair then points almost directly at the center of the line joining the two borons $B_3$ and $B_4$. Each of these borons forms a single external B—H bond and two bridge bonds, utilizing a total of three orbitals and two electrons. Because of the nearly trigonal symmetry of the hydrogens about these atoms, it seems reasonable to take the remaining orbital as pointing toward the unique atom $B_1$. The two orbitals from atoms $B_3$ and $B_4$ and the orbital of atom 1 lying in the symmetry plane may then be considered as forming a localized three-center orbital characterized by a parameter $\gamma/\beta \sim 1$ and concentrated in the triangular face $B_1B_3B_4$. The atoms $B_2$ and $B_5$ form two external B—H bonds and a single bridge bond. These bonds require the use of three orbitals and two and a half electrons and leave one orbital and one-half an electron for the framework. Again the nearly trigonal symmetry of the hydrogens about the boron atoms suggests strongly that the remaining orbital is pointed almost directly toward the unique atom 1. Hence, the pair of orbitals from $B_2$ and $B_5$ along with the single $p$ orbital contributed by atom 1 may be considered as forming a localized "open" three-center orbital

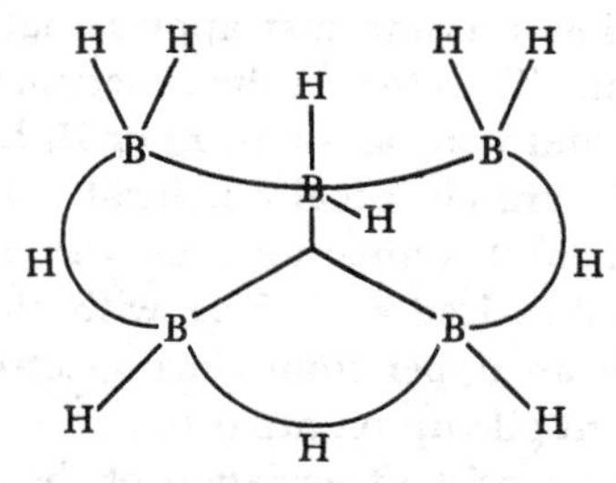

**Figure 2-7** *Three-center bonds in* $B_5H_{11}$. *These two descriptions have only the very small difference that the B—H bonds of the central* $BH_2$ *group require approximately* $sp^3$ *orbitals at the left, and* $sp^2$ *orbitals at the right from this B atom.*

lying essentially along the line joining the three atoms. Out of the six framework orbitals contributed by all the atoms, there are expected only two strongly bonding three-center orbitals, one nonbonding, and three antibonding three-center orbitals given essentially by the expressions

$$\psi_1(a') = \frac{1}{\sqrt{3}}[\phi_{1d} + \phi_3 + \phi_4] \qquad \text{bonding}$$

$$\psi_2(a'') = \frac{1}{\sqrt{2}}[\phi_3 - \phi_4] \qquad \text{antibonding}$$

$$\psi_3(a') = \frac{1}{\sqrt{6}}[2\phi_{1d} - \phi_3 - \phi_4] \qquad \text{antibonding}$$

$$\psi_4(a'') = \frac{1}{2}[\phi_2 + \sqrt{2}\,\phi_{1p} - \phi_5] \qquad \text{bonding}$$

$$\psi_5(a') = \frac{1}{\sqrt{2}}[\phi_2 + \phi_5] \qquad \text{nonbonding}$$

$$\psi_6(a'') = \frac{1}{2}[\phi_2 - \sqrt{2}\phi_{1p} - \phi_5] \qquad \text{antibonding}$$

Since there is a total of four electrons contributed to this framework, the two bonding orbitals are just filled and the resultant electronic configuration may be represented as $(a')^2(a'')^2 = {}^1A'$. Of course, as pointed out above, the choice of the forms for the two orbitals contributed by atom $B_1$ is entirely a matter of convenience, and hence the form of the wave functions is a matter of convenience. In essence, any linear combination of the three orbitals of symmetry $a'$ and the three orbitals of symmetry $a''$ will provide acceptable substitutes, but the general pattern of energy levels should not be greatly altered by such symmetrizing procedures since the forms chosen are such as to give qualitatively the largest resonance integrals and the least interaction integrals between localized orbitals. In particular, it is quite reasonable to choose two three-center bonds lying, respectively, in the triangles $B_1$, $B_2$, $B_3$ and $B_1$, $B_4$, $B_5$.

The coefficients given in the foregoing expressions for the bonding orbitals now allow a crude first approximation to the charge distribution and dipole moment. Thus the resultant net charge on atom 1 becomes $-\frac{2}{3}$ of an electronic charge and that on atoms $B_3$ or $B_4$ becomes $+\frac{1}{3}$ of an electronic charge; atoms $B_2$ or $B_5$ are electrically neutral. This charge distribution together with the experimental dimensions of the molecule lead to a dipole moment of $\frac{2}{3} \times 4.80 \times 1.48 = 4.75$ D with the negative end on atom $B_1$. Again this value is an upper limit since electron repulsion will reduce the tendency for charge to pile up on atom $B_1$.

Self-consistent variation of the Coulomb integrals similar to that utilized in discussing the dipole moment of $B_5H_9$ may be employed to estimate the reduction. In this case, both filled three-center orbitals must be polarized, since the charge will certainly tend to spread over the entire molecule in so far as possible. These calculations[69] lead to dipole moments of 1.1 and 1.9 D, corresponding to resonance integrals of $-2$ and $-3$ ev, respectively. The general features of the charge distribution are as expected intuitively from the simple treatment: the large negative charge of atom $B_1$ is reduced by spreading to the other B atoms. Thus atoms $B_2$ and $B_5$, which were originally neutral, become slightly negative, while atoms $B_3$ and $B_4$ become less positive.

It is interesting to note that the charge distribution suggested for this molecule should lead to asymmetry of the bridge hydrogens with the hydrogens closer to atoms $B_3$ and $B_4$, and farther from atoms $B_2$ and $B_5$. The crystal-structure studies do not give sufficient detail to confirm this effect.

### $B_6H_{10}$

This molecule has a plane of symmetry, which is also a symmetry element of the crystal itself. Since there are 14 electron pairs in the molecule, and 10 are used for the six B—H bonds and the four B—H—B bridges, there are four electron pairs remaining for the framework bonds. These four pairs are to be distributed among the 10 B atomic orbitals which are not involved in bonds to H atoms. Hence there are two three-center bonds and two two-center bonds

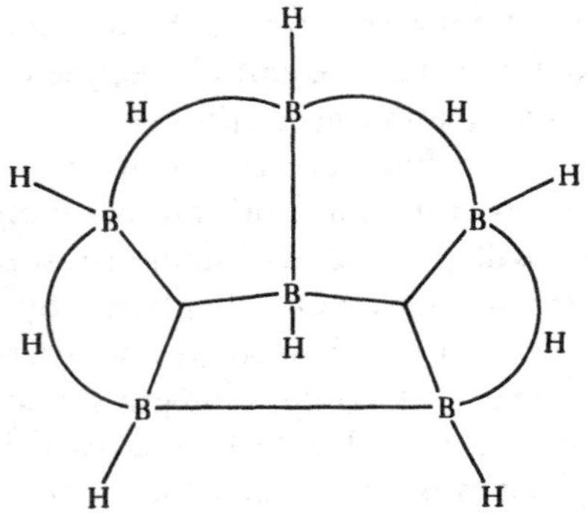


**Figure 2-8** *Three-center bonds in $B_6H_{10}$.*

in the B · · · B framework (Fig. 2–8). The most convenient choice of hybrids is clearest for atoms 2, 3, and 6, each of which have one approximately $sp^3$ orbital pointed toward atom 1, the apex B atom. In order to bond the apex atom to the other five one must require two three-center bonds and one single bond, and hence the symmetry of the molecule is best satisfied by placing the single bond from $B_1$ to $B_2$, and the two three-center bonds on either side of the mirror plane, between $B_1$, $B_3$, and $B_4$, and $B_1$, $B_5$, and $B_6$.

### $B_9H_{15}$

This structure strongly resembles $B_4H_{10}$ at the end where the $BH_2$ group occurs, and $B_5H_{11}$ at the other end, where the three contiguous bridge H atoms occur. The 21 pairs, less the 15 bonds involving the 10 B—H and 5 B—H—B bonds, leave six pairs for the B—B bonds. Since 16 orbitals remain for B · · · B bonding after formation of the bonds to H, there are two single B—B bonds and four three-center B—B—B bonds. Three of the ways of placing these bonds in the $B_9$ skeleton of the mirror plane of symmetry are shown in Fig. 2–9. Aside from small charges of hybridization toward some of the H atoms, all these structures represent the same electron-density function, and can be transformed into one another by performing various linear combinations of atomic orbitals in such a way that the bonding is left almost unchanged.

### $B_{10}H_{14}$

The structural data for this molecule are summarized in Chapter 1 and Fig. 1–7. The molecule exhibits $C_{2v}$ symmetry and has four different types of

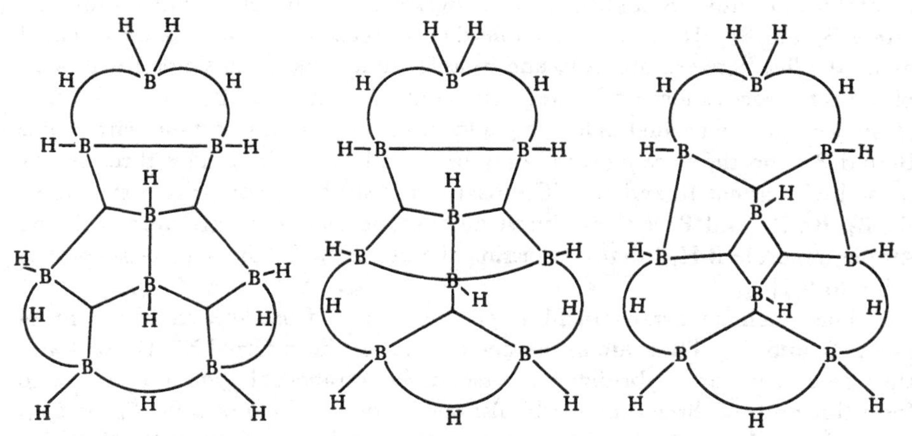

**Figure 2–9** *Bonding in $B_9H_{15}$. The two valence structures on the left most strongly resemble $B_4H_{10}$ near the top and $B_5H_{11}$ near the lower part of the molecule. The molecular symmetry is $C_s$, even though some of the H's are distorted for clarity in the drawings.*

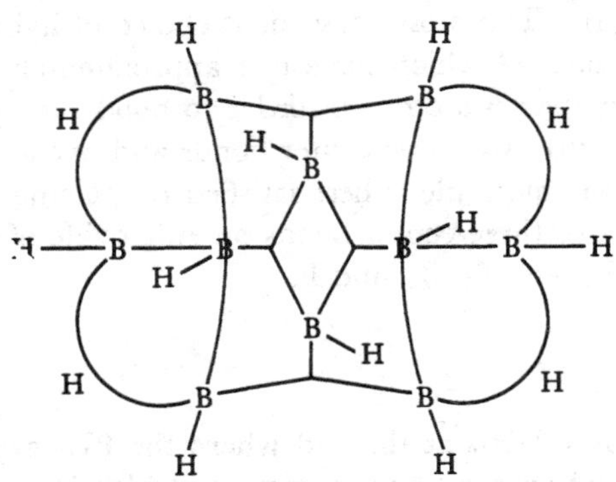


**Figure 2-10** *Bonding in $B_{10}H_{14}$. The molecular symmetry is $C_{2v}$.*

boron atoms, all of which form only a single normal B—H external bond. Each of the atoms $B_6$ and $B_9$ forms in addition two B—H—B bridge bonds which, together with the external B—H bond, consume three orbitals and two electrons. The remaining orbital probably points directly toward, respectively, $B_2$ and $B_4$, as suggested by the essentially trigonal symmetry of the hydrogens about $B_6$ and $B_9$ (Fig. 2-10). Each of the atoms $B_2$ and $B_4$ forms only the one normal B—H bond and hence contributes three other orbitals and two electrons to the framework. Again there is no unambiguous way of choosing these three orbitals, but consistent with the symmetry of the molecule, it is convenient to regard two as forming an $sp^2$ hybrid with the normal B—H bond orbital, all three lying in the vertical plane containing the $C_2$ axis and atoms $B_6$ and $B_2$. The fourth orbital is then a pure $p$ orbital normal to this plane.

It is evident from inspection that this form of hybridization directs one orbital toward atom $B_6$ and the second toward the center of the line joining the atoms $B_1$ and $B_3$. It seems reasonable then to regard the two orbitals directed along the line between atoms $B_6$ and $B_2$ as forming a localized single B—B bond of the same sort as formed between the atoms $B_1$ and $B_3$ in $B_4H_{10}$. The other $sp^2$ hybrid may be treated as forming a localized three-center orbital with atoms $B_1$ and $B_3$, and the pure $p$ orbital may be considered as forming a three-center orbital with atoms $B_5$ and $B_7$. The part of the structure comprising the atoms $B_5$, $B_7$, $B_2$, $B_1$, and $B_3$ is then almost identical in electronic structure with the simple molecule $B_5H_{11}$; that comprising the atoms $B_6$, $B_5$, $B_7$, and $B_2$ is comparable to $B_4H_{10}$.

That such an arrangement is possible results from the environment of atoms $B_1$ and $B_3$. These atoms use one orbital for the external B—H bond and the other three are hybridized in essentially tetrahedral symmetry so as to form three sets of three-center orbitals: one involving the atoms $B_7$, $B_8$, and $B_3$, or, equivalently, $B_5$, $B_{10}$, and $B_1$; the second involving $B_3$, $B_1$, and $B_2$; the third involving $B_3$, $B_1$, and $B_4$. Of these, the kind involving the atoms $B_2$ and $B_4$ is probably characterized by a parameter $\gamma/\beta \sim 1$, whereas those involving the atoms $B_5$, $B_7$, $B_8$, and $B_{10}$ may well be characterized by a smaller magnitude for this parameter. For ease in computing charge distribution, however, it will be assumed as a first approximation that these orbitals are all of similar character and that each atomic orbital contributes equally in the bonding molecular orbital. The $B_5$, $B_7$, $B_8$, and $B_{10}$ atoms form in addition to the external B—H bond one B—H—B bridge bond. These two bonds use two boron orbitals and one and a half electrons leaving two orbitals and one and a half electrons for the framework. These orbitals point in the general direction of atoms $B_2$ and $B_4$ and the center of the triangle formed by the neighboring $B_5$, $B_7$, $B_8$, $B_{10}$, $B_1$, and $B_3$ atoms. This orientation is again consistent with the localized three-center bond formulation previously outlined.

In summary, there are available 22 hybrid boron orbitals and 16 electrons from which to construct the framework. Normal single bonds are formed between atoms $B_6$ and $B_2$ and between atoms $B_4$ and $B_9$, using a total of four boron orbitals and giving two strongly bonding localized orbitals requiring four electrons. Four three-center orbitals are formed requiring 12 boron atomic orbitals and giving rise to four strongly bonding localized three-center orbitals concentrated in the triangles formed by atoms of the types $B_1$, $B_2$, $B_3$ and $B_1$, $B_5$, $B_{10}$. These four bonding orbitals require eight electrons. The remaining orbitals form weaker three-center orbitals lying along the line of atoms $B_5$, $B_2$, $B_7$ and the corresponding trio $B_8$, $B_4$, $B_{10}$. Each of these gives rise to one bonding orbital, one nonbonding orbital, and one antibonding orbital. Thus the remaining four electrons are required for filling the bonding orbitals. There are thus a total of eight bonding orbitals which are just filled by the available 16 electrons. The highest filled one of these orbitals is probably the weak three-center orbital of the type between atoms $B_5$, $B_2$, $B_7$, and excitation of an electron from this orbital to the first excited, nonbonding level is perhaps responsible for the absorption spectrum in the near ultraviolet. Since this orbital is characterized by the parameter $\gamma/\beta \sim 0$, the energy of the transition should be approximately $-\sqrt{2}\,\beta$ on the crude LCAO model and by comparison with the absorption spectrum[91,92,232] allows an estimate of $-2.5$ ev for the resonance integral $\beta$.

The molecular orbitals are given above in terms of simple prototype, localized three-center orbitals. Naturally, delocalization can be obtained by combining these according to the appropriate symmetry properties of the molecule, but the choice of orbitals indicated above is such as to minimize serious interaction between them, and hence little change in the energy-level pattern of coefficients of individual atomic orbitals is to be expected in this process. The coefficients may be determined immediately from the corresponding three-center prototypes and lead to the following charge distribution: atoms $B_6$ and $B_9$: neutral; atoms $B_5$, $B_7$, $B_8$, and $B_{10}$: $+\frac{1}{3}$ electronic charge; atoms $B_2$ and $B_4$: $-\frac{2}{3}$ electronic charge; atoms $B_1$ and $B_3$: neutral. Thus, the charge displacement appears between atoms $B_2$ and $B_5$ and gives rise to a dipole moment

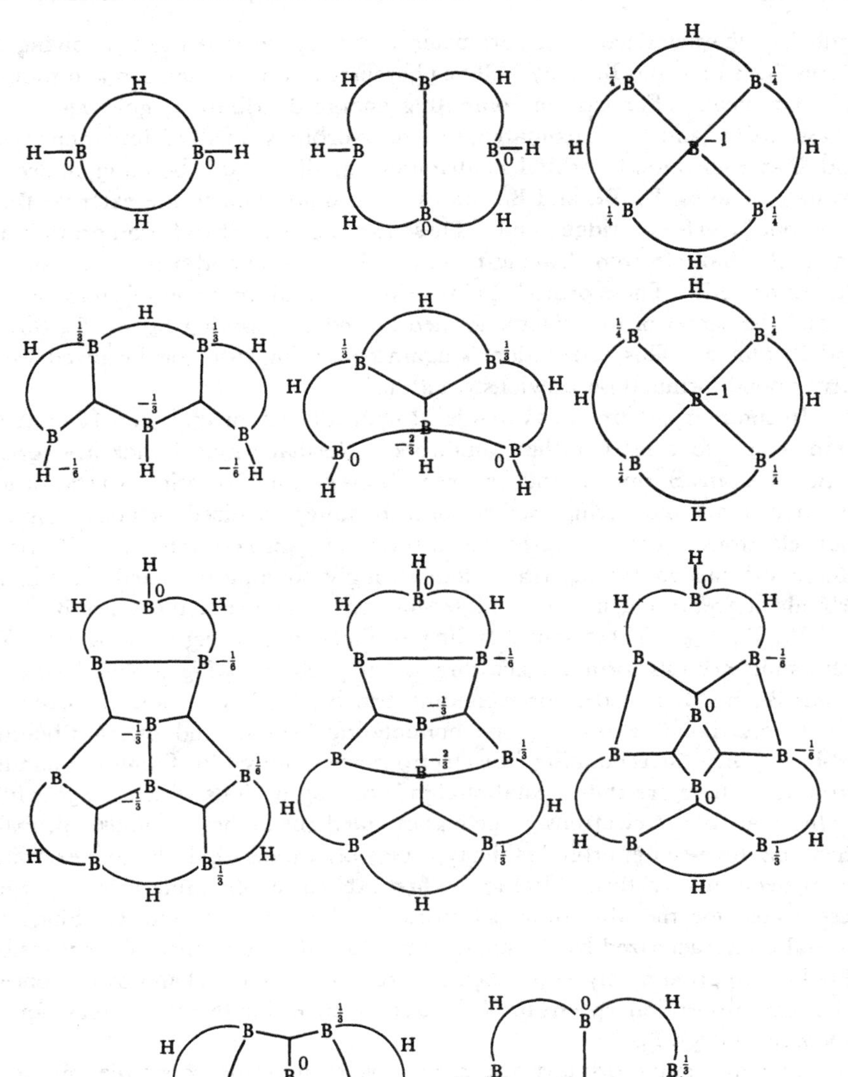

**Figure 2-11** *Zeroth approximation to the charge distribution in the known boron hydrides. The abbreviated notation omits all terminal hydrogens except when a* $BH_2$ *group is present. These results are obtained when the bond charge distribution in Fig. 2-2 is applied to the structures in Fig. 2-9. The third valence structure for* $B_9H_{15}$ *(extreme right) correlates relatively poorly with the observed bond distances and with the other hydride structures. Charges in* $B_5H_9$ *are averaged over the resonance hybrid.*

of $2 \times \frac{2}{3} \times 4.80 \times 0.92 = 6.2$ D, where the altitude of the triangle formed by $B_5$, $B_2$, and $B_7$ is estimated as 0.92 A. Again this value is to be taken as an upper limit which will be reduced by a redistribution of the coefficients of the weak three-center orbitals. Such a redistribution[69] leads to a resultant dipole moment of 1.6 to 2.2 D, which corresponds to resonance integrals of $-2$ to $-3$ ev, respectively. This value is to be compared with the experimental moment $3.52 \pm 0.02$ D determined by Laubengayer and Bottei[163] from dielectric-constant studies on solutions. Although the experimental value may be too high because of underestimation of atomic polarization, we shall be the first to point out that our calculation of the redistribution of charge is subject to significant improvement. Such improvements are, in fact, discussed in Chapter 3.

Although it is not within the three-center bond approximation to discuss the finer details of the structure, some of the gross features are of interest. The $B_2$—$B_6$ and the $B_4$—$B_9$ bonds are suggested to be essentially single bonds, and hence the boron-boron distance might be expected to be somewhat shorter than the other boron-boron distances. This effect is indeed indicated by the experimental data but there may be some doubt as to the magnitude, 1.72 A, as compared to the average B—B distance of 1.77 A (neglecting the long bonds).

#### The Method for Obtaining Three-Center Charge Distributions

In Fig. 2–11 are collected the very crude approximation to the charge distribution in these hydrides as indicated by the overlocalized three-center approximation. These charge distributions are more detailed than those given[170] by a free-electron model of charge distribution in a hemisphere or hemispherical shell, but both of these treatments agree in a general way. Negative charge tends to concentrate in the inner regions of the molecule because the wave functions vanish toward the exposed regions. If one superimposes this general effect on the LCAO results shown in Fig. 2–11, one may well expect that the two outer borons attached to bridge hydrogens in $B_{10}H_{14}$ would be more positive than the innermost pair of borons. If one adds the further conclusion that the bridge hydrogens carry about $-0.1e$ to $-0.2e$ of charge, these two outermost borons, labelled with 0 excess charge in Fig. 2–11, would become even more positive. It then becomes possible to indicate that this outer pair of borons or the next inner group of four equivalent borons is the more positive. On the other hand, the identity of the most negative pair of borons in $B_{10}H_{14}$ is reasonably certainly the one shown in Fig. 2–11, as has been confirmed by experiments described in Chapter 4.

## 2–5 The Equations of Balance

The relative numbers of orbitals, electrons, hydrogen, and boron atoms, and bonds of various types, can easily be expressed in a systematic way. The

**Table 2–1** *Tables of s, t, y, and x boron hydride structures,* $B_pH_{q+p}$*

| | | | | | | | | |
|---|---|---|---|---|---|---|---|---|
| $B_2H_2$ | $B_2H_4$ | $B_2H_6$ | | | | | | |
| 0200 | 2010 | 2002 | | | | | | |
| | 1101 | | | | | | | |
| $B_3H_3$ | $B_3H_5$ | $B_3H_7$ | $B_3H_9$ | | | | | |
| 0300 | 2110 | 3011 | 3003 | | | | | |
| | 1201 | 2102 | | | | | | |
| $B_4H_4$ | $B_4H_6$ | $B_4H_8$ | $B_4H_{10}$ | $B_4H_{12}$ | | | | |
| 0400 | 2210 | 4020 | 4012 | 4004 | | | | |
| | 1301 | 3111 | 3103 | | | | | |
| | | 2202 | | | | | | |
| $B_5H_5$ | $B_5H_7$ | $B_5H_9$ | $B_5H_{11}$ | $B_5H_{13}$ | $B_5H_{15}$ | | | |
| 0500 | 2310 | 4120 | 5021 | 5013 | 5005 | | | |
| | 1401 | 3211 | 4112 | 4104 | | | | |
| | | 2302 | 3203 | | | | | |
| $B_6H_6$ | $B_6H_8$ | $B_6H_{10}$ | $B_6H_{12}$ | $B_6H_{14}$ | $B_6H_{16}$ | $B_6H_{18}$ | | |
| 0600 | 2410 | 4220 | 6030 | 6022 | 6014 | 6006 | | |
| | 1501 | 3311 | 5121 | 5113 | 5105 | | | |
| | | 2402 | 4212 | 4204 | | | | |
| | | | 3303 | | | | | |
| $B_7H_7$ | $B_7H_9$ | $B_7H_{11}$ | $B_7H_{13}$ | $B_7H_{15}$ | $B_7H_{17}$ | $B_7H_{19}$ | $B_7H_{21}$ | |
| 0700 | 2510 | 4320 | 6130 | 7031 | 7023 | 7015 | 7007 | |
| | 1601 | 3411 | 5221 | 6122 | 6114 | 6106 | | |
| | | 2502 | 4312 | 5213 | 5205 | | | |
| | | | 3403 | 4304 | | | | |
| $B_8H_8$ | $B_8H_{10}$ | $B_8H_{12}$ | $B_8H_{14}$ | $B_8H_{16}$ | $B_8H_{18}$ | $B_8H_{20}$ | $B_8H_{22}$ | $B_8H_{24}$ |
| 0800 | 2610 | 4420 | 6230 | 8040 | 8032 | 8024 | 8016 | 8008 |
| | 1701 | 3510 | 5321 | 7131 | 7123 | 7115 | 7107 | |
| | | 2602 | 4412 | 6222 | 6214 | 6206 | | |
| | | | 3503 | 5313 | 5305 | | | |
| | | | | 4404 | | | | |

| $B_9H_9$ | $B_9H_{11}$ | $B_9H_{13}$ | $B_9H_{15}$ | $B_9H_{17}$ | $B_9H_{19}$ | $B_9H_{21}$ | $B_9H_{23}$ | $B_9H_{25}$ | $B_9H_{27}$ | |
|---|---|---|---|---|---|---|---|---|---|---|
| 0900 | 2710 | 4520 | 6330 | 8140 | 9041 | 9033 | 9025 | 9017 | 9009 | |
| | 1801 | 3611 | 5421 | 7231 | 8132 | 8124 | 8116 | 8108 | | |
| | | 2702 | 4512 | 6322 | 7223 | 7215 | 7207 | | | |
| | | | 3603 | 5413 | 6314 | 6306 | | | | |
| | | | | 4504 | 5405 | | | | | |

| $B_{10}H_{10}$ | $B_{10}H_{12}$ | $B_{10}H_{14}$ | $B_{10}H_{16}$ | $B_{10}H_{18}$ | $B_{10}H_{20}$ | $B_{10}H_{22}$ | $B_{10}H_{24}$ | $B_{10}H_{26}$ | $B_{10}H_{28}$ | $B_{10}H_{30}$ |
|---|---|---|---|---|---|---|---|---|---|---|
| 0.10.0.0 | 2810 | 4620 | 6430 | 8240 | 10.0.5.0 | 10.0.4.2 | 10.0.3.4 | 10.0.2.6 | 10.0.1.8 | 10.0.0.10 |
| | 1901 | 3711 | 5521 | 7331 | 9141 | 9133 | 9125 | 9117 | 9109 | |
| | | 2802 | 4612 | 6422 | 8232 | 8224 | 8216 | 8208 | | |
| | | | 3703 | 5513 | 7323 | 7315 | 7307 | | | |
| | | | | 4604 | 6414 | 6406 | | | | |
| | | | | | 5505 | | | | | |

| $B_{11}H_{11}$ | $B_{11}H_{13}$ | $B_{11}H_{15}$ | $B_{11}H_{17}$ | $B_{11}H_{19}$, etc. | | $B_{12}H_{12}$ | $B_{12}H_{14}$ | $B_{12}H_{16}$ | $B_{12}H_{18}$ | $B_{12}H_{20}$, etc. |
|---|---|---|---|---|---|---|---|---|---|---|
| 0.11.0.0 | 2910 | 4720 | 6530 | 8340 | | 0.12.0.0 | 2.10.1.0 | 4820 | 6630 | 8440 |
| | 1.10.0.1 | 3811 | 5621 | 7431 | | | 1.11.0.1 | 3911 | 5721 | 7531 |
| | | 2902 | 4712 | 6522 | | | | 2.10.0.2 | 4812 | 6622 |
| | | | 3803 | 5613 | | | | | 3903 | 5713 |
| | | | | 4704 | | | | | | 4804 |

---

* $s$ = number of hydrogen bridges, B⌒B.

$t$ = number of boron-boron three-center bonds, B–<(B, B) or B⌒(B)B.

$y$ = number of boron-boron single bonds, B—B.

$x$ = number of extra H atoms on B—H; i.e., $BH_2$ groups.

$s + x = q$, $s + t = p$, $p = t + y + q/2$ (see p. 47).

**Table 2–2** *Values of s, t, y, and x for* $B_pH_{p+q-1}^{-1}$ *ions* ($s + x = q - 1$, $s + t = p - 1$, $t + y = p + 1 - q/2$)

| $B_2H_3^-$ | $B_2H_5^-$ | $B_2H_7^-$ | |
|---|---|---|---|
| 1020 | 1012 | 1004 | |
| 0111 | 0103 | | |
| $B_3H_4^-$ | $B_3H_6^-$ | $B_3H_8^-$ | $B_3H_{10}^-$ |
| 1120 | 2021 | 2013 | 2005 |
| 0211 | 1112 | 1104 | |
| | 0203 | | |
| $B_4H_5^-$ | $B_4H_7^-$ | $B_4H_9^-$ | $B_4H_{11}^-$, etc. |
| 1220 | 3030 | 3022 | 3014 |
| 0311 | 2121 | 2113 | 2105 |
| | 1212 | 1204 | |
| | 0303 | | |
| $B_5H_6^-$ | $B_5H_8^-$ | $B_5H_{10}^-$ | $B_5H_{12}^-$, etc. |
| 1320 | 3130 | 4031 | 4023 |
| 0411 | 2221 | 3122 | 3114 |
| | 1312 | 2213 | 2205 |
| | 0403 | 1304 | |
| $B_6H_7^-$ | $B_6H_9^-$ | $B_6H_{11}^-$ | $B_6H_{13}^-$, etc. |
| 1420 | 3230 | 5040 | 5032 |
| 0511 | 2321 | 4131 | 4123 |
| | 1412 | 3222 | 3214 |
| | 0503 | 2313 | 2305 |
| | | 1404 | |
| $B_8H_9^-$ | $B_8H_{11}^-$ | $B_8H_{13}^-$ | $B_8H_{15}^-$, etc. |
| 1620 | 3430 | 5240 | 7050 |
| 0711 | 2521 | 4331 | 6141 |
| | 1612 | 3422 | 5232 |
| | 0703 | 2513 | 4323 |
| | | 1604 | 3414 |
| | | | 2505 |
| $B_9H_{10}^-$ | $B_9H_{12}^-$ | $B_9H_{14}^-$ | $B_9H_{16}^-$, etc. |
| 1720 | 3530 | 5340 | 7150 |
| 8011 | 2621 | 4431 | 6241 |
| | 1712 | 3522 | 5332 |
| | 0803 | 2613 | 4423 |
| | | 1704 | 3514 |
| | | | 2605 |

Table 2-2 (*continued*)

| $B_{10}H_{11}^-$ | $B_{10}H_{13}^-$ | $B_{10}H_{15}^-$ | $B_{10}H_{17}^-$, etc. |
|---|---|---|---|
| 1820 | 3630 | 5440 | 7250 |
| 0911 | 2721 | 4531 | 6341 |
| | 1812 | 3622 | 5432 |
| | 0903 | 2713 | 4523 |
| | | 1804 | 3614 |
| | | | 2705 |
| $B_{11}H_{12}^-$ | $B_{11}H_{14}^-$ | $B_{11}H_{16}^-$ | $B_{11}H_{18}^-$, etc. |
| 1920 | 3730 | 5540 | 7350 |
| 0.10.1.1 | 2821 | 4631 | 6441 |
| | 1912 | 3722 | 5532 |
| | 0.10.0.3 | 2813 | 4623 |
| | | 1904 | 3714 |
| | | | 2805 |
| $B_{12}H_{13}^-$ | $B_{12}H_{15}^-$ | $B_{12}H_{17}^-$ | $B_{12}H_{19}^-$ |
| 1.10.2.0 | 3830 | 5640 | 7450 |
| 0.11.1.1 | 2921 | 4731 | 6541 |
| | 1.10.1.2 | 3822 | 5632 |
| | 0.11.0.3 | 2913 | 4723 |
| | | 1.10.0.4 | 3814 |
| | | | 2905 |

procedure[69] can be simplified as follows. For a boron hydride, $B_pH_{p+q}$, containing $s$ H bridges, $x$ extra B—H groups, $t$ three-center B—B—B bonds, and $y$ two-center B—B bonds, the equations of balance may be written in a simplified form as follows. The hydrogen-atom balance is simply $s + x = q$. Since each boron supplies four orbitals but only three electrons, the total number of three-center bonds in the molecule is the same as the number of boron atoms, $s + t = p$. Finally, if we consider B—H as the bonding unit, each of which then supplies one electron pair, these $p$ pairs are used up as three-center B—B—B bonds, two-center B—B bonds, and in supplying half of each pair to each extra hydrogen; thus $p = t + y + q/2$.

Extension of these rules to ions of charge $c$ having the generalized formula $B_pH_{p+q+c}^c$ leads to the more general equations of balance,[174]

$$s + x = q + c$$
$$s + t = p + c$$
$$t + y = p - c - \frac{q}{2}$$

which reduce to the equations of the preceding paragraph if $c = 0$.

**Table 2-3** *Values of s, t, y, and x for* $B_pH^{=}_{p+q-2}$ *ions* $(s + x = q - 2, s + t = p - 2, t + y = p + 2 - q/2)$

| $B_2H_2^{=}$ | $B_2H_4^{=}$ | $B_2H_6^{=}$ | $B_2H_8^{=}$ | |
|---|---|---|---|---|
| 0030 | 0022 | 0014 | 0006 | |
| $B_3H_3^{=}$ | $B_3H_5^{=}$ | $B_3H_7^{=}$ | $B_3H_9^{=}$ | $B_3H_{11}^{=}$ |
| 0130 | 1031 | 1023 | 1015 | 1007 |
| | 0122 | 0114 | 0106 | |
| $B_4H_4^{=}$ | $B_4H_6^{=}$ | $B_4H_8^{=}$ | $B_4H_{10}^{=}$ | $B_4H_{12}^{=}$, etc. |
| 0230 | 2040 | 2032 | 2024 | 2016 |
| | 1131 | 1123 | 1115 | 1107 |
| | 0222 | 0214 | 0206 | |
| $B_5H_5^{=}$ | $B_5H_7^{=}$ | $B_5H_9^{=}$ | $B_5H_{11}^{=}$ | $B_5H_{13}^{=}$, etc. |
| 0330 | 2140 | 3041 | 3033 | 3025 |
| | 1231 | 2132 | 2124 | 2116 |
| | 0322 | 1223 | 1215 | 1207 |
| | | 0314 | 0306 | |
| $B_6H_6^{=}$ | $B_6H_8^{=}$ | $B_6H_{10}^{=}$ | $B_6H_{12}^{=}$ | $B_6H_{14}^{=}$, etc. |
| 0430 | 2240 | 4050 | 4042 | 4034 |
| | 1331 | 3141 | 3133 | 3125 |
| | 0422 | 2232 | 2224 | 2216 |
| | | 1323 | 1315 | 1307 |
| | | 0414 | 0406 | |
| $B_7H_7^{=}$ | $B_7H_9^{=}$ | $B_7H_{11}^{=}$ | $B_7H_{13}^{=}$ | $B_7H_{15}^{=}$, etc. |
| 0530 | 2340 | 4150 | 5051 | 5043 |
| | 1431 | 3241 | 4142 | 4134 |
| | 0522 | 2332 | 3233 | 3225 |
| | | 1423 | 2324 | 2316 |
| | | 0514 | 1415 | 1407 |
| | | | 0506 | |
| $B_8H_8^{=}$ | $B_8H_{10}^{=}$ | $B_8H_{12}^{=}$ | $B_8H_{14}^{=}$ | $B_8H_{16}^{=}$, etc. |
| 0630 | 2440 | 4250 | 6060 | 6052 |
| | 1531 | 3341 | 5151 | 5143 |
| | 0622 | 2432 | 4242 | 4234 |
| | | 1523 | 3333 | 3325 |
| | | 0614 | 2424 | 2416 |
| | | | 1515 | 1507 |
| | | | 0606 | |
| $B_9H_9^{=}$ | $B_9H_{11}^{=}$ | $B_9H_{13}^{=}$ | $B_9H_{15}^{=}$ | $B_9H_{17}^{=}$, etc. |
| 0730 | 2540 | 4350 | 6160 | 7061 |
| | 1631 | 3441 | 5251 | 6152 |
| | 0722 | 2532 | 4342 | 5243 |
| | | 1623 | 3433 | 4334 |
| | | 0714 | 2524 | 3425 |
| | | | 1615 | 2516 |
| | | | 0706 | 1607 |

Table 2-3 (*continued*)

| $B_{10}H_{10}^{=}$ | $B_{10}H_{12}^{=}$ | $B_{10}H_{14}^{=}$ | $B_{10}H_{16}^{=}$ | $B_{10}H_{18}^{=}$, etc. |
|---|---|---|---|---|
| 0830 | 2640 | 4450 | 6260 | 8070 |
| | 1731 | 3541 | 5351 | 7161 |
| | 0822 | 2632 | 4442 | 6252 |
| | | 1723 | 3533 | 5343 |
| | | 0814 | 2624 | 4434 |
| | | | 1715 | 3525 |
| | | | 0806 | 2616 |
| | | | | 1707 |
| $B_{11}H_{11}^{=}$ | $B_{11}H_{13}^{=}$ | $B_{11}H_{15}^{=}$ | $B_{11}H_{17}^{=}$ | $B_{11}H_{19}^{=}$, etc. |
| 0930 | 2740 | 4550 | 6360 | 8170 |
| | 1831 | 3641 | 5451 | 7261 |
| | 0922 | 2732 | 4542 | 6352 |
| | | 1823 | 3633 | 5443 |
| | | 0914 | 2724 | 4534 |
| | | | 1815 | 3625 |
| | | | 0906 | 2716 |
| | | | | 1807 |
| $B_{12}H_{12}^{=}$ | $B_{12}H_{14}^{=}$ | $B_{12}H_{16}^{=}$ | $B_{12}H_{18}^{=}$ | $B_{12}H_{20}^{=}$, etc. |
| 0.10.3.0 | 2840 | 4650 | 6460 | 8270 |
| | 1931 | 3741 | 5551 | 7361 |
| | 0.10.2.2 | 2832 | 4642 | 6452 |
| | | 1923 | 3733 | 5543 |
| | | 0.10.1.4 | 2824 | 4634 |
| | | | 1915 | 3725 |
| | | | 0.10.0.6 | 2816 |
| | | | | 1907 |

The results of these equations, applied to the $B_n$ units for $2 \leq n \leq 12$ and to charges 0, $-1$, and $-2$ are summarized in Tables 2-1, 2-2, and 2-3. No restriction has yet been assumed concerning the geometrical arrangements of atoms.

## 2-6 A Topological Theory of Boron Hydrides

Without the introduction of some formal arguments which abstract the common structural features of the known hydrides, the three-center bond approximations described above lead to a large variety and number of possible hydrides. The topological ideas to be described here are those of Dickerson and Lipscomb[49] and are capable of very considerable extension to other systems. The major principles are concerned with the ways in which various types of bonds are able to connect the framework B—H units that are adjacent to one another:

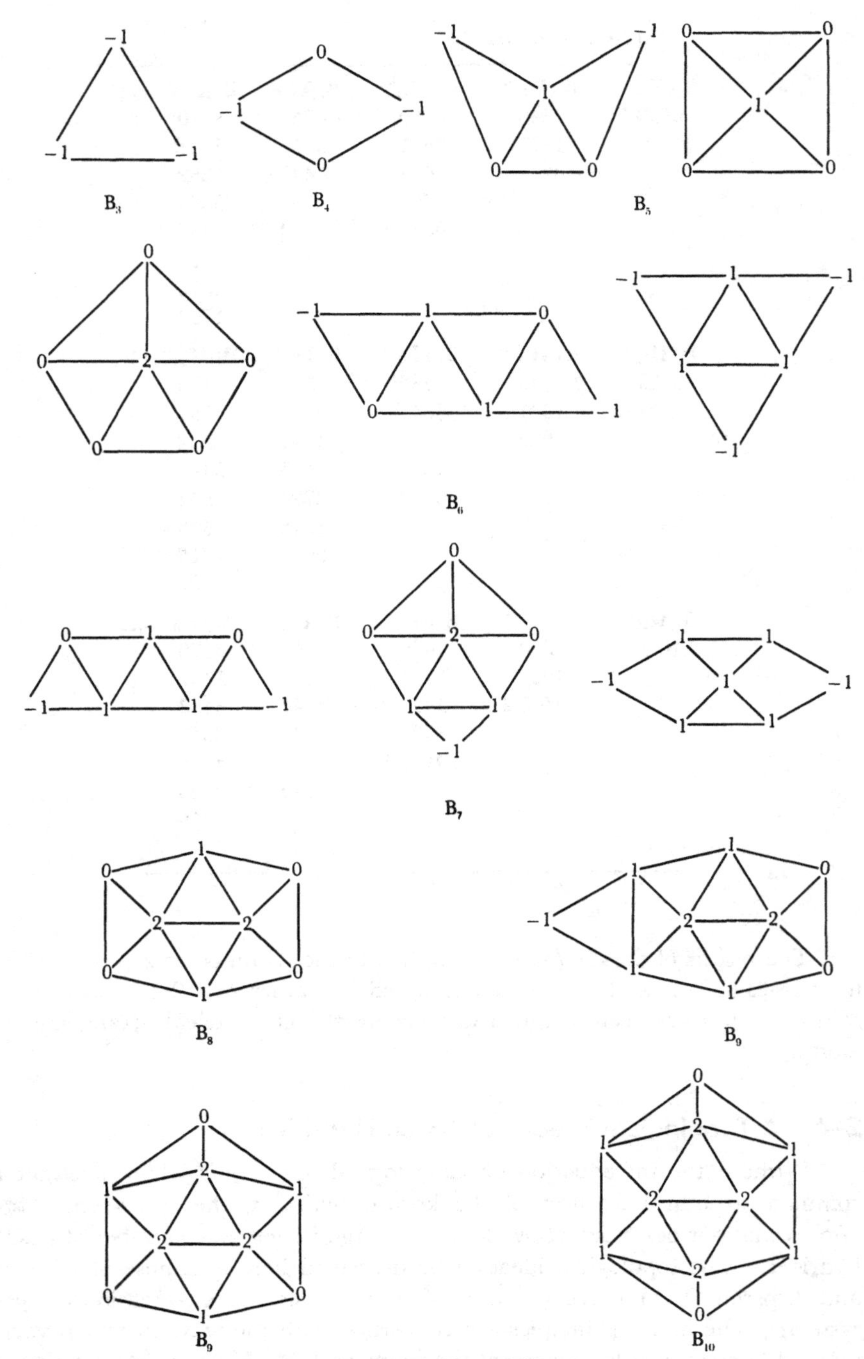


**Figure 2–12** *Excess connectivities of plausible boron frameworks.*

1. Assume that each boron must be connected to each adjacent boron atom by at least one two-center or three-center bond of some kind.

2. Define the connectivity of a boron atom as the number of adjacent boron atoms in the molecule. Define the excess connectivity (EC) as connectivity minus three.

3. Define the excess negative connectability of a bond arrangement at the site of a given boron atom as the number of other boron atoms that are connected to the given one by that bond arrangement. A bond arrangement may be several bonds considered as a unit.

Definition 2, applied to the more compact icosahedral and octahedral fragments to which our predictions are limited for convenience, is illustrated in Fig. 2–12. Definition 3, applied to various bonds and bond arrangements, leads to the excess negative connectabilities (ENC) shown in Fig. 2–13. Note that a B—B single bond or a B—H—B three-center bridge bond, when isolated from other bonding arrangements, connects two borons and uses one orbital from each so that addition of these bonds produces no change in the EC of a framework. Also note that if every boron were of connectivity 3, i.e., EC = 0, the three orbitals left after forming the B—H bond would suffice to form a framework of simple covalent bonds, provided enough electron pairs are available.

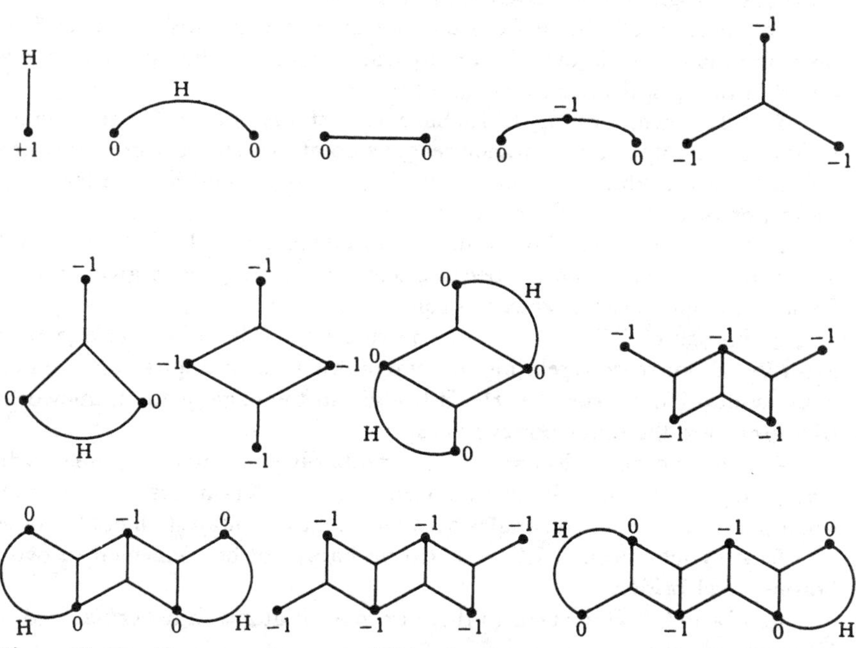

**Figure 2–13** *Excess negative connectabilities of various bonds. The extra B—H removes a boron orbital without connecting that B atom to any other B atom, and hence contributes +1 to this scheme.*

Hence, any EC $> 0$ must be removed by addition of three-center B—B—B bonds, and any EC $< 0$ must be removed by adding extra H's to form $BH_2$ groups. It is also permissible, as shown by the so-far unique example of a boron atom in $B_5H_{11}$, to add an extra H to a boron of EC $> -1$, thereby raising the EC of that boron atom by 1. Hence the procedure to be followed is:

1. Make a topological sketch of the assumed boron framework, indicating the EC at each boron site.

2. Add extra B—H bonds to every boron of EC $= -1$, and possibly to certain borons of EC $= 0$ or 1, but not to a boron of EC $= 2$.

3. Add three-center B—B—B bonds or arrangements of them, such as shown in Fig. 2–13, in such a way that the EC at every boron site is reduced to zero.

4. Complete the molecule with B—B and B—H—B bonds, subject to the equations of balance (Table 2–1).

It must be emphasized that this type of topological theory is not limited to three-center bonds. Its extension to other types of semilocalized multicentered molecular orbitals might well prove useful in an extensive study of valence in certain intermetallic compounds in which the nearest neighbors can easily be recognized, especially if the number of nearest neighbors is not too large. This type of theory is, moreover, not limited to finite molecules, single polyhedral fragments, or icosahedral fragments.

In order to eliminate from consideration certain structural features not present among the known boron hydrides, some further principles may be stated in order of decreasing reliability:

1. All known boron hydrides have at least a twofold element of symmetry. Low symmetry appears to provide reactive points which are averaged over when symmetry is introduced. A new hydride probably would have at least a plane, center, or twofold axis of symmetry.

2. No two adjacent boron atoms can be connected by both a central and an open three-center bond, because such an arrangement gives a very unfavorable angle between valence orbitals.

3. Borons of EC $= 2$ do not contribute to a B—H—B bridge or to an extra B—H, for reasons relating to unfavorable valence angles between orbitals. This assumption restricts B—H—B bridges to the edge of the framework, and $BH_2$ groups to the outer boron atoms.

4. Each boron of EC $= -1$ will probably be bonded by one hydrogen bridge to at least one of its two adjacent borons. Again this principle is based upon the requirement that angles between valence orbitals shall not be too small.

5. An edge boron of EC $= 1$ will probably not be connected to two other borons by H bridges.

6. If a $BH_2$ is connected to three or more B atoms, it never has two bridge H atoms joined to it.

7. The bridge H atom between two $BH_2$ groups, present in $B_2H_6$ and $B_3H_8$, is not found in the higher hydrides. This feature may become a steric

problem in a nonplanar B arrangement, and perhaps hydrides or ions with this feature would rearrange, when possible, to structures more stable on steric grounds. This is probably the weakest of our assumptions.

These further principles apply to bonding within a single polyhedral surface, and have, therefore, a slightly more complex form when two polyhedral surfaces meet, as they do in iso-$B_{18}H_{22}$. In each of the boron hydrides based upon a single polyhedral fragment there are two surfaces, the surface of the terminal H atoms, and a smaller, nearly spherical surface containing the B atoms, H bridges, and the extra H atoms of $BH_2$ groups. These structural principles are detailed formulations of the requirements that all H · · · H contacts must be greater than 2.0 A, and generally fill the inner surface with H · · · H contacts of about 2.2 to 2.6 A in regions not already occupied by B atoms.

## 2-7 Application to Icosahedral Fragments. Neutral Species

### $B_3$ Hydrides

Of the possible $B_3$ hydrides based upon a triangular arrangement, only the 3003 $B_3H_9$ structure is satisfactory (Fig. 2-14). As we shall see, this structure is related to the $B_3H_8^-$ structure, which may have unsymmetrical H bridges, by loss of $H^+$, leaving a single B—B bond in place of a former H bridge bond. The $B_3H_9$ molecule is suspected as an intermediate in the initial stage, $B_2H_6 + BH_3 \rightleftharpoons B_3H_9 \rightarrow B_3H_7 + H_2$, of the diborane pyrolysis, but it is not yet certain whether it is stable with respect to further decomposition to $B_3H_7$.

A 2102 $B_3H_7$ structure (Fig. 2-15) is not considered satisfactory because of unfavorable bond angles about the B atoms. All other $B_3$ structures are unsatisfactory. A suggested[227] linear structure for $B_3H_8^-$ has too many electron pairs, and in reality should be a possible $B_3H_8^+$ structure (Fig. 2-16), which, however, might readily lose a proton and undergo rearrangement and further reaction.

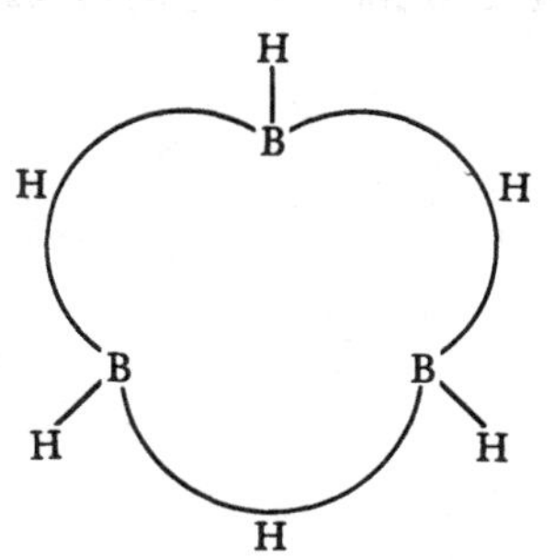


**Figure 2-14** *Hypothetical $B_3H_9$ structure.*

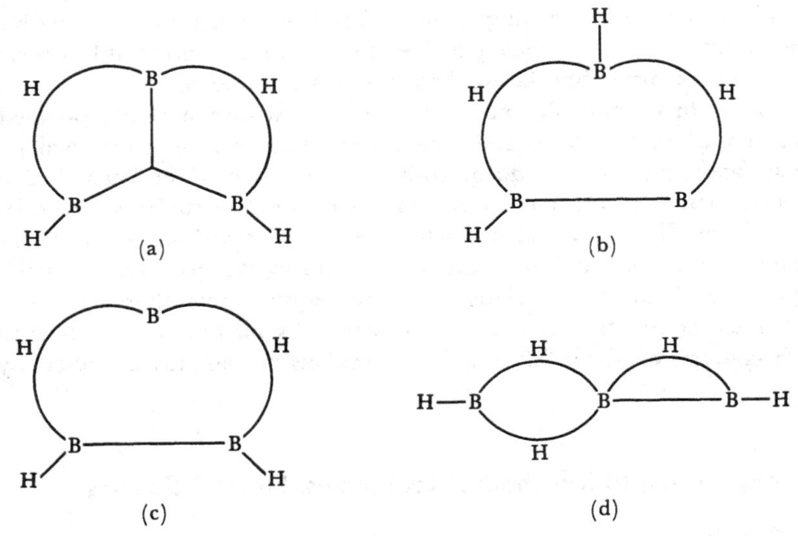


**Figure 2-15** *Unsatisfactory $B_3H_7$ structures.*

## $B_4$ Hydrides

Polyhedral hydrides and ions are discussed below, and hence we restrict this discussion to the $B_4$ arrangement in $B_4H_{10}$. There are two B atoms of excess negative connectability $-1$, and hence at least two $BH_2$ groups are required, and thus $x \geq 2$. The possibilities of 2202 $B_4H_8$, 4012 $B_4H_{10}$, and 3103 $B_4H_{10}$ (Fig. 2-17) leave only the known 4012 $B_4H_{10}$ as the only satisfactory structure if a plane of symmetry is required. In general, we shall not list low-symmetry possibilities, e.g., 3103 $B_4H_{10}$, but such structures may play some role as intermediates in intramolecular hydrogen rearrangements, indicated by tracer studies.[156] The relation between the 4012 and 3103 structures will be abstracted later as Fig. 2-29, so that these two structures will not be considered as distinct, and the left side will be favored in neutral molecules, whereas the right side will be favored in the negative ions.

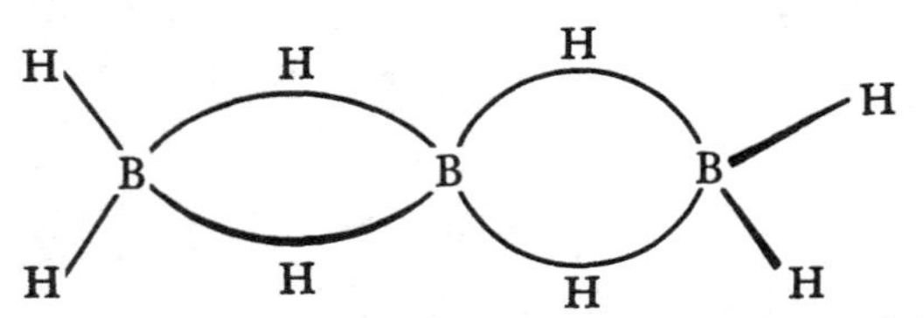


**Figure 2-16** *A structure for $B_3H_8^+$ analogous to $Be(BH_4)_2$.*

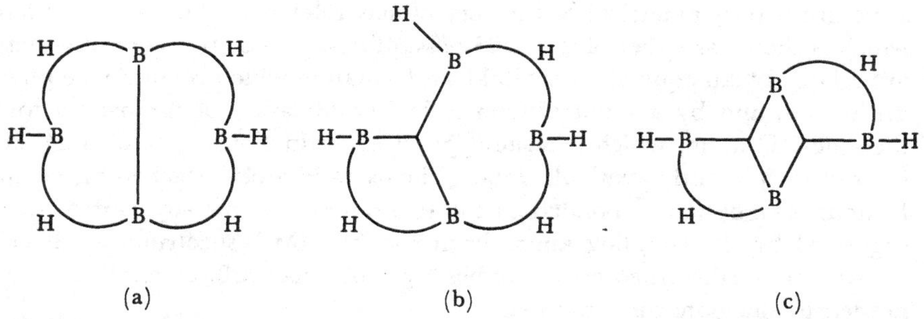

(a) (b) (c)

**Figure 2–17** *Possible structures of $B_4$ hydrides: (a) 4012 $B_4H_{10}$; (b) 3103 $B_4H_{10}$; (c) 2202 $B_4H_8$. The 4012 structure is by far the most satisfactory.*

### $B_5$ Hydrides

The two B atoms of excess negative connectability −1 require two $BH_2$ groups, so that $x \geq 2$. Also, not more than one bridge H atom may be joined to a B atom joined to four others, so that $s \leq 4$. The only possibilities are then 2302 $B_5H_9$, 4112 $B_5H_{11}$, 3203 $B_5H_{11}$, and 4104 $B_5H_{13}$ (Fig. 2–18). No reasonable formula seems possible for $B_5H_{13}$ without violating the rules based upon known compounds. The 2302 $B_5H_9$ structure would very likely be capable of collapsing easily to the tetragonal pyramidal $B_5H_9$ structure, accompanied by a simple rearrangement of the H atoms. The 4112 $B_5H_{11}$

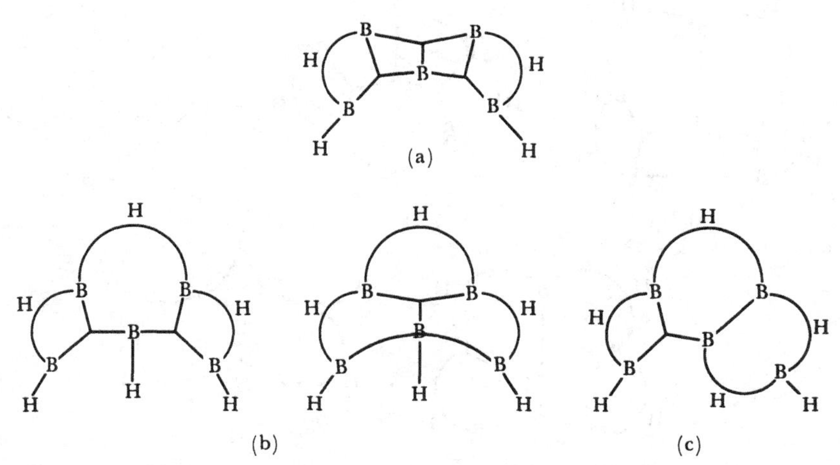


**Figure 2–18** *Possible $B_5$ hydrides. The $B_5H_9$ structure (a) may collapse to the known tetragonal pyramidal $B_5H_9$. The 4112 $B_5H_{11}$ (c) structure may be only an unstable intermediate in reactions. (b) is the 3203 $B_5H_{11}$ structure.*

structure is unsymmetrical but is very closely related to the correct structure, which is shown as either of the 3203 possibilities. These two correct structures are related by an equivalent orbital transformation which is not a true physical distinction, and by a slight difference in hybridization of the orbital toward the apical H atom, which is a slight physical distinction not resolvable on the basis of the presently available data. The extra H atom attached to the apical B atom may be partly bonded to the two outer B atoms adjacent to it, as is suggested by the coupling anomaly in the $B^{11}$ NMR spectrum, as described below, but further theoretical studies by molecular orbital methods may be needed to elucidate this situation.

### $B_6$ Hydrides

Here, two possible boron frameworks may be considered (Fig. 2–19). The more compact one, the icosahedral cap, has more B · · · B contacts and has a plane of symmetry, and hence may be regarded as more stable. The central B atom of excess negative connectability 2 requires two three-center bonds, so that $t \geq 2$. The bonding geometry of the central atom renders it impossible to have $t > 3$, so that $t = 2$ or 3, and if $t = 2$ there must also be at least one single B—B bond ($y \geq 1$) to bond the remaining orbital of the central B atom to the outer B atoms. The possibilities are therefore 4220 $B_6H_{10}$, 3311 $B_6H_{10}$, 4212 $B_6H_{12}$, and 3303 $B_6H_{12}$, as shown in Fig. 2–19. As we shall

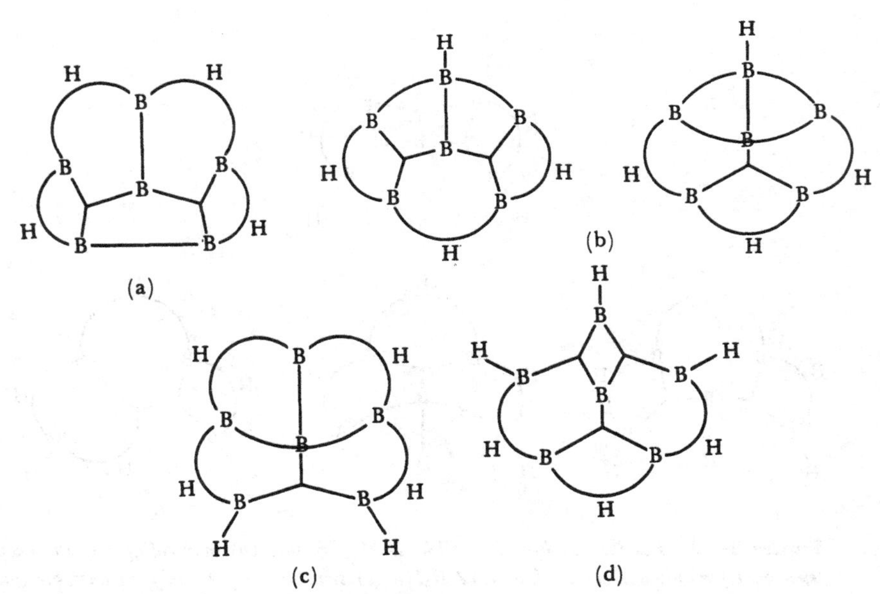


**Figure 2–19** *The $B_6$ hydrides based upon an icosahedral cap of B atoms. (a) 4220 $B_6H_{10}$; (b) 3311 $B_6H_{10}$; (c) 4212 $B_6H_{12}$; (d) 3303 $B_6H_{12}$. Note steric problems in (c) and (d).*

see, the 3311 $B_6H_{10}$ structure may be an intermediate in the hydrogen rearrangement of the 4220 structure. The compound $B_6H_{12}$ has not yet been isolated in pure form, but evidence concerning it is discussed on page 12.

Other possible structures for $B_6$ hydrides based upon the open frameworks shown in Fig. 2–19 may be derived. For the threefold symmetrical arrangement of $B_6$ in Fig. 2–12, no reasonable possible structures have been derived. The most obvious 6030 structure of $C_{3v}$ symmetry violates the valence rules. On the other hand, the even-less-compact arrangement (c) for $B_6$ yields the three possible structures shown in Fig. 2–20. In arrangement (a) there are two boron atoms of connectability −1, and therefore rather exposed, outer atoms. So far these have always been $BH_2$ groups, and hence we assign them as such to then give these atoms connectability zero. Such extra B—H's may also be assigned to atoms of connectability 0 or 1, in which case these atoms then become of connectivity 1 or 2. Then bonds that use fewer orbitals from a given boron than the number of immediate neighbors connected are employed; these bonds are placed so that the connectabilities of all atoms are reduced to zero. Then the remaining bonds, single B—B bonds and bridge hydrogens which leave the connectivities unchanged, are introduced, so that one set of *styx* numbers is satisfied.

Now in arrangement (c) of $B_6$ (Fig. 2–19) let us introduce two extra B—H's to obtain two $BH_2$ groups. This means that $x \geq 2$. In addition let us retain the twofold axis of the molecule. Hence $s$ and $x$ are even only. We are im-

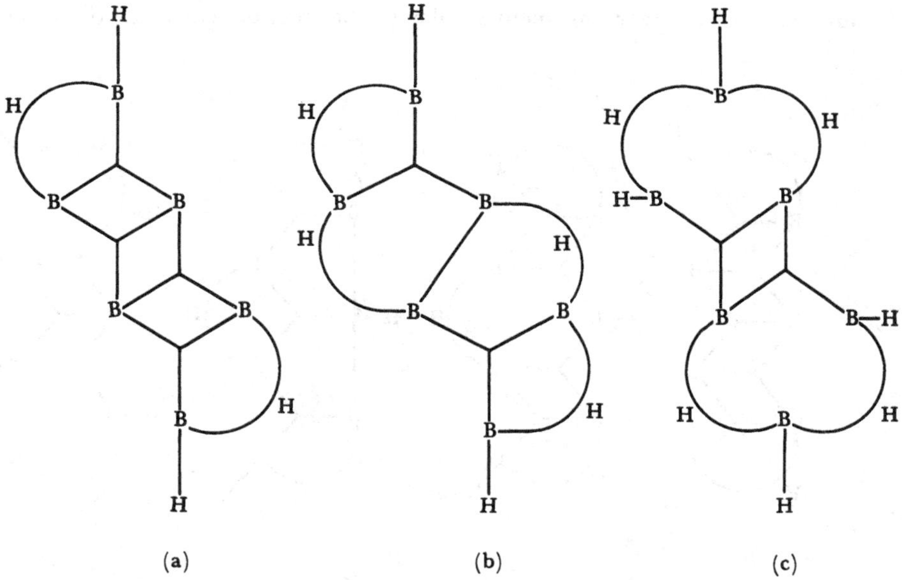

(a) (b) (c)

**Figure 2–20** *$B_6$ hydride formulas based upon other B arrangements. (a) 2402 $B_6H_{10}$; (b) 4212 $B_6H_{12}$; (c) 4204 $B_6H_{14}$.*

mediately reduced to the possibilities 2402, 4212, 6022, 4204, and 6014. Only one 6014 structure and only one 6022 structure can be sketched, and both are unsatisfactory in that they fail to connect $B_1$ with $B_1$ (Fig. 2-19a) for 6022, and $B_1$ with $B_0$ for 6014. The 2402 $B_6H_{10}$ possibility of Fig. 2-20 is also unique, because each $BH_2$ group must be connected by at least one H bridge. Also, the one single bond of the 4212 $B_6H_{12}$ structure of Fig. 2-20 must be at the molecular center, thus fixing the remaining bonds. Finally, the 4204 $B_6H_{14}$ structure uniquely requires the two three-center bonds connecting $B_1$ atoms, thus fixing its geometry. We leave open the question of their relative stabilities, but add that there is an already predicted $B_6H_{12}$ structure based upon the relatively compact pentagonal pyramidal arrangement of B—H groups known[114] to exist in $B_6H_{10}$.

#### $B_7$ Hydrides

The procedure is so nearly the same for a similar study of the possible $B_7$ hydrides that only the results are summarized (Fig. 2-21). No satisfactory structures were obtained from the second arrangement shown in Fig. 2-12. The general arguments leading to the structures shown also lead quite easily to the restrictions $x \geq 2$, $s \geq 2$, and $1 \leq t \leq 4$ at the earlier stages of the argument,[49] not given in detail here.

#### $B_8$ Hydrides

Again we summarize the results (Fig. 2-22). The three $B_8H_{12}$ structures are related by a rearrangement similar to that mentioned in the discussion of

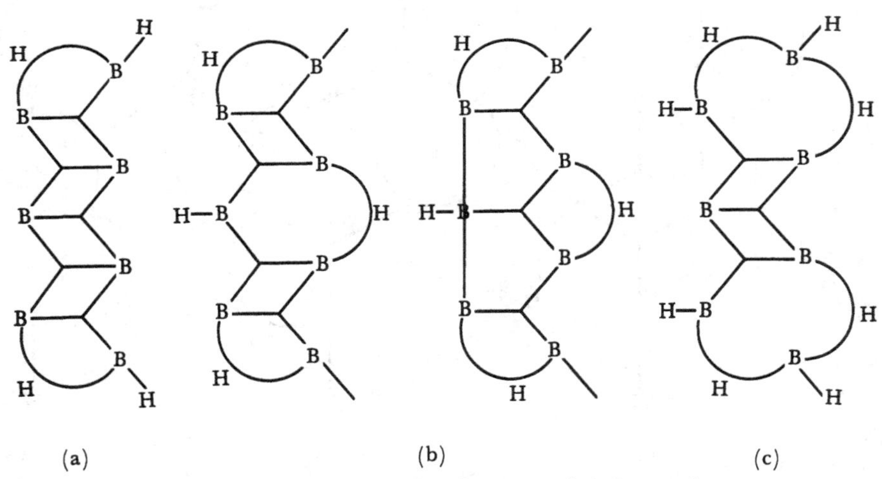


**Figure 2-21** *Possible $B_7$ hydrides, all based upon a relatively open boron arrangement and hence probably unstable.* (*a*) *2502 $B_7H_{11}$;* (*b*) *3403 $B_7H_{13}$;* (*c*) *4304 $B_7H_{15}$.*

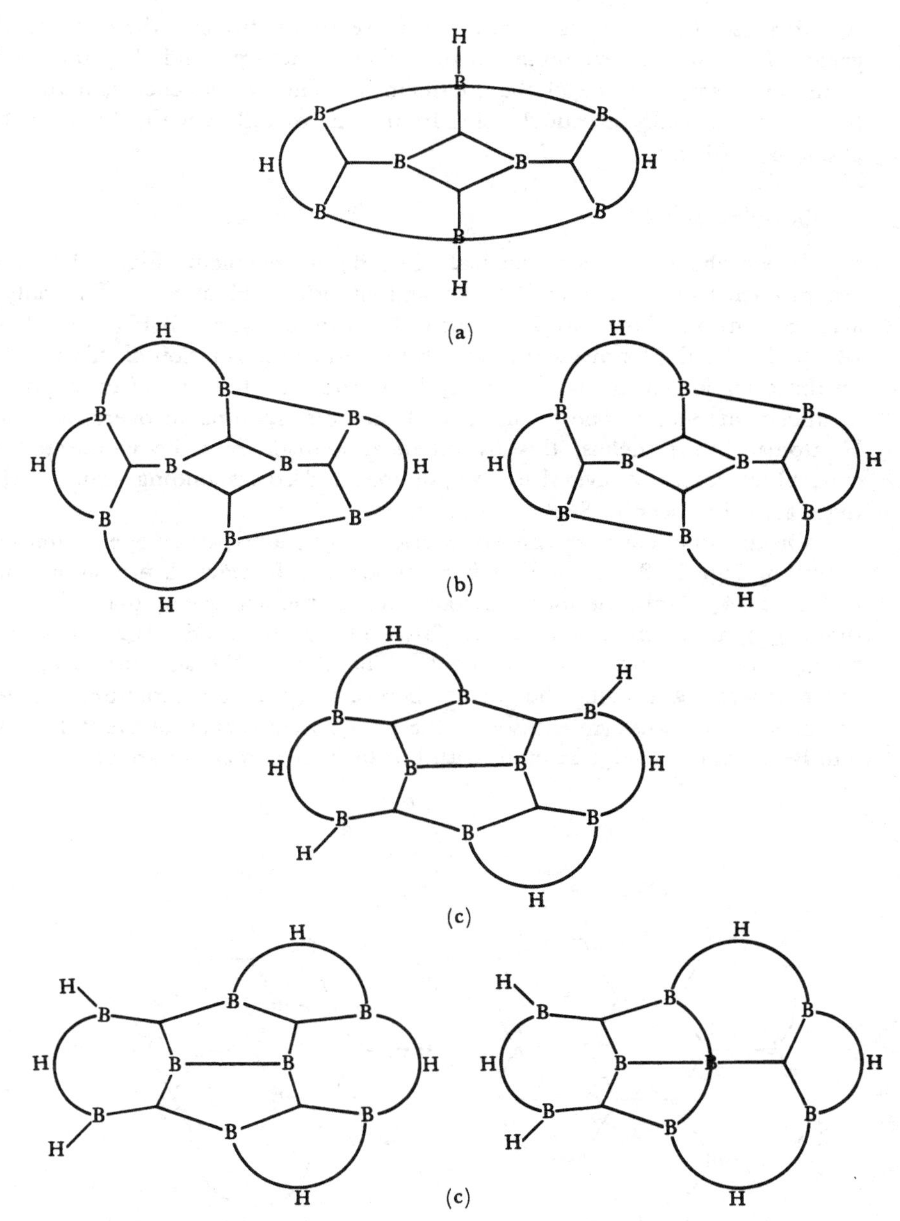


**Figure 2-22** *Possible $B_8$ hydrides. These structures for $B_8H_{12}$ are all related by the tautomeric shift shown in Fig. 2-29. (a) 2602 $B_8H_{12}$; (b) 4420 $B_8H_{12}$; (c) 4412 $B_8H_{14}$.*

$B_6$ hydrides. Later we shall collect these relations, because their recognition greatly facilitates the examination of various structures, and they may occur as tautomerisms in some of the compounds. The two valence structures for $B_8H_{14}$ are essentially identical, aside from a minor difference in hybridization of one B—H bond.

### $B_9$ Hydrides

If one chooses the symmetrical ($C_{3v}$) $B_9$ arrangement (Fig. 2–12), there are problems associated with placement of bridge H atoms. The only H arrangement and bond arrangement of $C_{3v}$ symmetry is a $B_9H_{15}$ (Fig. 2–23), which violates the topological rules. Although transformation of this structure by the rules shown in Fig. 2–23 may be written, all but one of these possible arrangements violate one or another of the rules relating to overcrowding of H atoms. Nevertheless, this hypothetical hydride may be considered the conceptual parent of several $B_9$ hydride ions and corresponding amine derivatives, as will be seen in Section 5–14.

On the other hand, the known $B_9H_{15}$ has the slightly more open framework shown in Fig. 2–12. Structures based upon this framework are summarized in Fig. 2–24. Only the most obvious parts of the argument, that $x \geq 1$ and that $t \geq 4$, are mentioned here. So far, we have not found a $B_9H_{17}$ possibility based upon an icosahedral fragment. The three 5421 structures represent different ways of drawing the valence bonds, and the 3605 structure may well be expected to transform easily into the 5421 by conversion of the $BH_2$ groups into B—H plus a bridge H atom with but little movement of atoms.

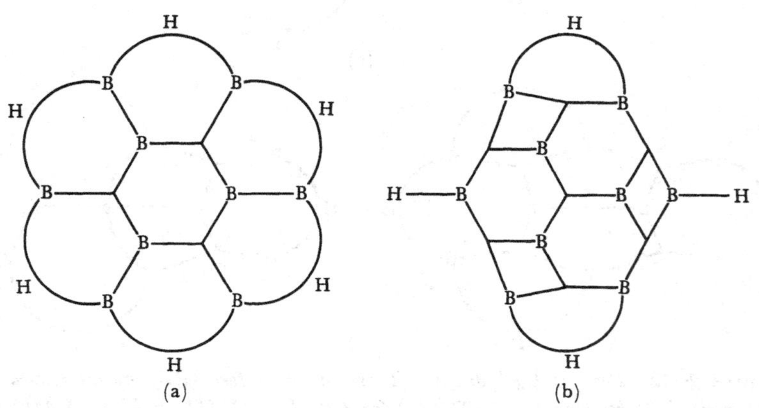


**Figure 2–23** *(a) A $B_9H_{15}$ structure which violates topological rule 5, but which forms the basis for several of the $B_9$ derivatives. (b) A $B_9H_{13}$ structure (2702) which is satisfactory. Dimerization by the replacement of one or two terminal B—H bonds, or further polymerization would seem to be a reasonable expectation.*

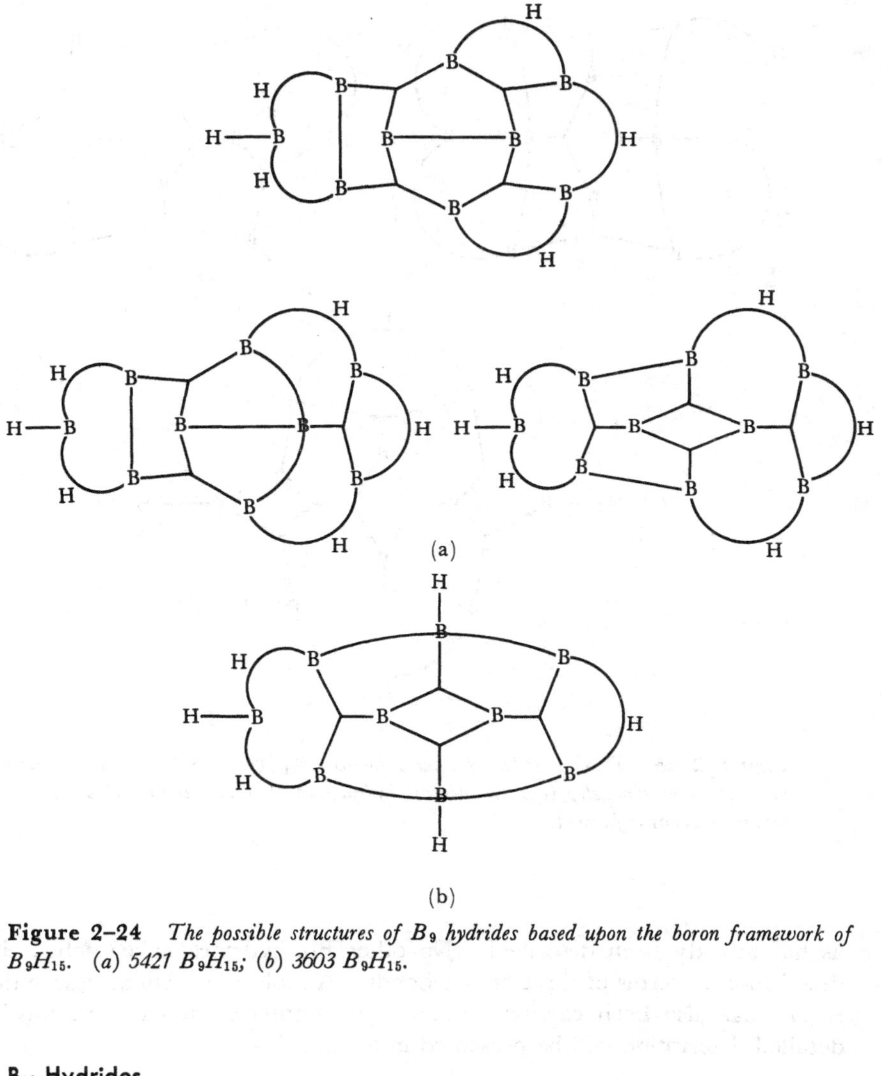

(a)

(b)

**Figure 2-24** *The possible structures of $B_9$ hydrides based upon the boron framework of $B_9H_{15}$. (a) 5421 $B_9H_{15}$; (b) 3603 $B_9H_{15}$.*

### $B_{10}$ Hydrides

Only the $B_{10}$ fragment shown in Fig. 2-12 seems to be a satisfactory choice for a $B_{10}$ hydride easily described in terms of three-center bonds. The known $B_{10}H_{16}$ formed by joining two $B_5H_8$ groups from $B_5H_9$ at the apical atoms is described in Section 1-1, and the hypothetical $B_{10}H_{16}$ formed geometrically from $B_{12}H_{12}^{=}$ by removing opposite apical $BH^-$ units and substituting $H_3$ groups is best described in terms of molecular orbitals (Chapter 3).

Only two H-atom arrangements seem reasonable for $B_{10}H_{14}$ (Fig. 2-25). The known structure is that on the left-hand side and has four bridge H atoms,

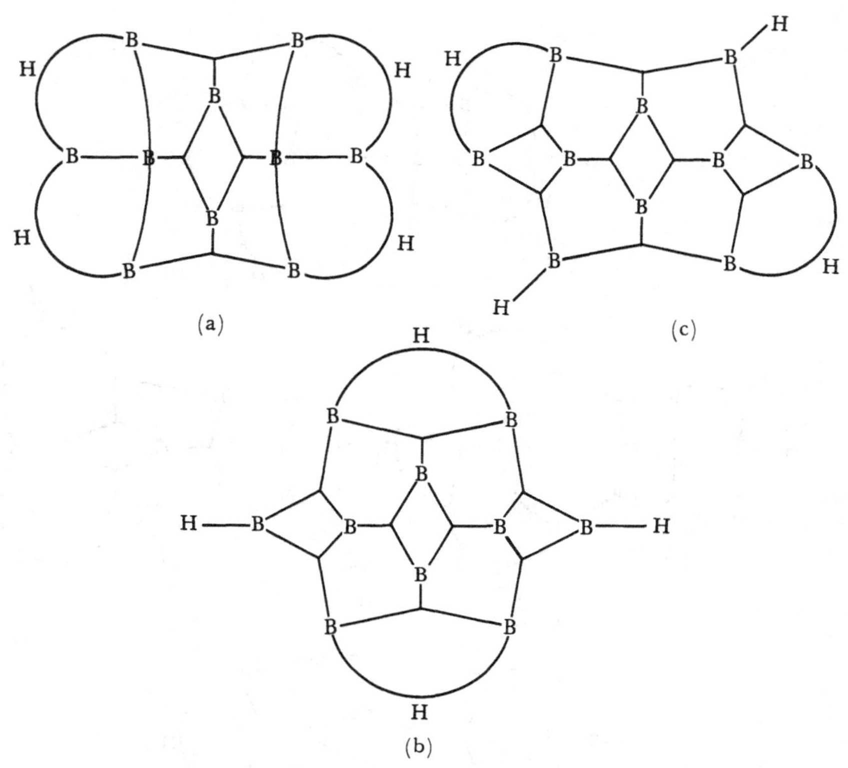


**Figure 2–25** *Two plausible structures for $B_{10}H_{14}$, (a) and (b), and an intermediate form of lower symmetry (c). In neutral hydrides the structure with the larger number of bridge H atoms is favored.*

as has already been described. No other $B_{10}$ hydrides have such a simple description in terms of three-center bonds. A molecular orbital description of $B_{10}H_{14}$ has also been carried out, and the results obtained from this more detailed description will be presented in Section 3–4.

## 2–8 Generalization and Resonance Structures of Neutral Compounds

### B Atoms without Terminal H

For $B_{10}H_{16}$, $B_{18}H_{22}$, and iso-$B_{18}H_{22}$ an extension of the orbital–electron–H atom balances is required. Write the hydride formula in the form $B_rB_pH^c_{p+q+c}$, where there are $r$ B atoms without terminal H atoms, and the charge is $c$ units. Then the orbital, electron, and H-atom balances are[179]

$$s + x = q + c$$

$$s + t = p + r + c$$

$$t + y = p + \frac{3}{2}r - c - \frac{q}{2}$$

where $s$ = B—H—B, $t$ = B—B—B, $y$ = B—B, and $x$ = extra H on B—H (i.e., $BH_2$) as defined previously. For $B_{10}H_{16}$, $r = 2$, $p = 8$, $q = 8$, so that $s = 8$, $t = 2$, $y = 5$, and $x = 0$.

#### Resonance in the Boron Hydrides

The three-center formalism has only recently[123] been explored in its relatively complete form, in which all resonance structures are written and a hybrid of these is taken as the basis for the charge distribution. Heretofore, we have written only the most symmetrical of the valence structures. In $B_{10}H_{14}$, for example, there is only one such structure, and it has the chemically disturbing feature that the 6,9 boron atoms (Fig. 2–26) are not the most positive. In principle there is certainly no objection to valence structures of lower symmetry than that of the nuclear framework, provided that a resonance hybrid is formed of all possible such structures. The problem of the $B_5H_9$ orbital structure makes this point clear: In the three-center approximation no structure having the molecular symmetry $C_{4v}$ can be written, and hence a hybrid of $C_{2v}$ valence structures must be used (Fig. 2–11).

We now distinguish between resonance and equivalent structures. The former, as for $B_5H_9$, are a set of symmetry equivalent structures which, when considered as a group, have the molecular symmetry. The latter, as the two usual structures drawn for $B_5H_{11}$, are distinct ways of showing relative arrangements of single bonds and three-center bonds in the boron framework. Although symmetry demands that all the resonance structures of a given equivalent form enter the wave function or the population analysis with equal coefficients, there are no such symmetry requirements on different equivalent forms. The fact that they are weighted equally in the subsequent discussion only indicates the symmetry of our ignorance.

The problem, then is to find *all* the equivalent and resonance structures of a molecule. Experience in this matter has shown that the human mind is both a fallible and extremely efficient mechanism. We have often thought that we had all the equivalent structures, only to find a week, or a year, later that we had missed some very obvious ones. The listing of boron hydride structures in Table 2–4 is, however, probably complete, inasmuch as we have finally succeeded in convincing a computer to exhaust all possible structures. However, we have noted that omission of a few structures does not affect greatly the ordering in the population analysis.

Figure 2–26 shows some of the boron hydride frameworks. The various structures derived for the neutral boron hydrides are listed in Table 2–4. Column 3 of Table 2–4 contains the number of resonance forms belonging to

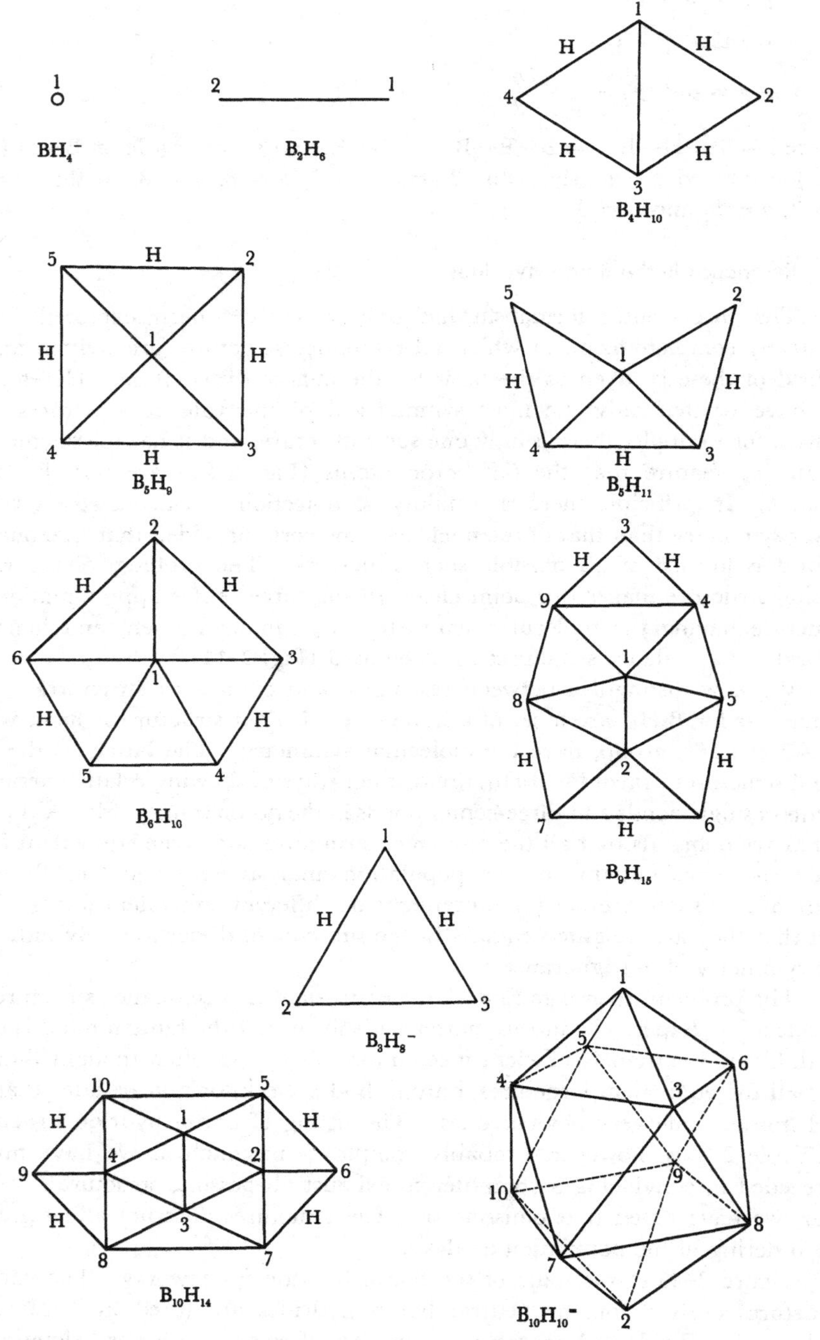


**Figure 2–26** *Numbering convention*[38, 39] *for the boron hydrides.*

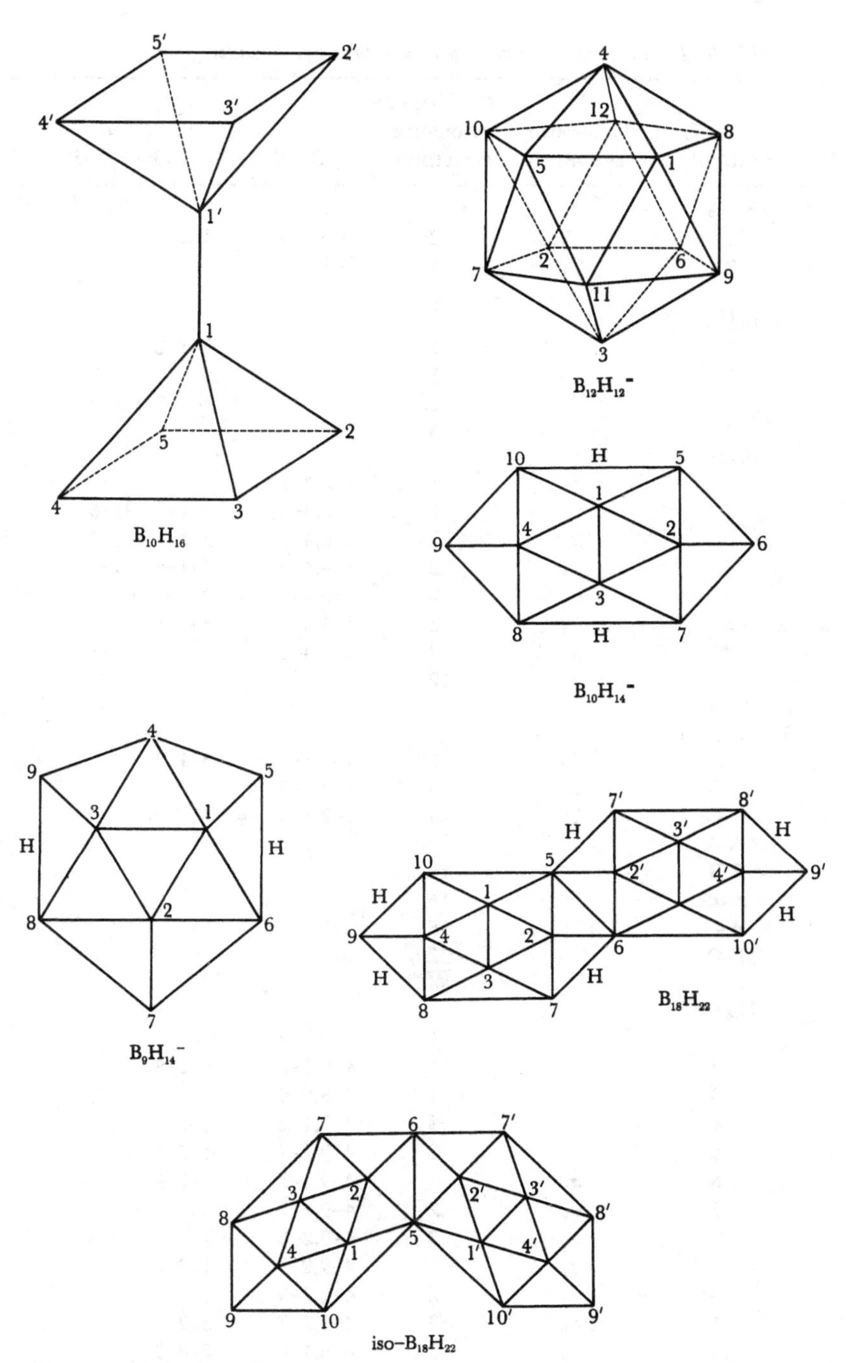


**Figure 2–26** (*continued*)

**Table 2-4** *Three-center bond structures in the boron hydrides*

| Molecule | Generating symmetry | Number of resonance structures | B—B | B⌒B (B over arc) |
|---|---|---|---|---|
| $B_5H_9$ | | | | |
| 1 | $C_2$ | 2 | 1–2,1–4 | 3–1–5 |
| 2 | $C_4$ | 4 | 1–2,1–3 | |
| | | 6 | | |
| $B_5H_{11}$ | | | | |
| 1 | | 1 | | 5–1–2 |
| 2 | | 1 | | |
| 3 | | 1 | | 5–1–3,2–1–4 |
| | | 3 | | |
| $B_6H_{10}$ | | | | |
| 1 | | 1 | 1–2,4–5 | |
| 2 | | 1 | 1–2,4–5 | 3–1–5,4–1–6 |
| 3 | $\sigma_v$ | 2 | 1–6,4–5 | 2–1–4,3–1–5 |
| 4 | $\sigma_v$ | 2 | 1–4,4–5 | 2–1–5,3–1–6 |
| 5 | $\sigma_v$ | 2 | 1–6,4–5 | 2–1–5 |
| 6 | $\sigma_v$ | 2 | 1–4,4–5 | 3–1–5 |
| 7 | $\sigma_v$ | 2 | 1–4,4–5 | |
| | | 12 | | |
| $B_{10}H_{16}$ | | | | |
| Half-structures | | | | |
| A | $C_2$ | 2 | 1–2,1–4 | 3–1–5 |
| B | $C_4$ | 4 | 1–2,1–3 | |
| C | $C_4$ | 4 | 1–2,1–3,1–4 | 1′–1–5 |
| A-A | | 4 | | |
| B-B | | 16 | | |
| A-B | | 16 | | |
| A-C | | 16 | | |
| B-C | | 32 | | |
| | | 84 | | |
| $B_9H_{15}$ | | | | |
| 1 | | 1 | 4–9,1–2 | |
| 2 | | 1 | 4–9,1–2 | 5–2–8 |
| 3 | | 1 | 4–5,8–9 | |
| 4 | | 1 | 4–9,1–2 | 5–2–7,6–2–8 |
| 5 | $\sigma_v$ | 2 | 4–9,2–7 | 1–2–6,5–2–8 |
| 6 | $\sigma_v$ | 2 | 4–9,2–7 | 6–2–8 |
| 7 | $\sigma_v$ | 2 | 4–9,2–7 | |
| 8 | $\sigma_v$ | 2 | 4–9,2–5 | 1–2–7,6–2–8 |
| 9 | $\sigma_v$ | 2 | 4–9,2–5 | 1–2–6 |
| 10 | $\sigma_v$ | 2 | 4–9,2–5 | |
| 11 | $\sigma_v$ | 2 | 4–5,2–7 | 6–2–8 |
| 12 | $\sigma_v$ | 2 | 4–5,1–8 | 2–8–9 |
| 13 | $\sigma_v$ | 2 | 1–4,2–7 | 9–4–5,6–2–8 |
| 14 | $\sigma_v$ | 2 | 1–4,2–8 | 9–4–5 |
| 15 | $\sigma_v$ | 2 | 1–4,2–6 | 9–4–5 |
| | | 26 | | |

**Table 2-4** (*continued*)

| Molecule | Generating symmetry | Number of resonance structures | B—B | B⌒B (B bridge) |
|---|---|---|---|---|
| $B_{10}H_{14}$ | | | | |
| 1 | | 1 | 4–9,2–6 | 5–2–7,10–4–8 |
| 2 | $\sigma_v'$ | 2 | 4–9,2–6 | 5–2–3,7–2–1,10–4–8 |
| 3 | $\sigma_v$ | 2 | 2–5,4–10 | |
| 4 | $\sigma_v$ | 2 | 2–7,4–10 | |
| 5 | $\sigma_v$ | 2 | 1–3,5–10 | 2–3–4 |
| 6 | $\sigma_v, \sigma_v'$ | 4 | 4–9,5–10 | |
| 7 | $\sigma_v, \sigma_v'$ | 4 | 4–9,5–10 | 1–3–7,3–2–6,2–7–8 |
| 8 | $\sigma_v, \sigma_v'$ | 4 | 4–9,1–5 | 10–4–8,1–2–6 |
| 9 | $\sigma_v, \sigma_v'$ | 4 | 4–9,1–5 | 10–4–8,1–2–7,3–2–6 |
| 10 | $\sigma_v, \sigma_v'$ | 4 | 4–9,1–5 | 10–4–8 |
| 11 | $\sigma_v, \sigma_v'$ | 4 | 4–9,1–2 | 10–4–8,5–2–7,3–2–6 |
| 12 | $\sigma_v, \sigma_v'$ | 4 | 4–9,1–2 | 10–4–8,5–2–3 |
| 13 | $\sigma_v, \sigma_v'$ | 4 | 4–9,1–2 | 10–4–8 |
| 14 | $\sigma_v, \sigma_v'$ | 4 | 4–9,1–5 | 10–5–2 |
| 15 | $\sigma_v, \sigma_v'$ | 4 | 4–9,2–5 | |
| 16 | $\sigma_v, \sigma_v'$ | 4 | 4–9,2–5 | 10–4–3,1–4–8 |
| 17 | $\sigma_v, \sigma_v'$ | 4 | 10–4,2–5 | 1–4–8,3–4–9 |
| 18 | $\sigma_v, \sigma_v'$ | 4 | 10–4,2–5 | 1–4–9 |
| 19 | $\sigma_v, \sigma_v'$ | 4 | 10–4,2–7 | 1–4–8,3–4–9 |
| 20 | $\sigma_v, \sigma_v'$ | 4 | 10–4,2–7 | 1–4–9 |
| 21 | $\sigma_v, \sigma_v'$ | 4 | 10–4,3–7 | 1–4–9,2–7–8 |
| 22 | $\sigma_v, \sigma_v'$ | 4 | 10–4,7–8 | 1–4–9 |
| 23 | $\sigma_v, \sigma_v'$ | 4 | 10–4,1–2 | 5–2–7,3–2–6 |
| 24 | $\sigma_v, \sigma_v'$ | 4 | 10–4,1–2 | 3–2–5,1–2–6 |
| 25 | $\sigma_v, \sigma_v'$ | 4 | 10–4,1–2 | |
| 26 | $\sigma_v, \sigma_v'$ | 4 | 10–4,2–3 | 5–2–7,1–2–6 |
| 27 | $\sigma_v, \sigma_v'$ | 4 | 10–4,2–3 | |
| 28 | $\sigma_v, \sigma_v'$ | 4 | 10–4,2–3 | 1–2–7 |
| 29 | $\sigma_v, \sigma_v'$ | 4 | 1–4,3–7 | 2–7–8 |
| 30 | $\sigma_v, \sigma_v'$ | 4 | 1–4,7–8 | |
| | | 109 | | |
| Additional $B_9H_{15}$ structures guessed by computer | | | | |
| 16 | $\sigma_v$ | 2 | 2–5,8–9 | |
| 17 | $\sigma_v$ | 2 | 2–7,8–9 | |
| | New total = | 30 | | |
| Additional $B_{10}H_{14}$ structures guessed by computer | | | | |
| 31 | $\sigma_v'$ | 2 | 2–6,4–9 | 10–4–8 |
| | New total = | 111 | | |

**Table 2-5** *Three-center charge distributions*

| Molecule | All structures | "Good" structures* | Symmetric structures | Most symmetric |
|---|---|---|---|---|
| $B_{10}H_{14}$ | 109 | 79 | 7 | 1 |
| 1,3 | −0.03 | −0.07 | −0.14 | 0 |
| 2,4 | −0.45 | −0.36 | −0.24 | −0.67 |
| 6,9 | 0.29 | 0.29 | 0.29 | 0 |
| 5,7,8,10 | 0.10 | 0.07 | 1.05 | 0.33 |
| $B_9H_{15}$ | 26 | 21 | 3 | |
| 1 | −0.10 | −0.11 | −0.22 | |
| 2 | −0.53 | −0.48 | −0.33 | |
| 3 | 0 | 0 | 0 | |
| 4,9 | −0.12 | −0.11 | −0.17 | |
| 5,8 | 0.12 | 0.09 | 0.11 | |
| 6,7 | 0.31 | 0.32 | 0.33 | |
| $B_6H_{10}$ | 12 | 9 | 1 | |
| 1 | −0.72 | −0.63 | −0.33 | |
| 2 | 0.36 | 0.37 | 0 | |
| 3,6 | 0.35 | 0.30 | 0.33 | |
| 4,5 | −0.17 | −0.17 | −0.17 | |
| $B_5H_{11}$ | 3 | 2 | | |
| 1 | −0.67 | −0.50 | | |
| 2,5 | −0.06 | −0.08 | | |
| 3,4 | 0.39 | 0.33 | | |
| $B_5H_9$ | 6 | | | 2 |
| 1 | −0.78 | | | −1.00 |
| 2,3,4,5 | 0.20 | | | 0.25 |
| $B_{10}H_{16}$ | 84 | | | |
| 1,1′ | −0.70 | | | |
| 2–5,2′–5′ | −0.18 | | | |
| $B_4H_{10}$ | 1 | | | |
| 1,3 | 0 | | | |
| 2,4 | 0 | | | |
| $B_2H_6$ | 1 | | | |
| 1,2 | 0 | | | |

* Structures with two open three-center bonds crossing eliminated.

**Table 2–6** *Three-center bond orders*

| Bond | Distance | All structures | "Good" structures* | Symmetric structures | Most symmetric |
|---|---|---|---|---|---|
| $B_{10}H_{14}$ | | 109 | 79 | 7 | 1 |
| 5–10 | 2.01 | 0.70 | 0.71 | 0.71 | 0.67 |
| 1–2 | 1.79 | 0.73 | 0.74 | 0.67 | 0.67 |
| 1–5 | 1.78 | 0.72 | 0.74 | 0.67 | 0.67 |
| 2–5 | 1.78 | 0.75 | 0.74 | 0.77 | 0.71 |
| 2–6 | 1.72 | 0.75 | 0.74 | 0.71 | 1.00 |
| 1–3 | 1.71 | 0.73 | 0.75 | 1.24 | 1.33 |
| 5–H–6 | 1.77 | 0.69 | 0.72 | 0.64 | |
| $B_6H_{10}$ | | 12 | 7 | 1 | |
| 1–4 | 1.80 | 0.77 | 0.77 | 0.67 | |
| 1–3 | 1.75 | 0.60 | 0.72 | 0.67 | |
| 1–2 | 1.74 | 0.63 | 0.73 | 1.00 | |
| 4–5 | 1.60 | 1.00 | 1.00 | 1.00 | |
| 2–H–6 | 1.74 | 0.61 | 0.69 | 0.50 | |
| 3–H–4 | 1.74 | 0.67 | 0.79 | 1.17 | |
| $B_{10}H_{16}$ | | 84 | | | |
| 1–2 | 1.74 | 0.86 | | | |
| 1–1′ | 1.66 | 0.83 | | | |
| 2–H–3 | 1.77 | 0.58 | | | |
| $B_9H_{15}$ | | 26 | | | |
| 4–5 | 1.98 | 0.71 | | | |
| 2–5 | 1.83 | 0.73 | | | |
| 1–4 | 1.83 | 0.73 | | | |
| 4–9 | 1.80 | 0.87 | | | |
| 1–5 | 1.77 | 0.76 | | | |
| 1–2 | 1.77 | 0.74 | | | |
| 2–6 | 1.76 | 0.76 | | | |
| 5–H–6 | 1.82 | 0.60 | | | |
| 5–H–7 | 1.78 | 0.71 | | | |
| 3–H–4 | 1.76 | 0.50 | | | |
| $B_5H_9$ | | 6 | | | |
| 1–2 | 1.66 | 0.84 | | | |
| 2–H–3 | 1.77 | 0.61 | | | |
| $B_5H_{11}$ | | 3 | | | |
| 1–2 | 1.87 | 0.69 | | | |
| 1–3 | 1.72 | 0.68 | | | |
| 3–H–4 | 1.77 | 0.72 | | | |
| 2–H–3 | 1.76 | 0.89 | | | |
| $B_2H_6$ | | 1 | | | |
| 1–H–2 | 1.77 | 1.00 | | | |

Table 2–6 (*continued*)

| | | |
|---|---|---|
| $B_4H_{10}$ | | 1 |
| 1–3 | 1.71 | 1.00 |
| 1–H–2 | 1.84 | 0.50 |

* Structures with two open three-center bonds crossing eliminated.

the particular equivalent structure. Only one of the resonance forms is described; the others may be generated by applying the symmetry operations of column 2 to the listed form. To simplify the compilation, only the position of single bonds and open three-center bonds are given. This procedure is sufficient since the central three-center bonds can be filled in uniquely to complete the particular valence pattern of the molecule. $B_2H_6$ and $B_4H_{10}$ allow only one structure each and are therefore absent from the listing.

These structures contain some in which two open three-center bonds cross and a single bond originates at the crossing. Such structures, although not explicitly barred, were not considered in the first three-center bond treatments, because the necessary hybridization at the central boron seemed difficult to achieve. Although less favorable, these structures are not impossible; to see their effect the population analysis has been performed with (All structures) and without them ("Good" structures).

Bond orders are defined[120] simply as $2C_iC_j$, summed over all orbitals. There are thus four possible component bond orders: $\frac{1}{2}$ for two borons bonded by an H bridge, 1 for a normal single bond, $\frac{2}{3}$ for two borons bonded by a central three-center bond, and $\frac{1}{2}$ for an open B—B—B three-center bond.

In Table 2–5 we compare charge distributions obtained from analyzing all three-center structures with those obtained from some limited subset. In Table 2–6 we see the three-center bond orders, together with a comparison of various subsets for $B_6H_{10}$ and $B_{10}H_{14}$.

For $B_{18}H_{22}$ we show in Fig. 2–27 a valence structure closely related to the most symmetrical valence structure of $B_{10}H_{14}$. There are 6506 known resonance structures for $B_{18}H_{22}$. However, a simplified approach is taken here as follows.

A cursory examination of the bonding situation about the common edge, 5–6, shows that the only centrosymmetric structures are those in Fig. 2–28; (a) leads to 6.12.4.0 $B_{18}H_{22}$ structures, while (b) yields only 6.14.1.0 $B_{18}H_{22}^{++}$ and 6.10.7.0 $B_{18}H_{22}^{=}$. Thus all the centrosymmetric $B_{18}H_{22}$ graphs may be derived from a complete listing of $B_{10}H_{14}$ structures. Fortunately we have compiled what we hope is an exhaustive, or at least nearly complete, list of the latter. From this we obtain the 30 centrosymmetric structures described in Table 2–7. An obvious derivation gives 870 unsymmetrical (no center of

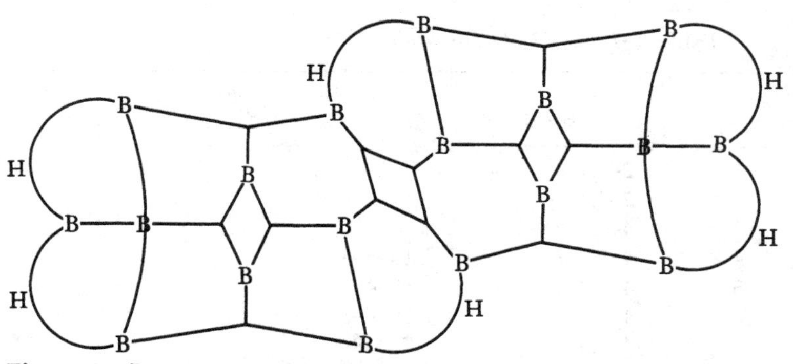


**Figure 2–27** *One of the many valence structures of $B_{18}H_{22}$. This bond arrangement shows the close relationship to the most symmetrical bonding arrangement in $B_{10}H_{14}$ (Fig. 2–10).*

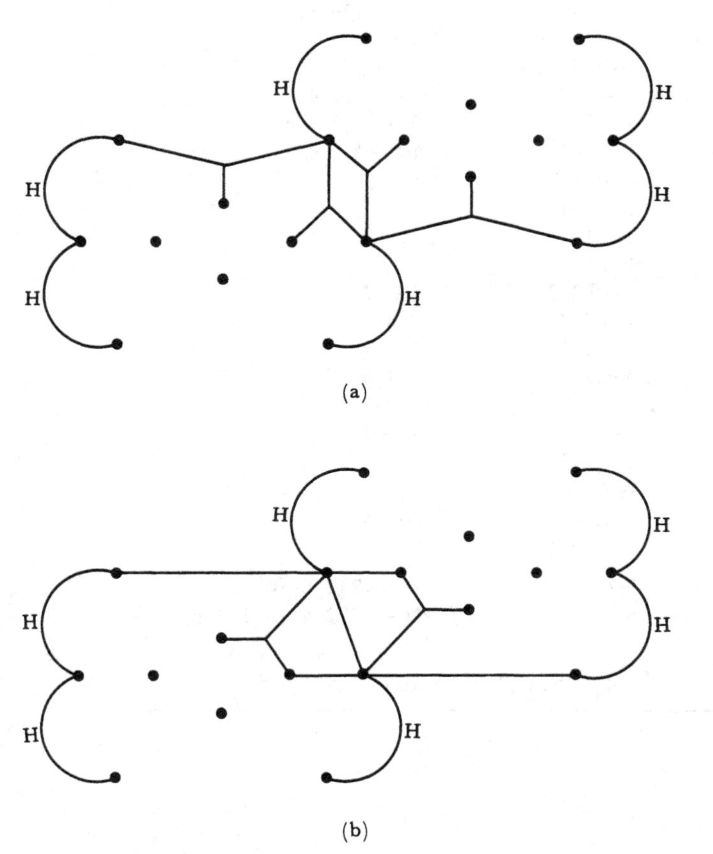


**Figure 2–28** *Centrosymmetric partial bonding arrangements in $B_{18}H_{22}$. These form the basis of the resonance structures which lead to the charge distribution in the text.*

Table 2-7 *$B_{18}H_{22}$ half-structures*

| | B—B | B⁀B (B above) |
|---|---|---|
| 1 | 2-7,4-8 | |
| 2 | 2-7,4-10 | |
| 3 | 4-9,7-8 | |
| 4 | 4-9,2-7 | 10-4-8 |
| 5 | 4-9,2-3 | 10-4-8,1-2,7 |
| 6 | 4-9,1-2 | 10-4-8 |
| 7 | 4-9,3-7 | 2-7-8 |
| 8 | 4-9,2-7 | |
| 9 | 4-8,2-7 | 10-4-3,1-4-9 |
| 10 | 4-8,2-7 | 9-4-3 |
| 11 | 4-10,2-7 | 1-4-8,3-4-9 |
| 12 | 4-10,3-7 | 1-4-9,2-7-8 |
| 13 | 4-10,7-8 | 1-4-9 |
| 14 | 4-10,2-7 | 1-4-9 |
| 15 | 3-4,2-7 | 1-4-9,10-4-8 |
| 16 | 4-9,2-7 | 3-4-10,1-4-8 |
| 17 | 1-4,2-7 | 10-4-8,3-4-9 |
| 18 | 4-10,1-2 | |
| 19 | 3-4,2-7 | |
| 20 | 4-8,1-2 | |
| 21 | 1-4,2-7 | |
| 22 | 4-10,2-3 | 1-2-7 |
| 23 | 1-4,2-7 | 3-4-10 |
| 24 | 4-8,2-3 | 1-2-7 |
| 25 | 3-4,2-7 | 1-4-8 |
| 26 | 1-4,3-7 | 2-7-8 |
| 27 | 1-2,3-8 | 4-8-7 |
| 28 | 1-4,7-8 | |
| 29 | 1-2,7-8 | |
| 30 | 1-3,7-8 | 2-1-4 |
| $B_{18}H_{22}^{++}$ | | |
| 1 | 5-6 | 10-5-2′,2-6-10′ |
| 2 | 5-6 | 10-5-2′,2-6-10′,1-3-8,7-8-4,9-4-3 |
| $B_{18}H_{22}^{=}$ | | |
| 1 | 5-6,4-9,1-3,7-8 | 10-5-2′,2-6-10′ |

**Table 2–8** *Charge distribution in $B_{18}H_{22}$ ($Q_1$ + $Q_1'$ + iso-$B_{18}H_{22}$)$Q_2$*

| Atom | 1 | 2 | 3 | 4 | 5 | 6 | 7 | 8 | 9 | 10 |
|---|---|---|---|---|---|---|---|---|---|---|
| $Q_1$ | −0.07 | −0.27 | −0.09 | −0.50 | 0.50 | 0.50 | −0.10 | 0.09 | 0.30 | 0.13 |
| $Q_1'$ | −0.05 | −0.35 | −0.06 | −0.47 | 0.42 | 0.42 | 0.02 | 0.09 | 0.29 | 0.12 |
| $Q_2$ | −0.04 | −0.33 | −0.06 | −0.47 | 0.16 | 0.62 | 0.01 | 0.09 | 0.27 | 0.13 |

inversion) structures for a grand total of 900. An equally facile examination of the 900 structures, equally weighted, yields the charge distribution in the first row of Table 2–8. In the second row we list the more nearly correct charges $Q_1'$ obtained by equal weighting of the complete list of 6506 resonance structures for $B_{18}H_{22}$. In the third row we list the charges $Q_2$ for corresponding atoms in the iso-$B_{18}H_{22}$ structure, in which all 6992 valence structures have been weighted equally. This method exaggerates the magnitude of these formal charges, but use of weights which are smaller, if the charges are larger in these resonance structures, has been shown to leave the order of the charge distributions unchanged in these two molecules.

Positions 5 and 6 appear to be the most positive ones in the molecule. They are, however, sterically inaccessible to nucleophilic reagents. Position 9 is next most positive, and it is probable that this is the region from which the hydrogen dissociates in the formation of $B_{18}H_{21}^-$, the existence of which has been reported.[234] In general the molecule should be more reactive toward electrophilic attacking groups and less reactive toward nucleophiles than decaborane. Positions 2 and 4 remain negative, but 4 is easily singled out as the most negative site in the molecule.

## 2–9 Resonance in $B_{10}C_2H_{12}$ Structures

In terms of the three-center bond formalism,[175] icosahedral $B_{12}H_{12}^{=}$ may be described as a composite of many resonance structures containing 2 normal B—B bonds and 10 central or open three-center B—B—B bonds. Although at first sight the number of resonance possibilities appears frightening, the topological bonding framework of the icosahedron is so tightly connected that the number of possibilities becomes quite reasonable. In exploring these resonance structures we have employed the following structural principles:

1. Each center has three orbitals with which to bond to five neighbors.
2. No two atoms can be connected by both a normal and a three-center bond.
3. No two atoms can be connected by both a central and an open three-center bond.
4. No atom may be the central member of two open three-center bonds.

One may easily see that there must be at least one pentagonal pyramid (6 atoms, 10 edges) in the icosahedron, in which single bonds are located on two edges. Proceeding from this pyramid we find that, conforming to the above structural principles, there are only five geometrically distinct bond arrangements (not including enantiomorphs) with only central three-center bonds, and nine with both open and central three-center bonds. Some of these also possess nonsuperimposable mirror images. The various arrangements possible are listed below.

a. 3 single bonds (S), 10 central three-center bonds (C), no open three-center bonds (O). Numbers refer to B atoms in Fig. 2–26.
   1. S: (9–11), (4–5), (2–7)
   2. S: (9–11), (4–5), (2–10)
   3. S: (9–11), (4–5), (2–12)
   4. S: (1–9), (4–5), (7–11)
   5. S: (1–9), (4–5), (2–12)

b. 3S; 7C; 3O
   1. S: (1, 9), (4–5), (7–11); O: (10–12–6), (8–6–2), (3–2–12)

c. 3S; 3C; 2O
   1. S: (4–5), (1–9), (2–10); O: (8–9–11), (12–2–7)
   2. S: (4–5), (1–9), (2–7); O: (8–9–11), (10–2–3)
   3. S: (4–5), (1–9), (2–6); O: (8–9–11), (12–2–3)
   4. S: (4–5), (9–11), (6–12); O: (10–4–1), (8–6–2)
   5. S: (4–5), (9–11), (2–3); O: (10–4–1), (6–2–7)

d. 3S; 9C; 1O
   1. S: (4–5), (9–11), (6–8); O: (10–4–1)
   2. S: (4–5), (9–11), (2–6); O: (10–4–1)
   3. S: (4–5), (9–11), (2–7); O: (10–4–1)

To study the carboranes we must bias the distribution of single bonds in order to concentrate charge on the more electronegative carbons. We have chosen to do so by fixing one of the single bonds between the two adjacent carbons in the $\alpha$(1–4) carborane and requiring that a single bond terminate

**Table 2–9**

| | OCS | CS |
|---|---|---|
| $\alpha$ | 10 | 27 |
| $\beta$ | 17 | 76 |
| $\gamma$ | 85 | 36 |

**Table 2–10** *Charge distribution in the icosahedral carboranes; comparison of three-center formalism and MO results*

| Position | OCS* | CS† | Three-center average‡ | MO |
|---|---|---|---|---|
| α 1,4 | −0.50¶ | −0.33 | −0.38 | −0.71 |
| 6,7 | −0.27 | −0.27 | −0.26 | −0.16 |
| 2,3 | −0.13 | −0.10 | −0.11 | −0.16 |
| 9,10,11,12 | −0.13 | −0.12 | −0.12 | −0.03 |
| 5,8 | 0.17 | −0.06 | 0.02 | 0.08 |
| β 1,3 | −0.67 | −0.33 | −0.39 | −0.85 |
| 10,12 | −0.29 | −0.19 | −0.21 | −0.16 |
| 5,6,7,8 | 0.05 | −0.14 | −0.11 | −0.03 |
| 2,4 | −0.14 | −0.09 | −0.10 | −0.03 |
| 9,11 | 0.00 | −0.11 | −0.09 | 0.10 |
| γ 1,2 | −0.50 | −0.33 | −0.45 | −0.85 |
| 3 to 12 | −0.10 | −0.13 | −0.11 | −0.03 |

* Open and central three-center bonds, single bonds.
† Central three-center bonds, single bonds.
‡ Weighted average with equal weights for each resonance structure (see text).
¶ $Q$ measured from 2 for both B and C.

on each of the two carbons in the β(1–3) and γ(1–2) carboranes. The total number of distinct resonance structures for central three-center bonds and single bonds (CS) and for open and central three-center bonds and single bonds (OCS) we find to be as shown in Table 2–9.

The charge distributions are easily calculated and are given in Table 2–10. The weighted average given is one which assigns equal weight to all resonance structures. The agreement is good, indicating an encouraging consistency in the two quite dissimilar approaches.

## 2–10 Ions of −1 and −2 Charge

It is hoped that in the near future the three-center bond logic can be programmed. Until the results of a more exhaustive study are available, the results of a preliminary survey of these ions based upon symmetrical bonding in icosahedral fragments is given in Appendix A in Figures A–1 and A–2. The number of structures consistent with the three-center resonance theory is much larger, and we do not therefore have a satisfactory criterion of stability or transformation to the most stable forms. Also, Fig. 2–29 contains a summary

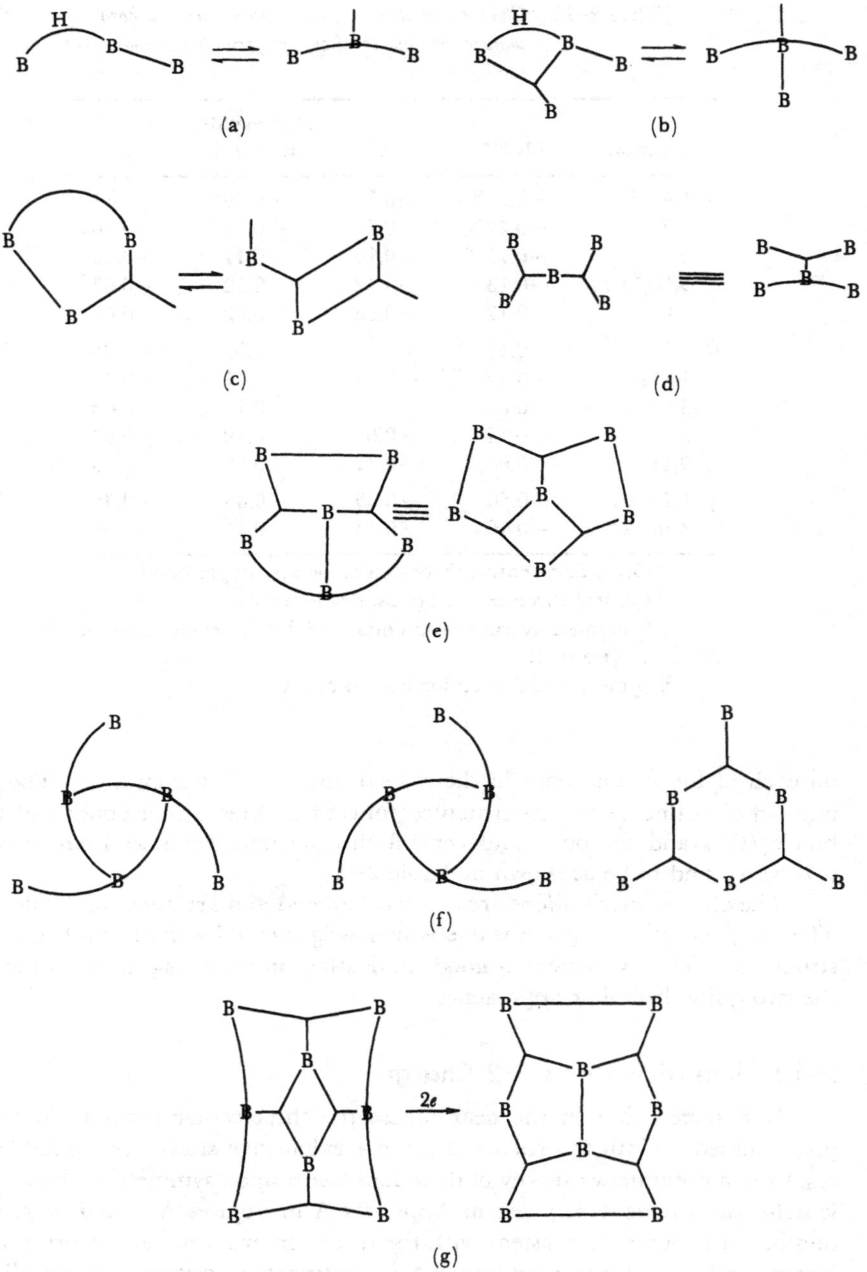

**Figure 2–29** *Summary of some of the more important relations among structural fragments in the three-center bond description.*

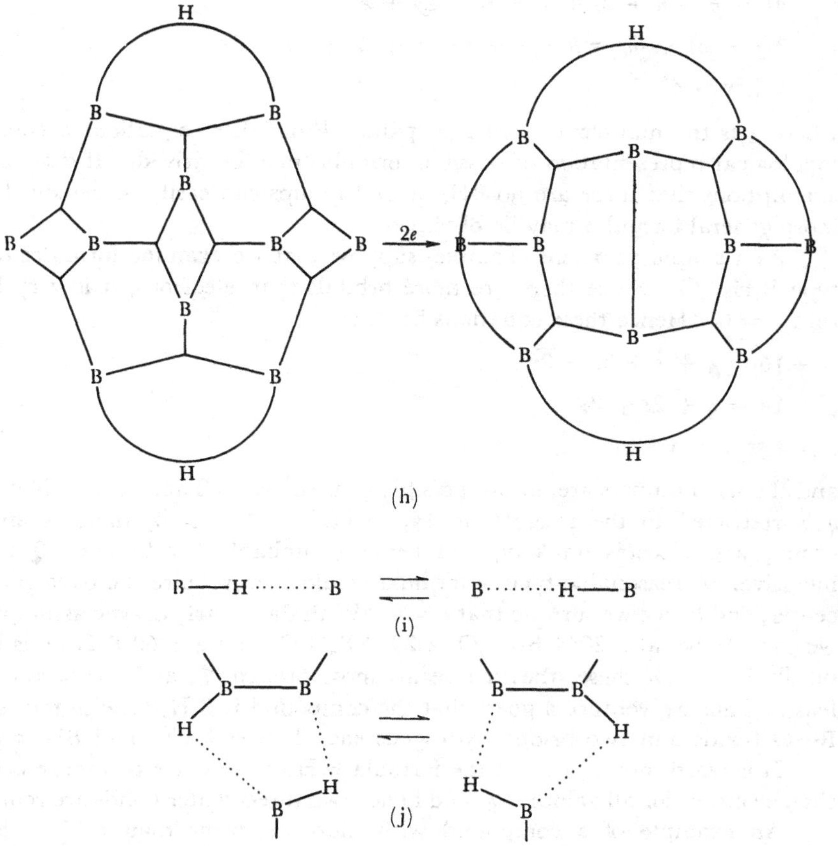

(h)

(i)

(j)

**Figure 2-29** (*continued*)

of most of the important relations among structural fragments in the three-center bond approximation to the valence theory.

## 2-11 Topology of $B_aH_bO_c$ Compounds

The report by W. H. Bauer and S. E. Wiberley[13] and by J. A. Hammond (private communication) of a possibly electron-deficient compound $B_4H_{12}O$ has suggested the investigation of valence rules for this and similar compounds. Assume that $B_{p-n}H_{p-n+q}O_n$ satisfies the same bonding rules as the boron hydrides. Then the equations of orbital balance, electron balance, and hydrogen balance[69] are

$$4p = p - n + 2s + x + 3t + 2y + z$$

$$3(p - n) + 6n = p - n + s + x + 2t + 2y + 2z$$

$$s + x = q$$

where $z$ is the number of unshared pairs. From these equations a complete topological representation of possible formulas may be drived. If desired, the assumptions that there are no $BH_3$ or OH groups can easily be removed, and more general formulas may be obtained.

As a simple pertinent example, suppose that we examine formulas of the type $B_4H_{4+q}O$. Since there are more orbitals than electrons, it is very likely that $z = 0$. Hence these equations become

$$16 = q + s + 3t + 2y$$

$$14 = q + 2t + 2y$$

$$s = q - x$$

and the *styx* numbers are, for all possible $q$, as shown in Table 2-11. Note that $q$ is restricted to the range 0 to 14, inclusive. If $x > 4$, there is an OH group, which seems unlikely, and hence $q$ probably lies between 0 and 6, inclusive. A reasonable type of formula would then require the oxygen at the center, and hence we surmise that $t = 0$. With these fairly drastic assumptions, we then arrive at a 2044 $B_4H_{10}O$, a 2052 $B_4H_8O$, and a 2060 $B_4H_6O$ as likely possibilities. Of these, the first seems most preferable, and the second, the least. Thus we venture a guess that the compound is $B_4H_{10}O$ with four single B—O bonds and two bridge hydrogens each between a pair of $BH_2$ groups.

It is worth noting that in the formula $B_4H_gO$ there are two more orbitals than electrons for all values of $g$, and hence two three-center bonds are required.

An example of a compound with more electrons than orbitals is the hypothetical $B_2H_2O_3$ (J. A. Hammond, private communication),[13,53,54] for which $n = 3$, $p = 5$, and $q = 0$. The equations give $s = x = 0$ and $z - t = 4$, but it is unlikely that there are both unshared pairs and three-center bonds, and so the most likely bond numbers are $s = 0$, $t = 0$, $y = 7$, $x = 0$, and $z = 4$. Two B—H groups each joined to three oxygen atoms are plausible, but there must also be one O—O bond; but formulas in which boron is three-coordinated are also possible.

**Table 2-11** *The styx numbers for* $B_4H_{4+q}O$

| $q$ | 14 | 12 | 10 | 8 | 6 | 4 | 2 | 0 |
|---|---|---|---|---|---|---|---|---|
| | 2.0.0.12 | 1.1.0.11 | 0.2.0.10 | 0218 | 0226 | 0234 | 0242 | 0250 |
| | | 2.0.1.10 | 1119 | 1127 | 1135 | 1143 | 1151 | |
| | | | 2028 | 2036 | 2044 | 2052 | 2060 | |

# 3

# Molecular Orbitals

The power of molecular orbital theory in predictions of charge distribution, large gaps between energies of highest filled orbitals and lowest unfilled molecular orbitals, relative stabilities of various molecular conformations, etc., is impressive compared with the relatively minor sacrifices, such as a tendency for each molecule to become a separate case, and a requirement that the investigator learn a bit of group representations. Nevertheless, there are patterns of molecular orbitals, such as the $\sigma$, $\pi$ situations, which can be carried from molecule to molecule. Moreover, a computer may be induced to produce the results even without the necessity for it to be taught how to do group theory. Indeed, it is not difficult to master the art of thinking in terms of symmetry orbitals, and we therefore outline this powerful method here, and give a number of applications to boron hydride molecular species in which these methods have increased depth of understanding and prediction.

## 3–1 Introduction to Symmetry and Representations. $B_5H_9$ Framework Orbitals

### Approximations

The boron hydrides and related negative ions are discussed here in the approximation that electron-electron repulsions and interactions of lower energy are neglected. Whether or not approximations are made, the presence of symmetry in a molecule allows one to draw conclusions which are sometimes of considerable power. For example, in this approximation the 1*s* levels of four interacting H atoms have the molecular energies shown in Fig. 3–1 for symmetries $C_1$, $C_{3v}$, and $T_d$, respectively. In the more usual approximation that the electron-nucleus interactions are the dominating interactions the conclusion that tetrahedral $H_4^{++}$ is a stable species can be suggested as worthy of experimental test. Thus the presence of symmetry, in particular axes of threefold or higher symmetry, produces relatively wide separations between energies, and these separations often occur between highest occupied molecular orbitals and lowest unoccupied molecular orbitals, as we saw in tetrahedral $H_4^{++}$. Inner-shell repulsions, electron-vibration interactions, poor overlap, and other molecular properties tend to lower symmetry, but the stronger tendency for boron to form more symmetrical molecular structures than, say, carbon is striking, and very likely is due to the relatively expanded valence orbitals, the underpopulated valence orbitals if only single bonds are formed, and the relatively small effect of the inner-shell repulsions in all the first row elements.

### Orbitals

We now consider the framework orbitals, left over after all bonds to H are formed, in $B_5H_9$ (Fig. 3–2). There are seven atomic orbitals, one nearly $sp^3$ hybrid on each of the outer B atoms, and on the apex (central) B atom one essentially *sp* orbital shown as a full circle and two $\pi$ orbitals. These are to be made into molecular orbitals in such a way that the resulting electron density (square of the wave function) must satisfy the molecular symmetry. For example, consider only the four atomic orbitals D, E, F, and G. Various linear combinations, e.g., D + E + F + G, form a molecular orbital (MO), but not all such molecular orbitals have an electron density which has the symmetry of the molecule. The (unnormalized) MO, D + E + F + G, does have the

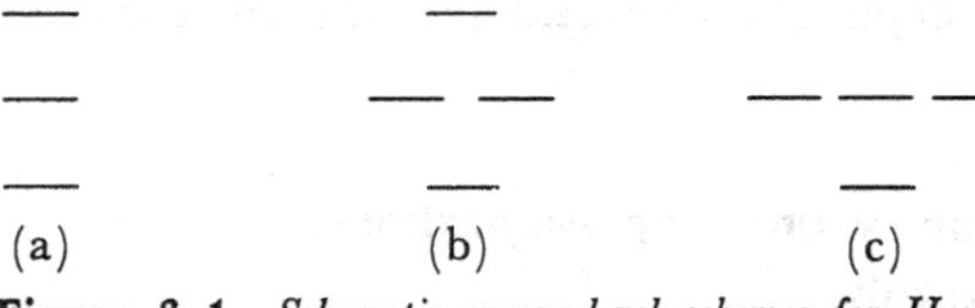


**Figure 3–1** *Schematic energy-level schemes for $H_4$ if the symmetry is (a) $C_1$ (identity element only, i.e., no axes, planes, centers, inversion axes of symmetry), (b) $C_{3v}$ (one threefold axis, three planes of symmetry each containing the threefold axis and 120° apart), and (c) $T_d$, the symmetry of a regular tetrahedron.*

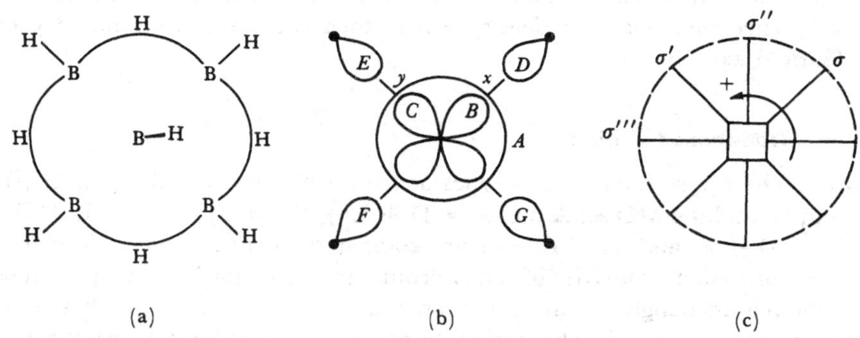


**Figure 3–2** *(a) View of $B_5H_9$ approximately along the fourfold axis, showing bonds to terminal and bridge H atoms. (b) Framework atomic orbitals available for B · · · B bonding, and x and y axes. (c) Symmetry planes, σ, etc., fourfold axis, and outline of diagram (dashed). This group of symmetry elements is designated by the symbol $C_{4v}$.*

molecular symmetry, since it tends to pile up* electron density equally between D and E, E and F, F and G, and G and D. Also the MO, D − E + F − G, shows molecular symmetry, because it requires a zero of electron density between every adjacent pair of atoms. However the electron density of the orbital D + E + F − G does not have the molecular symmetry, because the density is nonzero except between atoms F and G, and hence this orbital is not a proper molecular orbital. An orbital such as D − F, which would bond to a π orbital of the central atom also, does not have the symmetry of the molecule, but if one adds the densities of D − F and E − G, to obtain $D^2 - 2DF + F^2 + E^2 - 2EG + G^2$, the resulting density is invariant upon any of the rotations or reflections of the molecule and hence this pair of orbitals at the same energy, like a pair of π orbitals of a single atom, does yield a description of the electron density consistent with the molecular symmetry.

Thus for the six electrons needed for framework bonds in $B_5H_9$ the three bonding molecular orbitals are obtained by combining D + E + F + G with the axially symmetric *sp* orbital A, and the π-like pair D − F and E − G with, respectively, the π orbitals B and C on the central B in such a way that bonding occurs, i.e., with the same sign on the wave functions which overlap in the bonding regions. Aside from the easily handled question of normalization, we could equally well have chosen D + E − F − G and E + F − G − D for

* Consider the normalized molecular orbital, $\phi = (2+2S)^{-1/2}(A+B)$, where $S = \int AB\, d\tau$ is the overlap integral, and hence $\int \phi^2\, d\tau = 1$. If A and B are atomic orbitals on the same kind of atom, the electron density at the midpoint (A=B) is $(2+2S)^{-1}(A^2+2AB+B^2) = (1+S)^{-1}2A^2$. For all $S < 1$ this density is greater than the sum $A^2+B^2 = 2A^2$ for the midpoint if individual-atom electron densities are added.

this pair, and we could also have rotated the $\pi$ orbitals of the central atom by 45°, since their electron density would then still have the symmetry of the fourfold axis.

### Transformed Orbitals

The transformation properties of the set of atomic orbitals A,B,C,D,E,F, and G, and the MO set A,B,C,$\varphi_1 = D + E + F + G$, $\varphi_2 = D - E + F - G$, $\varphi_3 = D - F$, and $\varphi_4 = E - G$ are compared in Table 3–1. Note that the transformation properties of the A orbital are particularly simple; because this orbital is unchanged by any operation of the group it is called totally symmetric. It is only one step in abstraction to assign the numbers 1 to each symmetry element of the group with the meaning that as each operation is applied the symbol for that symmetry element multiplies the orbital in question. Such a set of numbers is called a *representation* of the group, and the numbers are called *characters*. They satisfy the "multiplication table" for the group (Table 3–2).

The rule for the symmetry table is (a) take an object, preferably an asymmetric one, (b) apply a chosen symmetry operation in the top row, and (c) follow this operation on the object by an operation of the left-hand column. If the symbol 1 is substituted for each symmetry element, the whole table is a set of 1's if the rule of combination is chosen as multiplication. When the number of symmetry elements is finite, the roots of unity are required for

**Table 3–1** *Transformation of atomic orbitals and of symmetry orbitals in $C_{4v}$*

| Rotation: | 0° | 180° | + 90° | − 90° | | | | |
|---|---|---|---|---|---|---|---|---|
| Orbital: | I | $C_2$ | $C_4^+$ | $C_4^-$ | $\sigma$ | $\sigma'$ | $\sigma''$ | $\sigma'''$ |
| A | A | A | A | A | A | A | A | A |
| B | B | −B | C | −C | B | −B | C | −C |
| C | C | −C | −B | B | −C | C | B | −B |
| D | D | F | E | G | D | F | E | G |
| E | E | G | F | D | G | E | D | F |
| F | F | D | G | E | F | D | G | E |
| G | G | E | D | F | E | G | F | D |
| $\varphi_1$ | $\varphi_1$ | $\varphi_1$ | $\varphi_1$ | $\varphi_1$ | $\varphi_1$ | $\varphi_1$ | $\varphi_1$ | $\varphi_1$ |
| $\varphi_2$ | $\varphi_2$ | $\varphi_2$ | $-\varphi_2$ | $-\varphi_2$ | $\varphi_2$ | $\varphi_2$ | $-\varphi_2$ | $-\varphi_2$ |
| $\varphi_3$ | $\varphi_3$ | $-\varphi_3$ | $\varphi_4$ | $-\varphi_4$ | $\varphi_3$ | $-\varphi_3$ | $\varphi_4$ | $-\varphi_4$ |
| $\varphi_4$ | $\varphi_4$ | $-\varphi_4$ | $-\varphi_3$ | $\varphi_3$ | $-\varphi_4$ | $\varphi_4$ | $\varphi_3$ | $-\varphi_3$ |
| $A_1$ | 1 | 1 | 1 | 1 | 1 | 1 | 1 | 1 |
| $A_2$ | 1 | 1 | 1 | 1 | −1 | −1 | −1 | −1 |
| $B_1$ | 1 | 1 | −1 | −1 | 1 | 1 | −1 | −1 |
| $B_2$ | 1 | 1 | −1 | −1 | −1 | −1 | 1 | 1 |
| E | 2 | −2 | 0 | 0 | 0 | 0 | 0 | 0 |

**Table 3–2** *"Multiplication table" for* $C_{4v}$

| First operation | | I | $C_4^+$ | $C_4^-$ | $C_2$ | $\sigma$ | $\sigma'$ | $\sigma''$ | $\sigma'''$ |
|---|---|---|---|---|---|---|---|---|---|
| Second operation | I | I | $C_4^+$ | $C_4^-$ | $C_2$ | $\sigma$ | $\sigma'$ | $\sigma''$ | $\sigma'''$ |
| | $C_4^+$ | $C_4^+$ | $C_2$ | I | $C_4^-$ | $\sigma''$ | $\sigma'''$ | $\sigma'$ | $\sigma$ |
| | $C_4^-$ | $C_4^-$ | I | $C_2$ | $C_4^+$ | $\sigma'''$ | $\sigma''$ | $\sigma$ | $\sigma'$ |
| | $C_2$ | $C_2$ | $C_4^-$ | $C_4^+$ | I | $\sigma'$ | $\sigma$ | $\sigma'''$ | $\sigma''$ |
| | $\sigma$ | $\sigma$ | $\sigma'''$ | $\sigma''$ | $\sigma'$ | I | $C_2$ | $C_4^-$ | $C_4^+$ |
| | $\sigma'$ | $\sigma'$ | $\sigma''$ | $\sigma'''$ | $\sigma$ | $C_2$ | I | $C_4^+$ | $C_4^-$ |
| | $\sigma''$ | $\sigma''$ | $\sigma$ | $\sigma'$ | $\sigma'''$ | $C_4^+$ | $C_4^-$ | I | $C_2$ |
| | $\sigma'''$ | $\sigma'''$ | $\sigma'$ | $\sigma$ | $\sigma''$ | $C_4^-$ | $C_4^+$ | $C_2$ | I |

nondegenerate representations, because successive applications of the operations must yield the identity, e.g., $(C_4^+)(C_4^+)(C_4^+)(C_4^+) = I$; hence $C_4^+$ must be represented by 1, $-1$, $i = \sqrt{-1}$, or $-i$. We now have the following group properties: (a) the identity element I is present; (b) two consecutive operations, whether different or not, are present as a distinct element of the group, e.g., $C_4^+\sigma = \sigma''$; (c) the associative law holds, e.g., $C_4^+(\sigma C_2) = (C_4^+\sigma)C_2$; and (d) every element has an inverse; e.g., the inverse of $C_4^+$ is $C_4^-$, and the inverse of $\sigma$ is $\sigma$. These properties are most useful to ensure that all symmetry elements have been found, a process aided by use of a multiplication table.

Other representations of the $C_{4v}$ group are also shown in Table 3–1. The numbers for the E representation are the sums of the diagonal elements of the matrices which correspond to the symmetry transformations

$$\begin{pmatrix} \cos\theta & -\sin\theta \\ \sin\theta & \cos\theta \end{pmatrix}\begin{pmatrix} x_1 \\ y_1 \end{pmatrix} = \begin{pmatrix} x_2 \\ y_2 \end{pmatrix}$$

for axes parallel to $z$, and

$$\begin{pmatrix} \cos 2\theta & \sin 2\theta \\ \sin 2\theta & -\cos 2\theta \end{pmatrix}\begin{pmatrix} x_1 \\ y_1 \end{pmatrix} = \begin{pmatrix} x_2 \\ y_2 \end{pmatrix}$$

for a plane containing the $z$ axis but oriented at an angle $\theta$ to the $x$ axis. Thus

$$I = \begin{pmatrix} 1 & 0 \\ 0 & 1 \end{pmatrix} \quad C_4^+ = \begin{pmatrix} 0 & -1 \\ 1 & 0 \end{pmatrix} \quad C_4^- = \begin{pmatrix} 0 & 1 \\ -1 & 0 \end{pmatrix} \quad C_2 = \begin{pmatrix} -1 & 0 \\ 0 & -1 \end{pmatrix}$$

$$\sigma = \begin{pmatrix} 1 & 0 \\ 0 & -1 \end{pmatrix} \quad \sigma' = \begin{pmatrix} -1 & 0 \\ 0 & 1 \end{pmatrix} \quad \sigma'' = \begin{pmatrix} 0 & 1 \\ 1 & 0 \end{pmatrix} \quad \sigma''' = \begin{pmatrix} 0 & -1 \\ -1 & 0 \end{pmatrix}$$

These problems arise because a rotation transforms $x$ into a linear combination of $x$ and $y$; e.g., $x_2 = x_1 \cos\theta - y_1 \sin\theta$, if one uses the rule for matrix multiplication. All the sequences of symbols for $A_1$, $A_2$, $B_1$, $B_2$ and the corresponding corners of the matrices behave as orthogonal vectors, with components

| | | | | | | | | |
|---|---|---|---|---|---|---|---|---|
| (1) | 1 | 1 | 1 | 1 | 1 | 1 | 1 | 1 |
| (2) | 1 | 1 | 1 | 1 | −1 | −1 | −1 | −1 |
| (3) | 1 | −1 | −1 | 1 | 1 | 1 | −1 | −1 |
| (4) | 1 | −1 | −1 | 1 | −1 | −1 | 1 | 1 |
| (5) | 1 | 0 | 0 | −1 | 1 | −1 | 0 | 0 |
| (6) | 0 | −1 | 1 | 0 | 0 | 0 | 1 | −1 |
| (7) | 0 | 1 | −1 | 0 | 0 | 0 | 1 | −1 |
| (8) | 1 | 0 | 0 | −1 | −1 | 1 | 0 | 0 |

a situation which gives rise to the orthogonality of these different representations and therefore to the orthogonality of wave functions constructed from use of these numbers, provided different representations are used. Finally, only the sum of diagonal terms of the matrices is independent of the coordinate system, and hence the E representation uses the symbols shown in Table 3–1.

To obtain the proper linear combinations of atomic orbitals one may either find by inspection the proper linear combinations, as we did above (note that orbital A transforms as $A_1$, B and C as E, $\varphi_1$ as $A_1$, $\varphi_2$ as $B_1$, and $\varphi_3$ and $\varphi_4$ as E), or better yet one can use the characters themselves to form these linear combinations. For example, the direct product (sum of products—corresponding symbol by corresponding symbol) of the $B_1$ characters with the atomic orbitals transformed starting from the D orbital are $(1)D + (1)F - (1)E - (1)G + (1)D + (1)F - (1)E - (1)G = 2(D - E + F - G)$, which is the correct (unnormalized) linear combination of $B_1$ symmetry. Since the characters of the $B_1$ representation are orthogonal to the characters of each of the other representations, it follows that this $B_1$ molecular orbital will be orthogonal to all other molecular orbitals of different symmetries. The other molecular orbitals described above may be obtained by a similar procedure. Note that one obtains zero if the $A_2$ or $B_2$ characters are used, but in a more complex problem this will not usually be true.

## 3–2 Energies of Molecular Orbitals

### $B_5H_9$ Framework Orbitals

If overlap integrals are neglected, the normalized molecular orbitals for $B_5H_9$ are

A symmetry: A and $(D + E + F + G)/2 = \varphi_1$

B symmetry: $(D - E + F - G)/2 = \varphi_2$

E symmetry: $\begin{cases} B \\ C \end{cases}$ and $\begin{cases} (D - F)/\sqrt{2} = \varphi_3 \\ (E - G)/\sqrt{2} = \varphi_4 \end{cases}$

Those within the same symmetry interact, whereas those of different symmetry do not. Only one orbital of $B_1$ symmetry is present, and it interacts with no other orbitals so that its energy is, from $H\psi = E\psi$,

$$E_{B_1} = \int \varphi_2 H \varphi_2 \, d\tau$$

Now substitute $\varphi_2 = (D - E + F - G)/2$, expand the 16 terms, define the Coulomb integrals $H_0 = \int D^2 \, d\tau = \int E^2 \, d\tau$, etc.; the resonance integrals $\gamma = \int DHE \, d\tau = \int EHF \, d\tau$, etc.; and neglect non-nearest-neighbor interactions. The result is $E = H_0 - 2\gamma$ for the orbital of $B_1$ symmetry. To the extent that $\gamma$ is $<<$ $H_0$ the energy becomes $H_0$, the same as the energy of the outer electron in the isolated atoms, and hence the orbital is called nonbonding.

A similarly simple calculation is not possible for the two orbitals of $A_1$ symmetry, because they can, and do, interact to form a strongly bonding combination; thus

$$\psi_1 = aA + b\varphi_1 = aA + \sqrt{1 - a^2}\,\varphi_1$$

where $b = \sqrt{1 - a^2}$ normalizes the function $\psi_1$, neglecting overlap. The energy is then

$$E_{A_1} = \int \psi_1 H \psi_1 \, d\tau = H_0 + a\sqrt{1 - a^2}\,4\alpha$$

where* $\alpha = \int AHD \, d\tau = \int DHA \, d\tau = \int AHE \, d\tau$, etc., and $H_0 = \int AHA \, d\tau = \int DHD \, d\tau$, etc. All these integrals are negative because the attractive potential between electron and nucleus dominates the total energy. To find the minimum $E_{A_1}$, set $dE_{A_1}/da = 0$, which yields $a = \pm 1/\sqrt{2}$, so that

$$\psi_1 = \frac{1}{\sqrt{2}}A + \frac{1}{\sqrt{2}}\varphi_1 = \frac{1}{\sqrt{2}}A + \frac{1}{2\sqrt{2}}(D + E + F + G)$$

and $E_{A_1} = H_0 + 2\alpha$. The antibonding orbital of $A_1$ symmetry,

$$-\psi_1' = \frac{1}{\sqrt{2}}A - \frac{1}{2\sqrt{2}}(D + E + F + G)$$

arises from the other choice of sign, and has energy $H_0 - 2\alpha$.

A similar calculation for the interaction of $B$ with $(D—F)/\sqrt{2}$ yields the orbitals and energies

$$\pm \frac{1}{\sqrt{2}}B + \tfrac{1}{2}(D - F) \qquad \text{at } E = H_0 \pm \sqrt{2}\,\beta$$

where $\beta = \int BHD \, d\tau = \int DHB \, d\tau = \int BHF \, d\tau$, etc. The + signs apply to the bonding orbital and the − signs to the antibonding orbital, and a completely similar set of orbitals and equations arise from orbitals C, E, and G. Here one must note that it is −F, not +F, which gives bonding overlap with the negative lobe of B. It is hoped that this introduction may provide a basis for

* The operator $H$ is hermitian.

the reader to follow the ideas of the calculations reported below, and perhaps the details in a recent reference[303] regarding the methods.

### $BH_5$ and $CH_5^+$

As a simple semiquantitative model we show in Fig. 3–3 the molecular orbital description of $BH_5$ or $CH_5^+$ of $D_{3h}$ symmetry. It is frequently helpful to separate initially one region of the molecule from another in order that the molecular orbitals and their interactions may be more easily seen. Here, the central B or $C^+$ is considered separately from the $H_5$ unit of $D_{3h}$ symmetry, and after molecular orbitals of the $H_5$ unit are found, those of appropriate symmetry to interact with the orbitals of B or $C^+$ are then given a bonding (and corresponding nonbinding) interaction proportional to a crude approximation of the amount of overlap of the wave functions. Three levels, such as the $\sigma_g$ set often, but not always, give approximately a strongly bonding, a strongly antibonding, and a nearly nonbonding orbital. Such a semiquantitative guess at the number and symmetries of the bonding levels is frequently sufficient in simple molecules, but is sometimes misleading in more complex molecules where better approximations are then required, as described in Section 3–5.

## 3–3 Polyhedral Molecules and Ions

### Theory

Molecular orbital descriptions have been given to the known $B_4Cl_4$ molecule,[69,185] the $B_8Cl_8$ molecule,[136] the $B_{10}H_{10}^{=}$ ion,[182,198] the $B_{12}H_{12}^{=}$

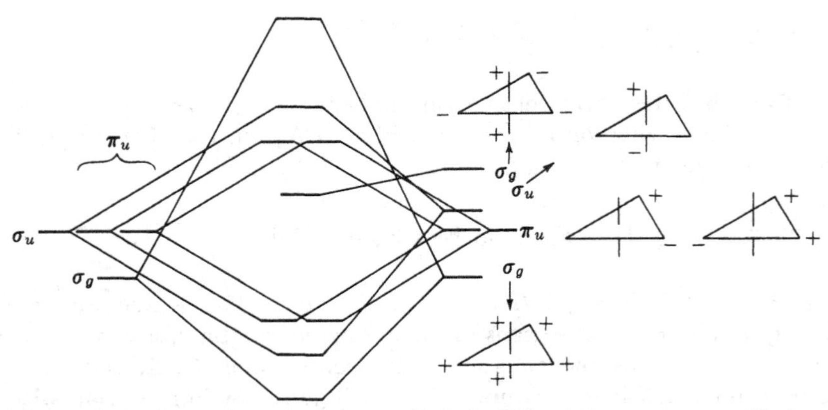


**Figure 3–3** *Interactions of symmetry orbitals in $BH_5$ of $D_{3h}$ symmetry. All three $\sigma_g$ orbitals interact, but the diagram has been simplified. Four electron pairs are needed to fill the bonding levels in $BH_5$ or in $CH_5^+$.*

ion,[69,188,198] and to the hypothetical $B_5H_5^=$ ion,[198] $B_6H_6^=$ ion,[69,179,187] and others.[179,198] Subsequent to these treatments, in which the predictions were at least uniform but the assumptions and approximations were quite varied, a unified treatment of all these polyhedral species,[120] as well as the hydrides themselves, has appeared.[123] The assumptions underlying this unified discussion are as follows.

The molecular orbital $\psi$ is built up as a linear combination of (real) atomic orbitals, $\phi$,

$$\psi = \sum_i c_i \phi_i$$

The coefficients $c_i$ satisfy the equations

$$(\alpha_r - ES_{rr})c_r - \Sigma'(\beta_{rs} - ES_{rs})c_s = 0$$

where $r = 1,2,3, \ldots, M$, $M$ is the number of orbitals considered, $E$ is the energy, $S_{rs} = \int \phi_r \phi_s \, d\tau$ is the overlap integral, $\alpha_r = \int \phi_r H \phi_r \, d\tau$ is the Coulomb integral, $\beta_{rs} = \int \phi_r H \phi_s \, d\tau$ is the resonance integral, and $H$ is the ubiquitous, but hardly ever specified, Hückel Hamiltonian.

All interactions between orbitals have been included, because inconsistencies arise if only nearest-neighbor interactions are included.[120] A different Coulomb integral is required[119,120] for the $2s$ orbital of B, as compared with the value for the $2p$ orbital. Normalized Slater orbitals with an exponent of 1.30 for B have been employed. The relation between resonance integrals and overlap integrals

$$\beta_{rs} = KS_{rs}$$

introduces the parameter $K$ and reduces the equations to

$$-xc_r + \Sigma' S_{rs} c_s = 0$$

where

$$-x = \frac{\alpha - E}{K - E}$$

in the approximation, later modified, that $\alpha(2s) = \alpha(2p)$. The order of the $x_i$'s is the same as the order of the $E_i$ for all $K > \alpha$, and hence we work with the $x_i$'s as energy parameters. The inverse relations are

$$E_i = \frac{\alpha + Kx_i}{1 + x_i}$$

Usually, we press the theory for no more than the order of energies and the charge distribution of the highest-filled or all-filled molecular orbitals.

This procedure has been programmed for a high-speed computer, which requires the atomic coordinates, the $s$ and $p$ Coulomb integrals (which may be different), the parameter $K$ (to which the results are not greatly sensitive), and parameters describing the unhybridized atomic orbitals. Only the $g$ or $u$

**Table 3–3** *Energy levels and symmetries in polyhedral species*

| $B_4H_4$ $T_d$,3$N$ | | $B_5H_5$ $C_{4v}$,3$N$ | | $B_5H_5^-$ $D_{3h}$,3$N$ | | $B_6H_6$ $C_{5v}$,3$N$ | | $B_6H_6^-$ $O_h$,3$N$ | | $B_{14}H_{14}^-$ $O_h$,4$N$ | |
|---|---|---|---|---|---|---|---|---|---|---|---|
| $f_2$ | −0.849 | $b_1$ | −0.881 | $e'$ | −0.863 | $a_1$ | −0.848 | $e_g$ | −0.884 | $a_{1g}$ | −19.66 |
| $f_1$ | −0.556 | $a_1$ | −0.853 | $a_2''$ | −0.858 | $e_2$ | −0.845 | $f_{1u}$ | −0.829 | $f_{1u}$ | −18.61 |
| $e$ | 0.033 | $e$ | −0.794 | $a_1'$ | −0.823 | $e_1$ | −0.801 | $f_{1g}$ | −0.671 | $e_g$ | −16.92 |
| $f_2$ | 0.685 | $b_2$ | −0.664 | $e''$ | −0.649 | $a_2$ | −0.726 | $f_{2u}$ | −0.416 | $f_{2g}$ | −16.55 |
| $a_1$ | 2.094 | $e$ | −0.542 | $a_2'$ | −0.556 | $e_1$ | −0.563 | $f_{2g}$ | 0.493 | $a_{1g}$ | −15.13 |
| | | $b_1$ | −0.409 | $e'$ | −0.301 | $e_2$ | −0.371 | $f_{1u}$ | 1.023 | $f_{2u}$ | −14.08 |
| | | $e$ | 0.106 | $e''$ | 0.335 | $e_1$ | 0.185 | $a_{1g}$ | 2.969 | $f_{1u}$ | −12.52 |
| | | $a_2$ | 0.492 | $e'$ | 0.758 | $e_2$ | 0.494 | | | $f_{1u}$ | −11.72 |
| | | $a_1$ | 0.711 | $a_2''$ | 1.108 | $a_1$ | 0.636 | | | $a_{2u}$ | −10.15 |
| | | $e$ | 0.816 | $a_1'$ | 2.567 | $e_1$ | 1.106 | | | $f_{1g}$ | −8.65 |
| | | $a_1$ | 2.433 | | | $a_1$ | 2.527 | | | $e_g$ | −8.35 |
| | | | | | | | | | | $f_{1g}$ | −6.56 |
| | | | | | | | | | | $a_{1g}$ | −5.86 |
| | | | | | | | | | | $f_{1u}$ | −0.92 |
| | | | | | | | | | | $e_u$ | 10.37 |
| | | | | | | | | | | $f_{1u}$ | 15.03 |
| | | | | | | | | | | $f_{2u}$ | 21.88 |
| | | | | | | | | | | $f_{2g}$ | 25.89 |
| | | | | | | | | | | $f_{1u}$ | 26.07 |
| | | | | | | | | | | $f_{1g}$ | 30.63 |
| | | | | | | | | | | $e_g$ | 30.88 |
| | | | | | | | | | | $a_{2u}$ | 31.34 |
| | | | | | | | | | | $f_{2g}$ | 50.20 |
| | | | | | | | | | | $a_{1g}$ | 62.79 |

(even or odd on inversion) symmetry separation is carried out, provided the molecule is centrosymmetric. No other symmetry factorization is made of the problem, inasmuch as it is difficult to teach group theory to the computer. If one terminal H orbital on each B is included, the order of the unfactored secular equations for the $x_i$ is 5$N$, where $N$ is the number of B atoms. If this H orbital is omitted, and only the $2s$, $2p_x$, $2p_y$, and $2p_z$ orbitals of B are included, this approximation is of order 4$N$. If $sp^3$ hybrids are formed at each B atom and the one (out) directed outward from the molecular center is dropped from the calculation (it forms a terminal, localized B—H bond), we are left with 3$N$ orbitals.

Even lower orders of approximation have been explored; for example, this 3$N$ set can be divided into a 2$N$ (surface) set of $\pi$ orbitals which interact on the surface of the polyhedron, and an $N$ (in) set, of largely $s$ character, pointing inward toward the molecular center. This last separation is not a good approximation[120] for description of the molecule, although the specific

**Table 3–3** (*continued*)

| $B_7H_7^{=}$ $D_{5h}$, $3\mathcal{N}$ | | $B_8H_8^{=}$ $D_{3d}$, $3\mathcal{N}$ | | $B_{10}H_{10}^{=}$ $D_{4d}$, $3\mathcal{N}$ | | $B_{12}H_{12}^{=}$ $I_h$, $3\mathcal{N}$ | | $B_{12}H_{12}^{=}$ $O_h$, $3\mathcal{N}$ | |
|---|---|---|---|---|---|---|---|---|---|
| $a_2''$ | −0.851 | $e_g$ | −0.887 | $e_2$ | −0.884 | $f_{2u}$ | −0.886 | $f$ | −0.885 |
| $a_1'$ | −0.847 | $a_{2u}$ | −0.831 | $e_3$ | −0.862 | $h_g$ | −0.856 | $e_g$ | −0.883 |
| $e_2'$ | −0.845 | $e_u$ | −0.829 | $b_2$ | −0.850 | $f_{1g}$ | −0.782 | $f_{2g}$ | −0.769 |
| $e_1'$ | −0.840 | $a_{2u}$ | −0.762 | $a_1$ | −0.812 | $f_{1u}$ | −0.773 | $f_{1g}$ | −0.756 |
| $a_2'$ | −0.726 | $a_{1g}$ | −0.736 | $b_2$ | −0.812 | $h_u$ | −0.678 | $a_{2g}$ | −0.739 |
| $e_1''$ | −0.669 | $e_g$ | −0.692 | $e_1$ | −0.776 | $g_g$ | −0.471 | $f_{1u}$ | −0.739 |
| $e_1'$ | −0.508 | $a_{2g}$ | −0.664 | $e_3$ | −0.747 | $g_u$ | 0.518 | $e_u$ | −0.669 |
| $e_2''$ | −0.371 | $e_u$ | −0.590 | $a_2$ | −0.739 | $h_g$ | 0.984 | $f_{2u}$ | −0.483 |
| $e_2'$ | 0.494 | $a_{1u}$ | −0.409 | $e_1$ | −0.615 | $f_{1u}$ | 1.907 | $f_{1g}$ | −0.232 |
| $e_1''$ | 0.524 | $e_g$ | −0.173 | $e_2$ | −0.598 | $a_g$ | 4.163 | $f_{1u}$ | 0.361 |
| $a_2''$ | 0.740 | $e_u$ | 0.114 | $b_1$ | −0.590 | | | $f_{2g}$ | 0.872 |
| $e_1'$ | 1.420 | $e_g$ | 0.698 | $e_3$ | −0.327 | | | $a_{2u}$ | 0.881 |
| $a_1'$ | 3.273 | $a_{1g}$ | 0.830 | $e_1$ | 0.328 | | | $e_g$ | 0.991 |
| | | $e_u$ | 1.090 | $e_2$ | 0.756 | | | $f_{1u}$ | 1.754 |
| | | $a_{2u}$ | 1.752 | $e_3$ | 0.760 | | | $a_{1g}$ | 3.616 |
| | | $a_{1g}$ | 3.362 | $a_1$ | 1.004 | | | | |
| | | | | $e_1$ | 1.447 | | | | |
| | | | | $b_2$ | 1.978 | | | | |
| | | | | $a_1$ | 3.877 | | | | |

interaction of the $2\mathcal{N}$ set of $\pi$ orbitals with a substituent is analogous to $\pi$ conjugation in aromatic carbon chemistry. In the following discussion these different procedures are referred to as $5\mathcal{N}$, $4\mathcal{N}$, $3\mathcal{N}$, and $2\mathcal{N}$.

### Examples of Filled Orbital Species

The main results are filled orbitals for the following neutral or negative species (Table 3–3). Back coordination situations will render the halogenated boron frameworks more stable,[185] and hence one might anticipate that $B_5Cl_5(C_{4v})$ and $B_6H_6(C_{5v})$ might be found, since the known chlorides are $BCl_3$, $B_2Cl_4$, $B_4Cl_4(T_d)$,[3] and $B_8Cl_8$ $(D_{2d})$.[185] However, subhalides having other features, such as terminal $BCl_2$ in place of Cl, polyhedra linked by B—B bonds, or B—Cl—B bridges, may also occur.

In Figs. 3–4, 3–5, and 3–6 are shown several polyhedra which are candidates for various possible boron or carbon compounds, summarized as follows (known species in italic):

Fig. 3–4a. *$B_4Cl_4$*, or tetrahedral $C_4H_4$(?)[177]
Fig. 3–4b. $C_8H_8$
Fig. 3–4c. $B_6H_6^{=}$, $CB_5H_6^{-}$, *$C_2B_4H_6$* (*two isomers*)

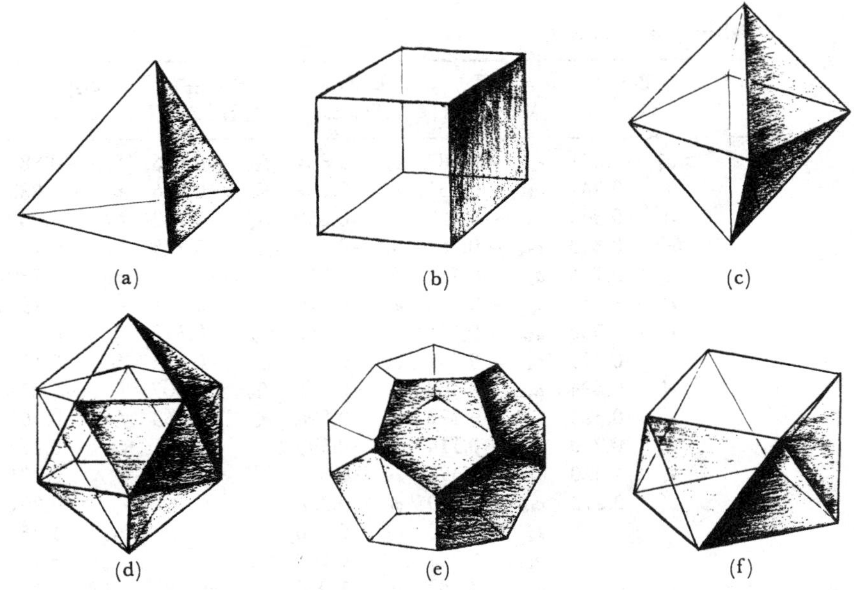


**Figure 3–4** *(a) Tetrahedron, (b) cube, (c) octahedron, (d) icosahedron, (e) dodecahedron, (f) Archimedean antiprism. (a), (b), and (e) are $C_NH_N$ possibilities.*

Fig. 3–4d. *$B_{12}H_{12}^{=}$*, $CB_{11}H_{12}^{-}$, $C_2B_{10}H_{12}$ (three isomers, at least one known)
Fig. 3–4e. $C_{20}H_{20}$
Fig. 3–4f. $B_8H_8^{=}$
Fig. 3–5a. $C_6H_6$
Fig. 3–5b. Collapses to 3*d*?
Fig. 3–5c. $C_{10}H_{10}$
Fig. 3–5d. Collapses to 1*d*?
Fig. 3–5e. Collapses to 2*f*?
Fig. 3–5f. *$B_{10}H_{10}^{=}$*, $CB_9H_{10}^{-}$ (two isomers), $C_2B_8H_{10}$ (7 isomers)
Fig. 3–6a. $B_{12}H_{12}^{=}$, collapses to 1*d*
Fig. 3–6b. $B_5H_5^{=}$, $CB_4H_5^{-}$ (two isomers), $C_2B_3H_5$ (three isomers, *one known*)
Fig. 3–6c. $B_7H_7^{=}$
Fig. 3–6d. $B_8H_8^{=}$ (poor electron distribution)
Fig. 3–6e. $C_{12}H_{12}$
Fig. 3–6f. $B_{14}H_{14}^{=}$

The known $B_8Cl_8$ polyhedron, of symmetry $D_{2d}$, is a significantly distorted Archimedian antiprism, in which each of the two square faces has a fold on one of its face diagonals.

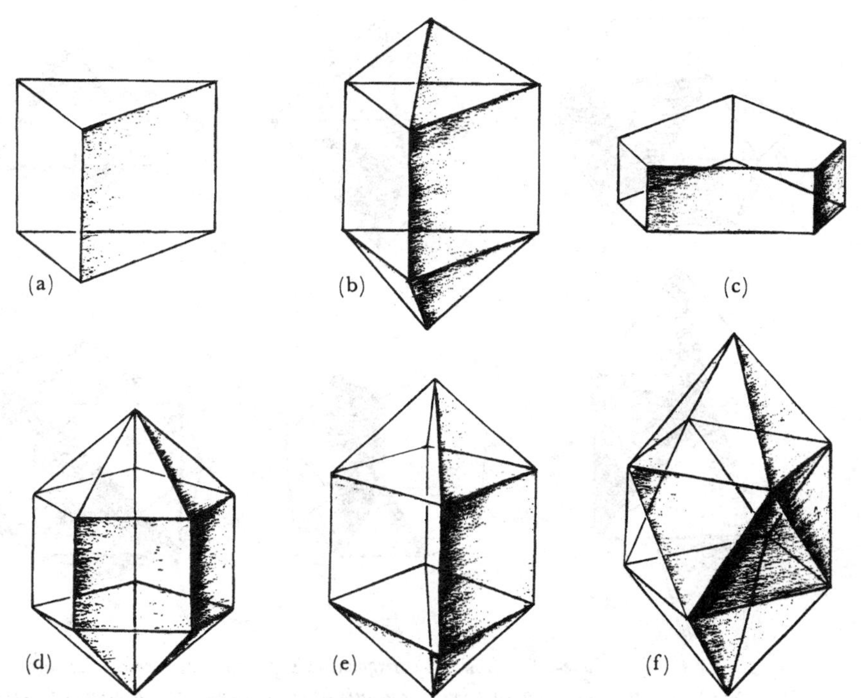


**Figure 3–5** *(a) Trigonal prism, (b) bicapped trigonal prism, (c) pentagonal prism, (d) bicapped pentagonal prism, (e) bicapped cube, (f) bicapped Archimedean antiprism. (a) and (c) are* $C_NH_N$ *possibilities.*

### Theory of Charge Distribution

A second set of results comes from analysis of the charge distributions[121] by Mulliken's method.[204] The overlap population in the $i$th molecular orbital (MO) between the $s$th basis orbital on atom $l$, and the $r$th basis orbital on atom $k$ is defined as

$$n(i;r_k,S_l) = 2\mathcal{N}(i)C_{r_k i}S_{r_k s_l}C_{s_l i}$$

where $\mathcal{N}(i)$ is the occupation number of the $i$th MO and $C_{s_l i}$ is the orbital coefficient. We shall be interested in the subtotal overlap population,

$$n(k,l) = \sum_i \sum_r \sum_s n(i;r_k,s_l)$$

The gross atomic population in MO $i$ on orbital $r_k$ is given by

$$\mathcal{N}(i;r_k) = \mathcal{N}(i)C_{r_k i}(SC)_{r_k i} = \mathcal{N}(i)C_{r_k i}\sum_l \sum_s S_{r_k s_l}C_{s_l i}$$

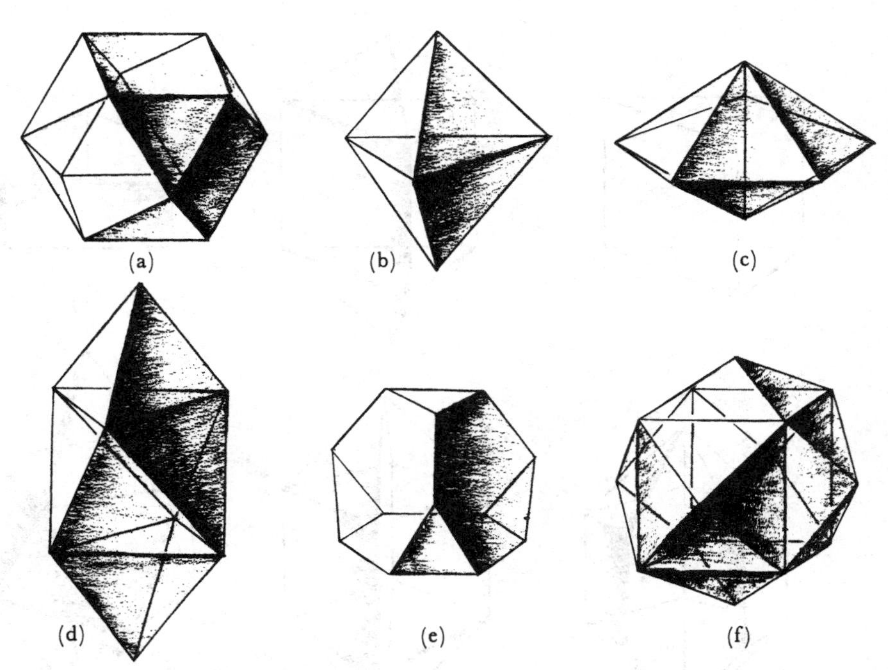


**Figure 3-6** *(a) Cube-octahedron, (b) trigonal bipyramid, (c) pentagonal bipyramid, (d) rhombic hexahedron-bicapped trigonal antiprism, (e) truncated tetrahedron, (f) rhombic dodecahedron-omnicapped cube. (e) is a* $C_NH_N$ *possibility.*

for a normalized orbital set. We shall also discuss the subtotal gross atomic population,

$$\mathcal{N}(k) = \sum_i \sum_r \mathcal{N}(i;r_k)$$

which we loosely call the total charge on atom $k$ [$\mathcal{N}(k)$ is thus different from $\mathcal{N}(i)$]. We shall also be interested in the perturbation in the charge distribution of a molecule due to a change in the Coulomb integral at a specified position. This perturbation is called the atom-atom polarizability, and is given by

$$\pi(k,l) = \frac{\partial \mathcal{N}(k)}{\partial \alpha(l)}$$

Using a perturbation procedure similar to that of Coulson,[40] we may derive an expression for $\pi(k,l)$, in which $C_{r_k i}$ has been contracted to $C_{ki}$,

$$\pi(k,l) = 2 \sum_{j=1}^{m} \sum_{i=m+1}^{n} \frac{C_{lj}C_{li}}{\epsilon_j - \epsilon_i} [C_{kj}(SC)_{ki} + C_{ki}(SC)_{kj}]$$

where the levels from 1 to $m$ are doubly occupied and all others are empty.

The $\epsilon_i$ are the zeroth-order energies. Note that $\pi(k,l) \neq \pi(l,k)$. Nevertheless, the perturbations are additive. A quantity $\bar{F}_{(k)}$, which we shall call the *bound valence*, may be defined, in analogy to a corresponding sum in the Hückel theory, as

$$\bar{F}(k) = \sum_i \sum_r \sum_{S_l}{}' n(i;r_k,S_l) = 2\mathcal{N}(k) - n(k,k)$$

### The Polyhedral Carboranes

The value for $K_2B_{10}H_{10}$ of 165 (in cubic centimeters per mole) for $-\chi_m \times 10^6$ is a remarkably high value.[118] When corrected[305] for $K^+$ (14), B (8.0), and H (2.0), the large value of 37 cm³/mole remains as a deviation from additivity, to be attributed primarily to molecular diamagnetism. This deviation is a strong indication of the "aromatic-like" character of $B_{10}H_{10}^=$, and presumably is a general characteristic of polyhedral molecules of this type.

The extraordinary stabilities of the polyhedral species (Fig. 3–7), as

**Table 3–4** *Population analysis for the trigonal bipyramid carboranes*

| | $k$ | $Q(k)$ | $\bar{F}(k)$ | $k$-$l$ | $n(k,l)$ |
|---|---|---|---|---|---|
| $B_5H_5^=$ | 1 | −0.251 | 2.05 | 1–2 | 0.316 |
| | 2 | −0.251 | 2.05 | 1–4 | 0.707 |
| | 3 | −0.251 | 2.05 | 4–5 | −0.135 |
| $\Sigma E = -184.44$ ev | 4 | −0.623 | 1.99 | | |
| | 5 | −0.623 | 1.99 | | |
| 1–2 | 1 | 0.171 | 1.83 | 1–2 | 0.154 |
| $B_3C_2H_5$ | 2 | 0.171 | 1.83 | 1–3 | 0.313 |
| | 3 | 0.050 | 1.84 | 1–4 | 0.680 |
| $\Sigma E = -194.32$ ev | 4 | −0.196 | 1.83 | 3–4 | 0.609 |
| | 5 | −0.196 | 1.83 | 4–5 | −0.140 |
| 4–5 | 1 | 0.264 | 1.78 | 1–2 | 0.174 |
| $B_3C_2H_5$ | 2 | 0.264 | 1.78 | 1–4 | 0.715 |
| | 3 | 0.264 | 1.78 | 4–5 | −0.218 |
| $\Sigma E = -197.04$ ev | 4 | −0.396 | 1.93 | | |
| | 5 | −0.396 | 1.93 | | |
| 3–4 | 1 | 0.130 | 1.83 | 1–2 | 0.273 |
| $B_3C_2H_5$ | 2 | 0.130 | 1.83 | 1–3 | 0.223 |
| | 3 | 0.271 | 1.85 | 1–4 | 0.661 |
| $\Sigma E = -195.42$ ev | 4 | −0.273 | 1.79 | 1–5 | 0.663 |
| | 5 | −0.256 | 1.89 | 3–4 | 0.642 |
| | | | | 3–5 | 0.746 |
| | | | | 4–5 | −0.177 |

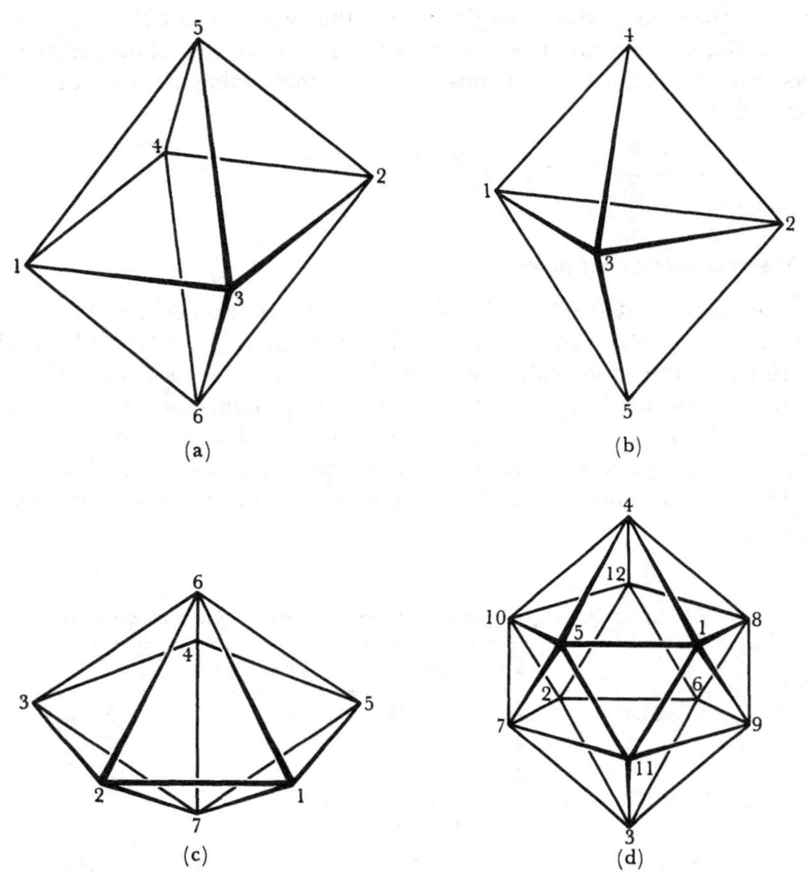


**Figure 3-7** *Labeling convention for the octahedron (a), trigonal bipyramid (b), pentagonal bipyramid (c), and icosahedron (d).*

compared with polyhedral fragments, is also strong evidence of aromatic-like stabilization, which can be described[121] either in resonance or molecular orbital languages. Therefore, one might expect a simple LCAO molecular orbital approach, such as that outlined above, to give at least as good an account of substituent chemistry in these polyhedral species as it does in the aromatic species, as given below. Also LCAO molecular orbital theory is applicable with reasonable results to heterocyclics, which are analogous to the carboranes, now to be discussed.

Parameters are 1.30 for the Slater exponent, $-15.36$ ev for boron $2s$, $-8.63$ ev for boron $2p$, $-21.01$ ev for carbon $2s$, $-11.17$ ev for carbon $2p$, $-23$ ev for K, and 1.75 A for all nearest distances. Use of 1.30 for orbital

Table 3–5 *Population analysis for the octahedral (tetragonal bipyramid) carboranes*

| | $k$ | $Q(k)$ | $\bar{F}(k)$ | $k$-$l$ | $n(k,l)$ |
|---|---|---|---|---|---|
| $B_6H_6^=$ | 1 | −0.333 | 2.11 | 1–2 | −0.179 |
| | 2 | −0.333 | 2.11 | 1–3 | 0.572 |
| | 3 | −0.333 | 2.11 | | |
| | 4 | −0.333 | 2.11 | | |
| $\Sigma E = -221.50$ | 5 | −0.333 | 2.11 | | |
| | 6 | −0.333 | 2.11 | | |
| 5–6 | 1 | 0.025 | 1.91 | 1–2 | −0.188 |
| $B_4C_2H_6$ | 2 | 0.025 | 1.91 | 1–3 | 0.449 |
| | 3 | 0.025 | 1.91 | 1–5 | 0.601 |
| | 4 | 0.025 | 1.91 | 4–5 | −0.364 |
| $\Sigma E = -231.96$ | 5 | −0.050 | 2.04 | | |
| | 6 | −0.050 | 2.04 | | |
| 4–5 | 1 | 0.003 | 1.93 | 1–2 | −0.180 |
| $B_4C_2H_6$ | 2 | 0.003 | 1.93 | 1–3 | 0.528 |
| | 3 | −0.114 | 2.02 | 1–4 | 0.525 |
| | 4 | 0.111 | 1.93 | 3–4 | −0.257 |
| $\Sigma E = -231.62$ | 5 | 0.111 | 1.93 | 3–5 | 0.604 |
| | 6 | −0.114 | 2.02 | 3–6 | 0.616 |
| | | | | 4–5 | 0.533 |

exponent of $C$ compensates in part for the use of 1.75 A for $B$—$C$, a value which is a bit large. A $3N$ type of calculation was made for all isomers of $B_NC_2H_{N+2}$ for $N = 3$ (trigonal bipyramid), $N = 4$ (octahedron), $N = 5$ (pentagonal bipyramid), and $N = 10$ (icosahedron).

In Tables 3–4 through 3–7 we list atomic charges $Q(k) = N(k) - A$ (where $A$ is 2 for boron and 3 for carbon), bound valences $\bar{F}(k)$, and all the important non-symmetry-equivalent overlap populations for the carboranes enumerated above. For purposes of comparison, the same parameters are also presented for the parent boron hydride ions, and the filled orbital energies are also given. In Table 3–8 the corresponding atom-atom polarizabilities matrices $\pi(k,e)$ are given; it should be noted that these have been reduced in a peculiar manner, to give the change in gross atomic population due to a perturbation of the Coulomb integral of the *surface* orbital.

It seems most likely that the "proved" reactivity indicators of aromatic theory,[303] such as the charge and free valence, might be significant here. We find further that varying the various parameters in the LCAO-MO computations leaves invariant the location of the most negative and most positive atoms, which would be the most likely candidates for electrophilic and nucleophilic

**Table 3-6** *Population analysis for the pentagonal bipyramid carboranes*

| | $k$ | $Q(k)$ | $\bar{F}(k)$ | $k$-$l$ | $n(k,l)$ |
|---|---|---|---|---|---|
| $B_7H_7^{=}$ | 1 | −0.418 | 2.11 | 1–2 | 0.698 |
| | 2 | −0.418 | 2.11 | 1–6 | 0.438 |
| | 3 | −0.418 | 2.11 | 6–7 | −0.166 |
| | 4 | −0.418 | 2.11 | | |
| | 5 | −0.418 | 2.11 | | |
| $\Sigma E = -252.84$ | 6 | 0.044 | 2.02 | | |
| | 7 | 0.044 | 2.02 | | |
| 6–7 | 1 | −0.161 | 1.98 | 1–2 | 0.595 |
| $B_5C_2H_7$ | 2 | −0.161 | 1.98 | 1–6 | 0.484 |
| | 3 | −0.161 | 1.98 | 6–7 | −0.554 |
| | 4 | −0.161 | 1.98 | | |
| | 5 | −0.161 | 1.98 | | |
| $\Sigma E = -260.90$ | 6 | 0.403 | 1.87 | | |
| | 7 | 0.403 | 1.87 | | |
| 3–5 | 1 | −0.179 | 2.01 | 1–2 | 0.714 |
| $B_5C_2H_7$ | 2 | −0.179 | 2.01 | 1–5 | 0.693 |
| | 3 | −0.171 | 2.04 | 1–6 | 0.396 |
| | 4 | 0.031 | 1.90 | 3–4 | 0.716 |
| | 5 | −0.171 | 2.04 | 3–6 | 0.444 |
| $\Sigma E = -263.78$ | 6 | 0.334 | 1.81 | 6–7 | −0.183 |
| | 7 | 0.334 | 1.81 | | |
| 4–5 | 1 | −0.165 | 2.00 | 1–2 | 0.704 |
| $B_5C_2H_7$ | 2 | 0.390 | 2.12 | 1–5 | 0.720 |
| | 3 | −0.165 | 2.00 | 1–6 | 0.385 |
| | 4 | 0.048 | 1.94 | 2–6 | 0.465 |
| | 5 | 0.048 | 1.94 | 6–7 | −0.175 |
| $\Sigma E = -263.30$ | 6 | 0.311 | 1.83 | 4–5 | 0.673 |
| | 7 | 0.311 | 1.83 | 4–6 | 0.384 |
| 5–7 | 1 | −0.087 | 1.95 | 1–2 | 0.650 |
| $B_5C_2H_7$ | 2 | −0.269 | 2.04 | 1–5 | 0.660 |
| | 3 | −0.269 | 2.04 | 1–6 | 0.391 |
| | 4 | −0.087 | 1.95 | 1–7 | 0.415 |
| | 5 | −0.046 | 1.95 | 2–3 | 0.646 |
| $\Sigma E = -262.98$ | 6 | 0.212 | 1.87 | 2–6 | 0.464 |
| | 7 | 0.546 | 1.91 | 2–7 | 0.483 |
| | | | | 5–6 | 0.451 |
| | | | | 5–7 | 0.411 |
| | | | | 6–7 | −0.293 |

Table 3-7 *Population analysis for the icosahedral carboranes*

| | $k$ | $Q(k)$ | $\bar{F}(k)$ | $k$-$l$ | $n(k,l)$ |
|---|---|---|---|---|---|
| $B_{12}H_{12}^{=}$ | 1 to | −0.1667 | 2.16 | 1–4 | 0.486 |
| $\Sigma E = -426.98$ | 12 | | | | |
| 1–4 ($\alpha$) | 1 | 0.294 | 2.08 | 1–4 | 0.506 |
| $B_{10}C_2H_{12}$ | 2 | −0.158 | 2.16 | 1–5 | 0.455 |
| | 3 | −0.158 | 2.16 | 1–9 | 0.522 |
| | 4 | 0.294 | 2.08 | 2–3 | 0.486 |
| | 5 | 0.081 | 2.00 | 2–6 | 0.485 |
| | 6 | −0.162 | 2.16 | 2–10 | 0.494 |
| | 7 | −0.162 | 2.16 | 5–7 | 0.502 |
| | 8 | 0.081 | 2.00 | 5–10 | 0.433 |
| | 9 | −0.027 | 2.08 | 6–9 | 0.493 |
| | 10 | −0.027 | 2.08 | 9–12 | 0.425 |
| $\Sigma E = -435.68$ | 11 | −0.027 | 2.08 | | |
| | 12 | −0.027 | 2.08 | | |
| 1–3 ($\beta$) | 1 | 0.148 | 2.15 | 1–4 | 0.504 |
| $B_{10}C_2H_{12}$ | 2 | −0.026 | 2.07 | 1–5 | 0.504 |
| | 3 | 0.148 | 2.15 | 1–9 | 0.517 |
| | 4 | −0.026 | 2.07 | 2–6 | 0.427 |
| | 5 | −0.030 | 2.08 | 2–10 | 0.494 |
| | 6 | −0.030 | 2.08 | 5–7 | 0.501 |
| | 7 | −0.030 | 2.08 | 5–10 | 0.494 |
| | 8 | −0.030 | 2.08 | 5–11 | 0.436 |
| | 9 | 0.101 | 1.99 | 9–11 | 0.364 |
| | 10 | −0.162 | 2.16 | 10–12 | 0.485 |
| $\Sigma E = -436.00$ | 11 | 0.101 | 1.99 | | |
| | 12 | −0.162 | 2.16 | | |
| 1–2 ($\gamma$) | 1 | 0.152 | 2.15 | 1–4 | 0.505 |
| $B_{10}C_2H_{12}$ | 2 | 0.152 | 2.15 | 3–6 | 0.427 |
| | 3 to | −0.031 | 2.08 | | |
| $\Sigma E = -436.00$ | 12 | −0.031 | 2.08 | | |

attack, respectively. Similarly, a low bound-valence probably indicates a susceptible position for free radical attack.

When derivatives of the carboranes are made, their reactivities will be of interest. By analogy with aromatic systems, it seems that substituents will be of greatest interest, e.g., atoms such as halogens or N, which have available $p$ orbitals that can interact with the surface orbitals of the polyhedron. It may be presumed, however, that these $\pi$ substituents will have a smaller effect on the properties of a carborane than $\sigma$ substituents. Consider as a first approxi-

**Table 3-8** *Atom-atom polarizabilities $\pi(k,e)$ for changes in surface Coulomb integrals in the carboranes (units of negative charge per $10^3$ ev)*

| 1-2 $B_3C_2H_5$ | | | | |
|---|---|---|---|---|
| −139 | 21 | 12 | 40 | 40 |
| 21 | −139 | 12 | 40 | 40 |
| 13 | 13 | −98 | 32 | 32 |
| 52 | 52 | 37 | −153 | 41 |
| 52 | 52 | 37 | 41 | −153 |

| 3-4 $B_3C_2H_5$ | | | | |
|---|---|---|---|---|
| −102 | 9 | 14 | 46 | 35 |
| 9 | −102 | 14 | 46 | 35 |
| 13 | 13 | −147 | 54 | 49 |
| 43 | 43 | 58 | −193 | 39 |
| 38 | 38 | 61 | 48 | −158 |

| $B_5H_5^{=}$ | | | | |
|---|---|---|---|---|
| −113 | 10 | 10 | 42 | 42 |
| 10 | −113 | 10 | 42 | 42 |
| 10 | 10 | −113 | 42 | 42 |
| 46 | 46 | 46 | −163 | 37 |
| 46 | 46 | 46 | 37 | −163 |

| 4-5 $B_3C_2H_5$ | | | | |
|---|---|---|---|---|
| −102 | 5 | 5 | 52 | 52 |
| 5 | −102 | 5 | 52 | 52 |
| 5 | 5 | −102 | 52 | 52 |
| 46 | 46 | 46 | −194 | 38 |
| 46 | 46 | 46 | 38 | −194 |

| $B_6H_6^{=}$ | | | | | |
|---|---|---|---|---|---|
| −116 | 12 | 26 | 26 | 26 | 26 |
| 12 | −116 | 26 | 26 | 26 | 26 |
| 26 | 26 | −116 | 12 | 26 | 26 |
| 26 | 26 | 12 | −116 | 26 | 26 |
| 26 | 26 | 26 | 26 | −116 | 12 |
| 26 | 26 | 26 | 26 | 12 | −116 |

| 5-6 $B_4C_2H_6$ | | | | | |
|---|---|---|---|---|---|
| −109 | 16 | 20 | 20 | 33 | 33 |
| 16 | −109 | 20 | 20 | 33 | 33 |
| 20 | 20 | −109 | 16 | 33 | 33 |
| 20 | 20 | 16 | −109 | 33 | 33 |
| 27 | 27 | 27 | 27 | −147 | 16 |
| 27 | 27 | 27 | 27 | 16 | −147 |

| 4-5 $B_4C_2H_6$ | | | | | |
|---|---|---|---|---|---|
| −108 | 14 | 22 | 30 | 30 | 22 |
| 14 | −108 | 22 | 30 | 30 | 22 |
| 21 | 21 | −112 | 18 | 30 | 26 |
| 26 | 26 | 14 | −146 | 38 | 27 |
| 26 | 26 | 27 | 38 | −146 | 14 |
| 21 | 21 | 26 | 30 | 18 | −112 |

| $B_7H_7^{=}$ | | | | | | |
|---|---|---|---|---|---|---|
| −124 | 43 | 4 | 4 | 43 | 18 | 18 |
| 43 | −124 | 43 | 4 | 4 | 18 | 18 |
| 4 | 43 | −124 | 43 | 4 | 18 | 18 |
| 4 | 4 | 43 | −124 | 43 | 18 | 18 |
| 43 | 4 | 4 | 43 | −124 | 18 | 18 |
| 16 | 16 | 16 | 16 | 16 | −93 | 2 |
| 16 | 16 | 16 | 16 | 16 | 2 | −93 |

| 6-7 $B_5C_2H_7$ | | | | | | |
|---|---|---|---|---|---|---|
| −116 | 35 | 4 | 4 | 35 | 24 | 24 |
| 35 | −116 | 35 | 4 | 4 | 24 | 24 |
| 4 | 35 | −116 | 35 | 4 | 24 | 24 |
| 4 | 4 | 35 | −116 | 35 | 24 | 24 |
| 35 | 4 | 4 | 35 | −116 | 24 | 24 |
| 19 | 19 | 19 | 19 | 19 | −122 | 3 |
| 19 | 19 | 19 | 19 | 19 | 3 | −122 |

| 4-5 $B_5C_2H_7$ | | | | | | |
|---|---|---|---|---|---|---|
| −120 | 43 | 3 | 8 | 51 | 14 | 14 |
| 41 | −124 | 41 | 4 | 4 | 18 | 18 |
| 3 | 43 | −120 | 51 | 8 | 14 | 14 |
| 7 | 3 | 45 | −156 | 56 | 18 | 18 |
| 45 | 3 | 7 | 56 | −156 | 18 | 18 |
| 12 | 16 | 12 | 19 | 19 | −86 | 2 |
| 12 | 16 | 12 | 19 | 19 | 2 | −86 |

| 3-5 $B_5C_2H_7$ | | | | | | |
|---|---|---|---|---|---|---|
| −121 | 42 | 4 | 5 | 52 | 15 | 15 |
| 42 | −121 | 52 | 5 | 4 | 15 | 15 |
| 3 | 44 | −153 | 43 | 4 | 20 | 20 |
| 5 | 5 | 52 | −116 | 52 | 12 | 12 |
| 44 | 3 | 4 | 43 | −153 | 20 | 20 |
| 13 | 13 | 21 | 10 | 21 | −85 | 3 |
| 13 | 13 | 21 | 10 | 21 | 3 | −85 |

**Table 3–8** (*continued*)

**5–7 $B_5C_2H_7$**

| | | | | | | |
|---|---|---|---|---|---|---|
| −116 | 39 | 4 | 6 | 48 | 15 | 21 |
| 37 | −120 | 39 | 4 | 5 | 18 | 23 |
| 4 | 39 | −120 | 37 | 5 | 18 | 23 |
| 6 | 4 | 39 | −116 | 48 | 15 | 21 |
| 41 | 4 | 4 | 41 | −151 | 20 | 28 |
| 13 | 16 | 16 | 13 | 20 | −89 | 3 |
| 16 | 18 | 18 | 16 | 26 | 3 | −120 |

**α 1–4 $B_{10}C_2H_{12}$**

| | | | | | | | | | | | |
|---|---|---|---|---|---|---|---|---|---|---|---|
| −125 | 2 | 1 | 27 | 18 | 1 | 1 | 18 | 20 | 2 | 20 | 2 |
| 3 | −98 | 18 | 1 | 1 | 18 | 18 | 1 | 1 | 18 | 1 | 18 |
| 1 | 18 | −98 | 3 | 1 | 18 | 18 | 1 | 18 | 1 | 18 | 1 |
| 27 | 1 | 2 | −125 | 18 | 1 | 1 | 18 | 2 | 20 | 2 | 20 |
| 21 | 1 | 1 | 21 | −93 | 2 | 18 | 2 | 2 | 15 | 15 | 2 |
| 2 | 18 | 18 | 2 | 2 | −98 | 1 | 18 | 18 | 1 | 1 | 18 |
| 1 | 18 | 18 | 1 | 18 | 1 | −98 | 2 | 1 | 18 | 18 | 1 |
| 21 | 1 | 1 | 21 | 2 | 18 | 2 | −93 | 15 | 2 | 2 | 15 |
| 23 | 1 | 18 | 2 | 2 | 18 | 1 | 15 | −95 | 2 | 16 | 1 |
| 2 | 18 | 1 | 23 | 15 | 1 | 18 | 2 | 2 | −95 | 1 | 16 |
| 23 | 1 | 18 | 2 | 15 | 1 | 18 | 2 | 16 | 1 | −95 | 2 |
| 2 | 18 | 1 | 23 | 2 | 18 | 1 | 15 | 1 | 16 | 2 | −95 |

**$B_{12}H_{12}^{=}$**

| | | | | | | | | | | | |
|---|---|---|---|---|---|---|---|---|---|---|---|
| −98 | 2 | 1 | 18 | 18 | 1 | 1 | 18 | 18 | 1 | 18 | 1 |
| 2 | −98 | 18 | 1 | 1 | 18 | 18 | 1 | 1 | 18 | 1 | 18 |
| 1 | 18 | −98 | 2 | 1 | 18 | 18 | 1 | 18 | 1 | 18 | 1 |
| 18 | 1 | 2 | −98 | 18 | 1 | 1 | 18 | 1 | 18 | 1 | 18 |
| 18 | 1 | 1 | 18 | −98 | 2 | 18 | 1 | 1 | 18 | 18 | 1 |
| 1 | 18 | 18 | 1 | 2 | −98 | 1 | 18 | 18 | 1 | 1 | 18 |
| 1 | 18 | 18 | 1 | 18 | 1 | −98 | 2 | 1 | 18 | 18 | 1 |
| 18 | 1 | 1 | 18 | 1 | 18 | 2 | −98 | 18 | 1 | 1 | 18 |
| 18 | 1 | 18 | 1 | 1 | 18 | 1 | 18 | −98 | 2 | 18 | 1 |
| 1 | 18 | 1 | 18 | 18 | 1 | 18 | 1 | 2 | −98 | 1 | 18 |
| 18 | 1 | 18 | 1 | 18 | 1 | 18 | 1 | 18 | 1 | −98 | 2 |
| 1 | 18 | 1 | 18 | 1 | 18 | 1 | 18 | 1 | 18 | 2 | −98 |

**γ 1–2 $B_{10}C_2H_{12}$**

| | | | | | | | | | | | |
|---|---|---|---|---|---|---|---|---|---|---|---|
| −126 | 3 | 1 | 20 | 20 | 1 | 1 | 20 | 20 | 1 | 20 | 1 |
| 3 | −126 | 20 | 1 | 1 | 20 | 20 | 1 | 1 | 20 | 1 | 20 |
| 1 | 23 | −95 | 2 | 1 | 16 | 16 | 1 | 18 | 2 | 18 | 2 |
| 23 | 1 | 2 | −95 | 16 | 1 | 1 | 16 | 2 | 18 | 2 | 18 |
| 23 | 1 | 1 | 16 | −95 | 2 | 18 | 2 | 2 | 18 | 16 | 1 |
| 1 | 23 | 16 | 1 | 2 | −95 | 2 | 18 | 18 | 2 | 1 | 16 |
| 1 | 23 | 16 | 1 | 18 | 2 | −95 | 2 | 1 | 16 | 18 | 2 |
| 23 | 1 | 1 | 16 | 2 | 18 | 2 | −95 | 16 | 1 | 2 | 18 |
| 23 | 1 | 18 | 2 | 2 | 18 | 1 | 16 | −95 | 2 | 16 | 1 |
| 1 | 23 | 2 | 18 | 18 | 2 | 16 | 1 | 2 | −95 | 1 | 16 |
| 23 | 1 | 18 | 2 | 16 | 1 | 18 | 2 | 16 | 1 | −95 | 2 |
| 1 | 23 | 2 | 18 | 1 | 16 | 2 | 18 | 1 | 16 | 2 | −95 |

**Table 3-8** *(continued)*

| $\beta$ 1-3 $B_{10}C_2H_{12}$ | | | | | | | | | | | |
|---|---|---|---|---|---|---|---|---|---|---|---|
| −126 | 2 | 1 | 20 | 20 | 1 | 1 | 20 | 20 | 1 | 20 | 1 |
| 3 | −95 | 23 | 1 | 1 | 16 | 16 | 1 | 1 | 18 | 1 | 18 |
| 1 | 20 | −126 | 2 | 1 | 20 | 20 | 1 | 20 | 1 | 20 | 1 |
| 23 | 1 | 3 | −95 | 16 | 1 | 1 | 16 | 1 | 18 | 1 | 18 |
| 24 | 1 | 1 | 16 | −95 | 2 | 18 | 1 | 2 | 18 | 16 | 1 |
| 1 | 16 | 24 | 1 | 2 | −95 | 1 | 18 | 16 | 1 | 2 | 18 |
| 1 | 16 | 24 | 1 | 18 | 1 | −95 | 2 | 2 | 18 | 16 | 1 |
| 24 | 1 | 1 | 16 | 1 | 18 | 2 | −95 | 16 | 1 | 2 | 18 |
| 23 | 1 | 23 | 1 | 1 | 16 | 1 | 16 | −92 | 2 | 13 | 1 |
| 1 | 18 | 1 | 18 | 18 | 1 | 18 | 1 | 2 | −98 | 1 | 18 |
| 23 | 1 | 23 | 1 | 16 | 1 | 16 | 1 | 13 | 1 | −92 | 2 |
| 1 | 18 | 1 | 18 | 1 | 18 | 1 | 18 | 1 | 18 | 2 | −98 |

mation the effect of a $\pi$ substituent as a change in the corresponding surface Coulomb integral. The alteration of the charge distribution is then easily computed with the $\pi(k,l)$ tabulated. For instance, suppose we have a substituent on atom 1 in 3-4 $B_3C_2H_5$. The appropriate polarizabilities are then found in the first column of the 3-4 $B_3C_2H_5$ matrix in Table 3-8. If the effect of the $\pi$ substituent is approximated by lowering $\alpha(1, \text{surface})$ by 2.0 ev, i.e., $\Delta\alpha = -2.0$, then $\Delta Q(k) = -2.0\pi(k,l)$, and the new charge distribution will be approximately −0.074, +0.148, +0.297, −0.187, −0.180, for atoms 1 through 5, respectively.

### Sequential Substitution[122] in $B_{10}H_{10}^{=}$ and $B_{12}H_{12}^{=}$

As a further study of the pseudoaromatic nature of polyhedral $B_{10}H_{10}^{=}$ of $D_{4d}$ symmetry[141,182] and $B_{12}H_{12}^{=}$ of icosahedral symmetry,[336] we consider the paths of sequential halogen replacement of H atoms. These compounds form a very interesting series of substitution derivatives,[141,149] as yet incompletely identified, of the type $B_{10}H_{10-n}X_n^{=}$, where $0 \leq n \leq 10$, and X = F, Cl, Br, or I. This indication that a large number of sequentially substituted derivatives can be prepared has led us to make predictions including resonance and inductive effects in an LCAO-MO description.

Formulated in terms of a specific example, the question investigated here is the following. Suppose that attack by a halogen is electrophilic and that successive halogen atoms attack sequentially; what can be said theoretically about the positions that successive halogen atoms substitute on the boron framework? To answer this question we first perform an LCAO-MO calculation of the $3N$ type on the $B_N$ framework, choosing the three hybrids as one, mostly *s*, pointing *in*, and two, pure *p*, directed along the polyhedron surface.

As a result of a Mulliken population analysis we then obtain a charge distribution for the molecule. We then assume that electrophilic attack will be most likely to take place at the most negative site of the polyhedron, this being an apex in $B_{10}H_{10}^{=}$ [$Q$ (apex) = $-0.535$, $Q$ (equatorial) = $-0.116$], or any position in $B_{12}H_{12}^{=}$. Once the substituent is on we inquire about the new charge distribution. A more complex calculation is then avoided by simulating the effect of the substituted boron. For this purpose the required atom-atom polarizabilities, i.e., changes in charge with respect to variation of an *in* or *surface* $\alpha$, have been computed. These polarizabilities are given in Table 3–9, except for $(\partial q/\partial \alpha)_{\text{surface}}$ for $B_{12}H_{12}^{=}$, which have already been written down elsewhere.[120]

We now make the following association with aromatic reaction theory. We presume we have substituents which act as sources or sinks of electronic charge by either a resonance ($R$) or an inductive ($I$) mechanism. The resonance interaction, probably not predominant, stabilizes structures in which substituents have $\pi$ orbitals which are available for extension of polyhedral delocalization. We simulate a resonance effect by a change in the *surface* Coulomb integral of the substituted site, an inductive effect by a change in the corresponding *in* Coulomb integral. For $+R$, $+I$ the substituent is a source of electrons; this effect is approximated by making the substituted site positive, i.e., $\Delta\alpha > 0$. For $-R$, $-I$ the substituent is a sink and we therefore make the substituted site negative by setting $\Delta\alpha < 0$.

The first clear result is that in apex-substituted $B_{10}H_9X$ electrophilic substitution should proceed at the other apex regardless of the nature of the substituent. Subsequently there will be eight equivalent sites left. From this point on we consider each molecule separately.

**Case I:** $+R$, $+I$; $\Delta\alpha_i > 0$

$B_{12}H_{12}^{=}$ (Fig. 3–7): Neighboring atoms become most negative and substitution propagates around the first substituted site. If site 1 is attacked first, the next substituent should go on at position 4,5,8,9, or 11. If it goes at 4, positions 8,10,11 succumb next, etc. The notation we shall use to indicate the process is a sequence of 12 numbers which give, in order from left to right, one symmetry equivalent pathway for sequential attack at the enumerated sites. Thus for the case discussed above, the sequence is

1,4,5,8,9,11,3,6,7,2,10,12

$B_{10}H_{10}^{=}$ (Fig. 3–7): As stated above, the sequence is apex 1, apex 2, followed by

3,7,4,9,6,8,5,10

**Case II:** $-R$, $-I$; $\Delta\alpha_i < 0$

$B_{12}H_{12}^{=}$: Here there are 15 symmetry nonequivalent pathways to fill the polyhedron, listed in Table 3–10; relative multiplicities of the paths are also

**Table 3–9** *Atom-atom polarizabilities ($10^{-3}$ units of electronic charge/ev)*

| $B_{12}H_{12}^{=}$ *in* | | | | | | | | | | | |
|---|---|---|---|---|---|---|---|---|---|---|---|
| −52 | 0 | 0 | 11 | 11 | 0 | 0 | 11 | 11 | 0 | 11 | 0 |
| 0 | −52 | 11 | 0 | 0 | 11 | 11 | 0 | 0 | 11 | 0 | 11 |
| 0 | 11 | −52 | 0 | 0 | 11 | 11 | 0 | 11 | 0 | 11 | 0 |
| 11 | 0 | 0 | −52 | 11 | 0 | 0 | 11 | 0 | 11 | 0 | 11 |
| 11 | 0 | 0 | 11 | −52 | 0 | 11 | 0 | 0 | 11 | 11 | 0 |
| 0 | 11 | 11 | 0 | 0 | −52 | 0 | 11 | 11 | 0 | 0 | 11 |
| 0 | 11 | 11 | 0 | 11 | 0 | −52 | 0 | 0 | 11 | 11 | 0 |
| 11 | 0 | 0 | 11 | 0 | 11 | 0 | −52 | 11 | 0 | 0 | 11 |
| 11 | 0 | 11 | 0 | 0 | 11 | 0 | 11 | −52 | 0 | 11 | 0 |
| 0 | 11 | 0 | 11 | 11 | 0 | 11 | 0 | 0 | −52 | 0 | 11 |
| 11 | 0 | 11 | 0 | 11 | 0 | 11 | 0 | 11 | 0 | −52 | 0 |
| 0 | 11 | 0 | 11 | 0 | 11 | 0 | 11 | 0 | 11 | 0 | −52 |

| $B_{10}H_{10}^{=}$ *surface* | | | | | | | | | |
|---|---|---|---|---|---|---|---|---|---|
| −179 | 24 | 40 | 40 | 40 | 40 | 4 | 4 | 4 | 4 |
| 24 | −179 | 4 | 4 | 4 | 4 | 40 | 40 | 40 | 40 |
| 35 | 4 | −121 | 17 | 3 | 17 | 20 | 0 | 20 | 0 |
| 35 | 4 | 17 | −121 | 17 | 3 | 20 | 0 | 0 | 20 |
| 35 | 4 | 3 | 17 | −121 | 17 | 0 | 20 | 0 | 20 |
| 35 | 4 | 17 | 3 | 17 | −121 | 0 | 20 | 20 | 0 |
| 4 | 35 | 20 | 20 | 0 | 0 | −121 | 3 | 17 | 17 |
| 4 | 35 | 0 | 0 | 20 | 20 | 3 | −121 | 17 | 17 |
| 4 | 35 | 20 | 0 | 0 | 20 | 17 | 17 | −121 | 3 |
| 4 | 35 | 0 | 20 | 20 | 0 | 17 | 17 | 3 | −121 |

| $B_{10}H_{10}^{=}$ *in* | | | | | | | | | |
|---|---|---|---|---|---|---|---|---|---|
| −87 | 0 | 12 | 12 | 12 | 12 | 0 | 0 | 0 | 0 |
| 0 | −87 | 0 | 0 | 0 | 0 | 12 | 12 | 12 | 12 |
| 21 | 1 | −80 | 15 | −1 | 15 | 19 | 0 | 19 | 0 |
| 21 | 1 | 15 | −80 | 15 | −1 | 19 | 0 | 0 | 19 |
| 21 | 1 | −1 | 15 | −80 | 15 | 0 | 19 | 0 | 19 |
| 21 | 1 | 15 | −1 | 15 | −80 | 0 | 19 | 19 | 0 |
| 1 | 21 | 19 | 19 | 0 | 0 | −80 | −1 | 15 | 15 |
| 1 | 21 | 0 | 0 | 19 | 19 | −1 | −80 | 15 | 15 |
| 1 | 21 | 19 | 0 | 0 | 19 | 15 | 15 | −80 | −1 |
| 1 | 21 | 0 | 19 | 19 | 0 | 15 | 15 | −1 | −80 |

given. Enantiomeric pathways are not distinguished. For $-R$ the matter is complicated by the presence often of two possible pathways of similar but unequal probabilities. The source of the difficulty is that for $-R$ if, say, $\Delta\alpha_1 < 0$, then the most negative sites are 3,6,7,10,12 (a) but site 2(b) is almost as negative and may react also. The final column of Table 3–10 gives a sequence of *a*'s and *b*'s which indicate how many "decisions" of this type are confronted

**Table 3–10** *Pathways for electrophilic substitution on $B_{12}H_{12}^{=}$, $B_{10}H_{10}^{=}$*

| Pathway | | | | | | | | | | | | Multiplicity | Weight |
|---|---|---|---|---|---|---|---|---|---|---|---|---|---|
| | | | | | | **$B_{12}H_{12}^{=}$** | | | | | | | |
| 1 | 3 | 10 | 6 | 7 | 12 | 4 | 9 | 5 | 2 | 8 | 11 | $133\frac{1}{3}$ | *aaa* |
| 1 | 3 | 10 | 6 | 7 | 4 | 11 | 8 | 2 | 5 | 9 | 12 | $133\frac{1}{3}$ | *aaba* |
| 1 | 3 | 10 | 6 | 7 | 4 | 8 | 11 | 2 | 5 | 9 | 12 | $133\frac{1}{3}$ | *aabb* |
| 1 | 3 | 10 | 6 | 4 | 7 | 11 | 8 | 2 | 5 | 9 | 12 | 100 | *aba* |
| 1 | 3 | 10 | 6 | 4 | 7 | 8 | 11 | 2 | 5 | 9 | 12 | 100 | *abb* |
| 1 | 2 | 3 | 5 | 8 | 10 | 11 | 9 | 12 | 7 | 4 | 6 | 48 | *baa* |
| 1 | 2 | 3 | 5 | 8 | 10 | 6 | 11 | 4 | 7 | 9 | 12 | 48 | *bab* |
| 1 | 2 | 3 | 4 | 5 | 8 | 9 | 12 | 7 | 6 | 10 | 11 | 4 | *bbaaa* |
| 1 | 2 | 3 | 4 | 5 | 8 | 9 | 7 | 12 | 6 | 10 | 11 | 8 | *bbaab* |
| 1 | 2 | 3 | 4 | 5 | 8 | 6 | 10 | 11 | 7 | 9 | 12 | 4 | *bbaba* |
| 1 | 2 | 3 | 4 | 5 | 8 | 6 | 7 | 9 | 10 | 12 | 11 | 1 | *bbabba* |
| 1 | 2 | 3 | 4 | 5 | 8 | 6 | 7 | 9 | 10 | 11 | 12 | 1 | *bbabbb* |
| 1 | 2 | 3 | 4 | 5 | 6 | 7 | 9 | 12 | 8 | 10 | 11 | 4 | *bbba* |
| 1 | 2 | 3 | 4 | 5 | 6 | 7 | 8 | 9 | 12 | 10 | 11 | 1 | *bbbba* |
| 1 | 2 | 3 | 4 | 5 | 6 | 7 | 8 | 9 | 10 | 11 | 12 | 1 | *bbbbb* |
| | | | | | | | | | | | | 720 | |
| | | | | | | **$B_{10}H_{10}^{=}$** | | | | | | | |
| | | 1 | 2 | 3 | 5 | 7 | 8 | 4 | 6 | 9 | 10 | 1 | *aa* |
| | | 1 | 2 | 3 | 5 | 7 | 8 | 4 | 9 | 6 | 10 | 1 | *ab* |
| | | 1 | 2 | 3 | 8 | 4 | 5 | 9 | 7 | 6 | 10 | 2 | *ba* |
| | | 1 | 2 | 3 | 8 | 4 | 9 | 5 | 7 | 6 | 10 | 2 | *bb* |
| | | | | | | | | | | | | 6 | |

in following a given path. To find the multiplicity of a given route being traveled, the geometric multiplicity must be multiplied by an appropriate weighting factor $a^n b^m$, where $a/2$ is the probability of choosing $a$ when confronted with one $a$ and one $b$; $b = 2 - a$, $1 \leq a \leq 2$. For $-I$, $a = b = 1$, and the criterion of the distinguishability of pathways is purely geometrical.

$B_{10}H_{10}^{=}$: For $-R$ one sequence is predominant; apex 1, apex 2, followed by

$$3,8,4,9,5,10,6,7$$

For $-I$ an ambiguity similar to the one for $-R$ in $B_{12}H_{12}^{=}$ arises. There are four alternate pathways, which are listed, along with their multiplicities and weights in Table 3–10.

From the above considerations we may thus attempt to make a prediction of the distribution of geometrical isomers for a given $n$ in $B_N H_{N-n} X_n^{=}$.

In criticism of the above treatment there are factors related to the validity of our basic assumptions which may require an extension of this relatively simple theory, if experimental results so demand: (a) An assumption, such as sequential, instead of coordinated, attack by halogen may have to be modified.

**Table 3-11** *Isomer ratios in $B_{12}X_nH_{12-n}^{=}$ and $B_{10}X_nH_{10-n}^{=}$*

| $n$ | Isomer type | Relative frequency |
|---|---|---|
| | | $B_{12}H_{12}^{=}$ |
| 2 | 1,3 | $600a$ |
| | 1,2 | $120b$ |
| 3 | 1,3,10 | $600a$ |
| | 1,2,3 | $120b$ |
| 4 | 1,3,10,6 | $600a$ |
| | 1,2,3,5 | $96ba$ |
| | 1,2,3,4 | $24b^2$ |
| 5 | 1,3,10,6,7 | $400a^2$ |
| | 1,3,10,6,4 | $296ab$ |
| | 1,2,3,4,5 | $24b^2$ |
| 6 | 1,3,10,6,7,12 | $133\frac{1}{3}a^3$ |
| | 1,3,10,6,7,4 | $266\frac{2}{3}a^2b + 296ab$ |
| | 1,2,3,4,5,8 | $18b^2a$ |
| | 1,2,3,4,5,6 | $6b^3$ |
| 7 | 1,3,10,6,7,12,4 | $133\frac{1}{3}a^3 + 133\frac{1}{3}a^3b + 148a^2b$ |
| | 1,3,10,6,7,4,8 | $133\frac{1}{3}a^2b^2 + 148ab^2$ |
| | 1,2,3,4,5,8,9 | $12b^2a^2$ |
| | 1,2,3,4,5,6,7 | $6b^3a + 6b^3$ |
| 8 | 1,3,4,6,7,9,10,12 | $133\frac{1}{3}a^3 + 133\frac{1}{3}(a^3b + a^2b^2) + 148a^2b + 100b^2a$ |
| | 1,2,3,5,8,9,10,11 | $48a^2b + 4b^2a^3$ |
| | 1,2,3,4,5,7,8,9 | $12b^3a^2 + 4b^3a$ |
| | 1,2,3,4,5,6,7,8 | $2b^4a + 2b^4$ |
| 9 | 1,3,4,5,6,7,9,10,12 | $133\frac{1}{3}(a^3 + a^3b + a^2b^2) + 148(a^2b + b^2a) + 4b^2a^3 + 12b^3a^2 + 4b^3a$ |
| | 1,2,3,4,5,6,7,8,9 | $b^4a + b^5 + b^4a^2 + b^5a$ |
| 10 | 1,2,3,4,5,6,7,9,10,12 | $133\frac{1}{3}(a^3 + a^3b + a^2b^2) + 148(a^2b^2 + b^2a) + 4b^2a^3 + 12b^3a^2 + 4b^3a + b^4a^2 + b^4a$ |
| | 1,2,3,4,5,6,7,8,9,10 | $b^3a^2 + b^5$ |
| | | $B_{10}H_{10}^{=}$ |
| 4 | 1,2,3,5 | $2a$ |
| | 1,2,3,8 | $4b$ |
| 5 | 1,2,3,5,7 | $2a$ |
| | 1,2,3,8,4 | $4b$ |
| 6 | 1,2,3,5,7,8 | $2a$ |
| | 1,2,3,8,4,5 | $2ba$ |
| | 1,2,3,8,4,9 | $2b^2$ |
| 7 | 1,2,3,5,7,8,4 | $2a$ |
| | 1,2,3,8,4,5,9 | $2ba + 2b^2$ |
| 8 | 1,2,3,4,5,6,7,8 | $a^2$ |
| | 1,2,3,4,5,7,8,9 | $3ab + 2b^2$ |

(b) There may be competing rearrangements of the framework, such as those discussed[141] for $B_{10}H_{10}^{=}$. For example, a motion of only about 0.35 A of B atoms will convert an icosahedron into a cube-octahedron. Two such consecutive transformations may lead to isomerization of 1,4-$B_{12}X_2H_{10}^{=}$ (ortho) to 1,10-$B_{12}X_2H_{10}^{=}$ (meta), but the 1,2 isomer (para) cannot arise from either the ortho or meta by this mechanism.[124]

Nevertheless, we think it worthwhile to compute expected isomer ratios for $B_NX_nH_{N-n}^{=}$. For $+R$, $+I$ only one sequence is predicted for both $B_{10}H_{10}^{=}$ and $B_{12}H_{12}^{=}$, thus no direct isomerism. For $-$R there is also one predominant sequence for $B_{10}H_{10}^{=}$ but a variety of pathways for $B_{12}H_{12}^{=}$. Expected ratios are 1,3-$B_{12}X_2H_{10}^{=}$/1,2-$B_{12}X_2H_{10}^{=}$ = $5a/b$. Other ratios are given in Table 3–11. For $+I$, the $B_{12}H_{12}^{=}$ ratios are the same as for $-R$, but with $a = b = 1$. The $B_{10}H_{10}^{=}$ $-I$ ratios are also given in Table 3–11.

It is probably too early to extend these results into a more generally valid substituent theory of these polyhedral molecules in a way comparable with the present theory of aromatic molecules. However, such an extension is clearly possible, and will be made if these preliminary theoretical findings are closely verified by experiment. The preliminary comparisons on $B_{10}H_{10}^{=}$ are at least encouraging, and hence the area of polyhedral molecules may form an ideal test for LCAO-MO theories of reactivity.

## 3–4 Molecular Orbitals in the Boron Hydrides and Their Ions

### Outline of the Theory and Calculations

A computer program described previously[120,121] has been extended to calculate the energy levels and wave functions of the known boron hydrides and borohydride ions.[123] The input parameters consist of the cartesian coordinates of $N$ arbitrarily situated boron atoms and $M$ hydrogens ($4N + M \leq 68$, but this limitation may be removed), boron $2s$ and $2p$ and hydrogen $1s$ Coulomb integrals, and the proportionality constant $K$, in the relation $\beta_{rs} = KS_{rs}$, where $S$ is the overlap between Slater orbitals and $K$ is taken equal for all interactions. The set of equations $\Sigma_s(H_{rs} - ES_{rs})C_s = 0$ is solved with all interactions or overlaps included. The program also performs a Mulliken population analysis on the resulting molecular orbitals, calculating overlap populations and gross atomic populations (charges). The boron Coulomb integrals were rather arbitrarily set equal to the corresponding valence-state ionization potentials, $\alpha(2s) = -15.36$ ev and $\alpha(2p) = -8.63$ ev. The value of $K$ was set equal to $-21$ ev throughout, primarily to compare with earlier calculations.[198]

The input coordinates for the known boron hydrides were in some cases ($B_{10}H_{14}$, $B_9H_{15}$, $B_{10}H_{10}^{=}$, $B_{12}H_{12}^{=}$) adjusted to yield the apparent symmetry planes; in the others the original parameters were retained and the symmetriza-

**Table 3–12**

| $n$ | Ours | Yamazaki's |
|---|---|---|
| $H_t$—B | 0.86 | 0.85 |
| $H_b$—B | 0.45 | 0.39 |
| B—B | 0.38 | 0.34 |

tion, when appropriate, applied to the results of the population analysis. For $B_5H_9$, considerations of consistency led us to use the X-ray-determined distances rather than the presumably more accurate microwave values. The ions $B_{10}H_{14}^{=}$ and $B_9H_{14}^{-}$ were included, even though the structure determinations have been performed only on their acetonitrile analogues, $B_{10}H_{12}(NCCH_3)_2$ and $B_9H_{13}NCCH_3$. The substituted hydrogens were replaced in these molecules at a distance of 1.21 A from the substituted boron, in the direction of the acetonitrile N, as found in the corresponding X-ray study. The framework projections and the labeling convention used in this book are given in Fig. 2–26.

In the first calculation on $B_2H_6$ the Coulomb integral for all hydrogens was set equal to $-13.61$ ev. The resulting charge distribution was: B, 0.233; $H_t$, $-0.156$; $H_b$, 0.080. These charges are in disagreement with the approximate, but nonempirical, SCF calculation of Yamazaki,[337] who obtains: B, 0.22; $H_t$, 0.01; $H_b$, $-0.24$.

There is, however, better agreement for the overlap populations (Table 3–12). No reasonable variation of any Coulomb integral or of $K$ was found which would make the terminal hydrogens more positive than those on the bridge, while preserving actual B—H distances. When the computation was repeated omitting the terminal hydrogens and the four $sp^2$ boron hybrids pointing toward them, the bridge hydrogens showed up negative, a result to be expected in view of the greater magnitude of the H Coulomb integral. The procedure of dropping the terminal hydrogens was deemed inadvisable for the higher hydrides, since certain boron orbitals would have to be omitted as well, and no consistent procedure for doing so was discovered.

The alternative chosen was to perform each calculation twice; once with all hydrogen Coulomb integrals equal to $-13.61$ ev (method I) and again after forcing the bridge hydrogens to be negative by using a different Coulomb integral for bridge and terminal hydrogens. If in diborane we take $\alpha(H_t) = -11.60$ ev, $\alpha(H_b) = -15.75$ ev (method II), we obtain the charge distribution: B, 0.147; $H_t$, $-0.004$; $H_b$, $-0.135$. This same choice of $\alpha(H)$ kept the charge on the bridge hydrogens between $-0.08$ and $-0.17$, and that on the terminal hydrogens between $-0.03$ and $+0.05$ for all the hydrides, with the exception of the unique terminal hydrogen in $B_5H_{11}$, which showed up very positive ($+0.30$). We optimistically anticipated that the boron charge distribution would not depend drastically on either choice of the hydrogen Coulomb integral.

It may well be that after reexamination and correction of errors in the LCAO-SCF calculation[337] of $B_2H_6$ that we shall reopen the question of the charge on the bridge H atoms, but at the present time it would appear that these bridge H atoms are indeed more negative than the terminal H, and therefore the results of method II are somewhat to be preferred over those of method I.

Large energy gaps for the known structures are obtained, a surprising additivity is encountered for the electronic energies, and the charge distributions appear realistic. We discuss these results in order.

### Energy Values

Energy gaps and ionization potentials are given in Table 3–13. Both the energies of the top occupied orbital and the magnitude of the gap to the lowest unfilled orbital are given. These gaps are nicely large, with no striking discrepancy between calculations I and II. It is notoriously difficult to predict electronic absorption spectra from calculations that do not explicitly incorporate electronic interactions. Experimentally the situation is not entirely adequate, but it leads to the following list of boron hydrides in estimated order of increasing wavelength of the first intense absorption band: $B_2H_6$,[243] $B_5H_9$,[238] $B_{12}H_{12}^{=}$,[123] $B_{10}H_{10}^{=}$,[123] $B_{10}H_{14}$.[232] This order is in fair agreement with the order of the corresponding calculated energy gaps.

**Table 3–13** *Highest occupied orbital energies and gaps to lowest unfilled orbitals in the boron hydrides*

| | Real | | | | | |
|---|---|---|---|---|---|---|
| | *E* top filled, ev | | Gap, ev | | Number of filled | Experimental |
| Molecule | I | II | I | II | orbitals | I.P., ev* |
| $B_2H_6$ | −14.08 | −19.95 | 9.21 | 8.18 | 6 | 12.1,11.9,11–12 |
| $B_4H_{10}$ | −13.04 | −13.05 | 12.59 | 12.48 | 11 | |
| $B_5H_9$ | −12.21 | −11.90 | 9.81 | 9.75 | 12 | 10.8,10.4,10.9 |
| $B_5H_{11}$ | −12.62 | −12.26 | 8.86 | 8.39 | 13 | |
| $B_6H_{10}$ | −12.37 | −12.08 | 7.93 | 7.23 | 14 | |
| $B_9H_{15}$ | −11.84 | −11.48 | 6.59 | 6.07 | 21 | |
| $B_{10}H_{14}$ | −12.83 | −12.38 | 6.75 | 6.32 | 22 | 11.0,10.7,10.26 |
| $B_{10}H_{16}$ | −11.38 | −11.24 | 8.64 | 8.73 | 23 | |
| $B_{10}H_{10}^{=}$ | −11.68 | −11.56 | 9.13 | 9.22 | 21 | |
| $B_{10}H_{14}^{=}$ | −11.34 | −11.16 | 9.57 | 9.73 | 23 | |
| $B_{12}H_{12}^{=}$ | −12.60 | −11.98 | 14.11 | 13.49 | 25 | |
| $BH_4^-$ | −14.67 | −13.75 | 26.54 | 29.28 | 4 | |
| $B_3H_8^-$ | −13.58 | −12.63 | 15.24 | 13.59 | 9 | |
| $B_9H_{14}^-$ | −11.89 | −11.32 | 6.59 | 6.67 | 21 | |

* For original sources see Ref. 123.

**Table 3-14** *Total electronic and binding energies*

| Molecule | Three-center bond parameters* | | | | Total energy $\Sigma E_i$, ev | | Binding energy* $\Sigma E_i^\infty - \Sigma E_i$ | | Calculated $\Sigma E_i$ | |
|---|---|---|---|---|---|---|---|---|---|---|
| | $s$ | $t$ | $y$ | $x$ | I | II | I | II | I† | II‡ |
| $B_2H_6$ | 2 | 0 | 0 | 4 | −191.6 | −188.7 | 44.7 | 45.5 | −190.5 | −187.1 |
| $B_4H_{10}$ | 4 | 0 | 1 | 6 | −347.0 | −344.4 | 80.4 | 81.3 | −346.8 | −343.7 |
| $B_5H_9$ | 4 | 1 | 2 | 5 | −375.3 | −374.4 | 89.7 | 90.3 | −374.8 | −373.6 |
| $B_5H_{11}$ | 3 | 2 | 0 | 8 | −408.2 | −401.4 | 95.3 | 98.2 | −410.4 | −401.6 |
| $B_6H_{10}$ | 4 | 2 | 2 | 6 | −437.5 | −434.8 | 105.7 | 106.5 | −437.0 | −434.0 |
| $B_9H_{15}$ | 5 | 4 | 2 | 10 | −655.5 | −648.4 | 157.8 | 160.0 | −656.2 | −648.4 |
| $B_{10}H_{14}$ | 4 | 6 | 2 | 10 | −686.1 | −676.9 | 169.4 | 171.7 | −685.6 | −675.7 |
| $B_{10}H_{16}$ | 8 | 2 | 5 | 8 | −715.4 | −716.7 | 171.5 | 171.7 | −715.4 | −716.7 |
| $B_{10}H_{10}^=$ | 0 | 8 | 3 | 10 | −643.6 | −628.5 | 181.3 | 186.3 | −643.6 | −628.5 |
| $B_{12}H_{12}^=$ | 0 | 10 | 3 | 12 | −768.4 | −750.6 | 213.7 | 220.0 | −767.9 | −749.4 |
| $B_{10}H_{14}^=$ | 2 | 6 | 3 | 12 | −706.9 | −691.8 | 190.2 | 194.9 | −709.8 | −694.8 |
| $B_3H_8^-$ | 2 | 0 | 1 | 6 | −280.7 | −275.0 | 74.0 | 76.0 | −280.7 | −275.0 |
| $B_9H_{14}^-$ | 2 | 6 | 1 | 12 | −649.7 | −635.3 | 165.6 | 171.0 | −653.7 | −637.5 |
| $BH_4^-$ | 0 | 0 | 0 | 4 | −124.6 | −118.4 | 37.6 | 39.4 | −124.4 | −118.4 |

* For negative ions, $\Sigma E_i^\infty$ is for neutral core only, and does not include $E(e^-)$ at $\infty$. $E_i^\infty(B) = 2\alpha(2sB) + \alpha(2pB)$; $E_i^\infty(H) = \alpha(1sH)$.

† $s = 33.04$, $t = 31.06$, $y = 28.04$, $x' = 31.10$.

‡ $s = 34.37$, $t = 30.82$, $y = 28.66$, $x' = 29.60$.

Ionization potentials have fared poorly experimentally. To our knowledge only three have been measured by various workers and are reproduced in the last column of Table 3-13. This number is too small to make a comparison, but it is clear that the calculated ionization potentials, i.e., the energies of the top filled orbitals, are somewhat too large, a situation that may be remedied by an appropriate adjustment of parameters.

Electronic energies (Table 3-14) do not include nuclear repulsion. In the Hückel theory the one-electron energies are the eigenvalues of a hamiltonian matrix, the matrix elements of which are integrals with respect to an unspecified hamiltonian $H_{eff}$. Now in constructing this hamiltonian one may or may not subtract out nuclear repulsions beforehand. The choice is purely formal, since one never carried out any quadratures with $H_{eff}$, but only guesses at its matrix elements. From many as-yet-unpublished calculations on the conformations of simple molecules, we have amassed a considerable body of evidence showing that it is unnecessary to add nuclear repulsions to the energies obtained, as the repulsions are largely simulated by our choice of the matrix elements of $H_{eff}$. Thus we believe that the effect of nuclear repulsions is largely included in the parametric assumptions when this $H_{eff}$ is employed.

A calculation has also been carried out for various $BH_3$ geometries. The molecule seems to prefer a planar equilateral triangle conformation with $\Sigma E_i = -95.63$ ev for all B—H = 1.21 A (calculation of type I), and $\Sigma E_i = -95.02$ ev for a geometry of half of $B_2H_6$. Thus diborane is calculated with our parameters to be 1.53 ev more stable than $2BH_3$ of diborane fragment geometry and 0.31 ev more stable than the most favorable $BH_3$ conformation.[24f]

### Heats of Formation

Heats of formation of most of the boron hydrides have been measured, and from them the heats of atomization are listed in the second column of Table 3–15. In the third column these heats are given as multiples of the heat of atomization of $B_2H_6$. It is clear that our calculated bonding energies are too large. In this type of calculation the binding energies can be made of arbitrary size by varying the overlap proportionality constant $K$; since our calculated energies are too large, we conclude that $K$ should be decreased, a change that would also move the ionization potentials in the desired direction.

But a more interesting comparison, assuming a rough proportionality of binding energy to $K$, is the *relative* calculated binding energy, which again is given for calculation I in multiples of the $B_2H_6$ binding energy, in the fourth column of Table 3–15. The agreement is good, and it would not be so if we had chosen our bonding energies with B at infinity in the $(2s)^2(2p)$ configuration. The agreement is not good enough to allow a calculated estimate of the heats of formation. However, we are encouraged to think that calculations with lower $K$ may even yield predictions of these relatively small (less than 1 ev) energies. The last column of Table 3–15 shows the ratios of the total energies relative to diborane, to be compared with values in column 2!

There are other astonishingly good additivities among these energies. In

**Table 3–15** *Heats of atomization and binding energies for the boron hydrides*

| Molecule | $\Delta H_a$, ev | $\frac{\Delta H_a}{\Delta H_a(B_2H_6)}$ | $\frac{B.E.}{B.E.(B_2H_6)}$ | $\frac{\Sigma E}{\Sigma E(B_2H_6)}$ |
|---|---|---|---|---|
| $B_2H_6$ | 24.94 | 1 | (1) | (1) |
| $B_4H_{10}$ | 45.43 | 1.822 | 1.800 | 1.811 |
| $B_5H_9$ | 48.98 | 1.964 | 2.008 | 1.959 |
| $B_5H_{11}$ | 53.18 | 2.132 | 2.134 | 2.131 |
| $B_6H_{10}$ | 56.90 | 2.281 | 2.366 | 2.284 |
| $B_{10}H_{14}$ | 90.10 | 3.613 | 3.792 | 3.582 |
| $BH_3$ | 11.85 | 0.475 | 0.497* | 0.499* |

* For most stable, i.e., planar, $BH_3$ configurations.

**Table 3-16** *Charge distributions in the boron hydrides*

| Molecule | Atoms | Molecular-orbital* I† | Molecular-orbital* II‡ | Three-center bonds¶ |
|---|---|---|---|---|
| $B_2H_6$ | 1,2 | 0.23 | 0.14 | 0 |
| $B_4H_{10}$ | 1,3 | 0.04 | 0.14 | 0 |
| | 2,4 | 0.15 | 0.03 | 0 |
| $B_5H_9$ | 1 | −0.23 | −0.36 | −0.78 |
| | 2,3,4,5 | 0.19 | 0.25 | 0.20 |
| $B_5H_{11}$ | 2,5 | 0.09 | −0.14 | −0.06 |
| | 3,4 | 0.19 | 0.24 | 0.33 |
| | 1 | −0.10 | −0.30 | −0.67 |
| $B_6H_{10}$ | 1 | 0.02 | −0.12 | −0.72 |
| | 4,5 | −0.03 | −0.07 | −0.17 |
| | 3,6 | 0.24 | 0.30 | 0.35 |
| | 2 | 0.20 | 0.28 | 0.36 |
| $B_9H_{15}$ | 3 | 0.09 | −0.05 | 0.00 |
| | 4,9 | 0.05 | 0.03 | −0.12 |
| | 1 | 0.11 | −0.02 | −0.10 |
| | 5,8 | 0.11 | 0.07 | 0.12 |
| | 2 | −0.01 | −0.17 | −0.53 |
| | 6,7 | 0.18 | 0.24 | 0.31 |
| $B_{10}H_{14}$ | 5,7,8,10 | 0.17 | 0.12 | 0.10 |
| | 2,4 | 0.04 | −0.10 | −0.45 |
| | 6,9 | 0.29 | 0.33 | 0.29 |
| | 1,3 | 0.10 | −0.04 | −0.03 |
| $B_{10}H_{16}$ | 1,1′ | −0.35 | −0.32 | −0.70 |
| | 2–5,2′–5′ | 0.18 | 0.24 | 0.18 |
| $B_{10}H_{10}^{=}$ | 1,2 | −0.15 | −0.29 | −0.33 |
| | 3–10 | −0.01 | −0.17 | −0.17 |
| $BH_4^-$ | 1 | −0.25 | −0.92 | −1.00 |
| $B_{12}H_{12}^{=}$ | 1→2 | −0.02 | −0.17 | −0.17 |
| $B_3H_8^-$ | 1 | 0.12 | −0.05 | 0 |
| | 2,3 | −0.16 | −0.030 | −0.50 |
| $B_{10}H_{14}^{=}$ | 1,3 | −0.04 | −0.19 | −0.33 |
| | 2,4 | 0.07 | −0.07 | −0.33 |
| | 5,7,8,10 | 0.00 | −0.05 | 0.33 |
| | 6,9 | −0.27 | −0.56 | −1.00 |
| $B_9H_{14}^-$ | 7 | −0.20 | −0.50 | −1.00 |
| | 2 | 0.06 | −0.08 | −0.33 |

Table 3-16 (*continued*)

| Molecule | Atoms | Molecular-orbital* I† | Molecular-orbital* II‡ | Three-center bonds¶ |
|---|---|---|---|---|
| $B_9H_{14}^-$ | 4 | 0.15 | −0.08 | 0.00 |
| | 1,3 | 0.02 | −0.14 | 0.00 |
| | 6,8 | 0.05 | 0.00 | 0.33 |
| | 5,9 | −0.01 | −0.16 | −0.17 |

* Averaged over the apparent symmetry planes for $B_4H_{10}$, $B_5H_{11}$, $B_{10}H_{14}$, $B_{10}H_{10}^=$, $B_{12}H_{12}^=$.

¶ For all structures weighted equally except as follows: for $B_{10}H_{14}^=$, $B_9H_{14}^-$, one structure only. For $B_{10}H_{10}^=$ only structures with central three-center bonds.

† $\alpha(H) = -13.61$ ev for all H's.

‡ $\alpha(H) = -15.75$ ev for bridge H's, $-11.60$ ev for terminal H's.

view of the small values of the heats of formation relative to the heats of atomization one might expect that the bonding and total energies for the neutral boron hydrides could be fitted well with a set of two parameters, corresponding to the energy per boron and hydrogen, respectively. This is indeed true. Even better is the fit where the 14 total energies and binding energies are calculated with four energy parameters corresponding to the four variables of the three-center bond theory (energy per hydrogen bridge, three-center B—B—B bond, single B—B bond, normal B—H bond). The average deviation from the observed energy in this fit is less than an electron volt per molecule. A set of such computed total energies may be examined in the last columns of Table 3-14.

### Charge Distributions

Charge distributions are tabulated in Table 3-16. As mentioned above, we had hoped that the difference between calculations I and II, i.e., different Coulomb integrals for bridge and terminal hydrogens, would not affect the boron charge distributions. This has largely proved true with the following exceptions: atoms 1,3 vs. 2,4 in $B_4H_{10}$; atom 1 vs. 4,5 in $B_6H_{10}$; atoms 4,9 in $B_9H_{15}$; atoms 2,4 vs. 5,7,8,10 in $B_{10}H_{14}^=$; atom 4 vs. 6,8 in $B_9H_{14}^-$. Note that the charge order in $B_{10}H_{14}$ is in agreement with that calculated earlier,[198] and is in accordance with experimental evidence.

In Table 3-16 are also listed charges derived from an analysis of, for the main part, all three-center structures, as discussed in Chapter 2. If we define "improvement" as increasing similarity to both experimentally inferred and

LCAO-MO computed charge distributions, then the inclusion of unsymmetrical structures clearly improves the three-center bond theoretical results. Decaborane is here the best example.

With the conviction that the LCAO-MO charge distributions are more realistic than those derived previously, we earnestly implore that they be tested in the interpretation of experimental facts such as reactivities. Where ambiguities exist, we hope that, although better calculations are not immediately forthcoming, a choice may be made on empirical grounds between methods I and II. No clear choice can be made at present.

## 3–5 Simplified Molecular-Orbital Descriptions

In view of the ease with which at least the one-electron approximation can be applied with the aid of a digital computer, as described in the previous sections, it would seem hardly worthwhile to make the even cruder approximations to be described here. These further approximations are (1) neglect of the orbitals localized to give bonding to either terminal or bridge hydrogen (where it occurs) or both; (2) artificial separation of pure $p$ orbitals in a plane tangential to the polyhedral surface from the hybrid orbital pointing inward toward the polyhedral center; (3) assumption of small interactions between molecular orbitals at chosen or obvious apices and equatorial orbitals, e.g., in the tetragonal bicapped antiprism, $B_{10}H_{10}^{=}$; and (4) ring-polar separations in the $\mathcal{N} = 5$, 6, and 7 bipyramids ($\mathcal{N} = 6$ corresponds to the octahedral $B_6H_6^{=}$) and in the $\mathcal{N} = 4$, 5, and 6 pyramids ($B_4H_7^-{:}C_{3v}$; $B_5H_9{:}C_{4v}$; and $B_6H_{11}^+{:}C_{5v}$). A discussion of these separations has been given by Hoffmann and Lipscomb.[120] These separations are not required by symmetry, but are invoked because of either real or imagined partial or nearly complete localization of bonding electrons in the corresponding molecular regions. These separations offer a description intermediate between a full molecular-orbital treatment and a localized bond treatment without the introduction of the resonance hybrids required if only three-center bonds are employed.[121] There are three main advantages to these approximations. First, they form a simplified description similar in many respects to molecular orbitals of $\pi$-electron systems of aromatic molecules, already familiar to most chemists. Second, they form descriptions in which molecular-orbital features may be carried over from one molecule to the next closely related one. Third, they provide the only yet satisfactory way of describing the nature of binding in an almost complete polyhedron, such as the predicted,[198] and recently described[2] $B_{11}H_{14}^-$ (icosahedral fragment), where localized bonds are appropriate in the open part while delocalized molecular orbitals are most appropriate for the boron-boron bonding.

### Polyhedra

First consider the $B_8H_8^{=}$ ($D_{3d}$ unknown), $B_{10}H_{10}^{=}$ ($D_{4d}$ known), and $B_{12}H_{12}^{=}$ ($I_h$ known, but considered as $D_{5d}$ symmetry here). First, form all

B—H terminal bonds, and then at the two opposite apices form framework bonds of the $\sigma$, $\pi$ type in the framework orbitals of $B_5H_9$. These bonds leave two hybrid orbitals on each of the 6, 8, or 10 "equatorial" B atoms, those closest to the equatorial plane of the species. In $B_8H_8^=$ this leaves 17 (total) $-8$ (for BH) $-2 \times 3$ (apex) $= 3$ pairs for equatorial bonding molecular orbitals to be constructed from 12 atomic orbitals. A group-theoretical argument, and rough evaluation of energies based upon nearest-neighbor interactions then gives[198] molecular orbitals $a_{1g}$ at $H_0 + 2\beta$ (filled), $e_u$ at $H_0 + \sqrt{3}\beta$ (filled), $e_g$ at $H_0 + \beta$ (back bonding from apical molecular orbitals of the same symmetry), $a_{1u}$ at $H_0$, $a_{2u}$ at $H_0$, $e_g$ at $H_0 - \beta$, $e_u$ at $H_0 - \sqrt{3}\beta$, and $a_{2g}$ at $H_0 - 2\beta$. The electron-orbital ratios, 6/12 for the equatorial region and 6/6 for each apical region, are so different that this $B_8H_8^=$ of $D_{3d}$ symmetry may not be stable. The corresponding ratios of 10/16 and 6/7 for $B_{10}H_{10}^=$ and of 14/16 and 6/8 for $B_{12}H_{12}^=$ are very much more reasonable. In $B_{10}H_{10}^=$ there are $21 - 10 - 2 \times 3 = 5$ pairs for bonding among the 16 equatorial orbitals which are[182] $a_1$ at $H_0 + 2\beta$ (filled), $e_3$ at $H_0 + \sqrt{2 + \sqrt{2}}\,\beta$ (filled), $e_1$ at $H_0 + \sqrt{2}\,\beta$ (filled), $e_2$ at $H_0 + \sqrt{2 - \sqrt{2}}\,\beta$, $b_1$ at $H_0$, $b_2$ at $H_0$, $e_2$ at $H_0 - \sqrt{2 - \sqrt{2}}\,\beta$, $e_1$ at $H_0 - \sqrt{2}\,\beta$, $e_3$ at $H_0 - \sqrt{2 + \sqrt{2}}\,\beta$, and $a_1$ at $H_0 - 2\beta$. In the subgroup $D_{5d}$ of $I_h$ there are in $B_{12}H_{12}^=$ $25 - 12 - 2(3) = 7$ pairs for equatorial bonding among 20 hybrid B atomic orbitals which form molecular orbitals of symmetry $a_{1g}$ at $H_0 + 2\beta$ (filled), $e_{1g}$ at $H_0 + 1.45\beta$ (filled), $e_{1u}$ at $H_0 + 1.40\beta$ (filled), $e_{2g}$ at $H_0 + 1.38\beta$ (filled), $e_{2u}$ at $H_0 + 1.12\beta$, $a_{2g}$ at $H_0$, $a_{2u}$ at $H_0$, $e_{2u}$ at $H_0 - 1.12\beta$, $e_{2g}$ at $H_0 - 1.38\beta$, $e_{1u}$ at $H_0 - 1.40\beta$, $e_{1g}$ at $H_0 - 1.45\beta$, and $a_{2g}$ at $H_0 - 2\beta$. We shall refer to these orbitals and energies below, where fragments of these polyhedra are discussed.

The ring-polar separations are applicable to the $N = 5,6,7$ bipyramids: $D_{3h}$, $B_5H_5^=$; $O_h$ in the subgroup $D_{4h}$, $B_6H_6^=$; and $D_{5h}$, $B_7H_7^=$. Carborane analogues isoelectronic with these ions have been found since the initial predictions[69,179,187,198] regarding probable stabilities of these ions. The method of description localizes the equatorial belt as a planar set of B—H units joined by $\sigma$ bonds, and forming also a $\pi$-electron set of molecular orbitals which are combined with the appropriate combination of orbitals formed from the additional B—H units above and below the equatorial plane. $B_5H_5^=(D_{3h})$, $B_6H_6^=(O_h)$, and $B_7H_7^=(D_{5h})$ ions may be used to illustrate these points. Remove the apical B—H groups, thus leaving planar $B_3H_3^=$, $B_4H_4^=$, and $B_5H_5^=$ units, which are treated by the usual molecular-orbital methods employed for aromatic hydrocarbons.

The in-plane bonding of each unit has 3, 4, and 5 single bonds, respectively, and there are two extra electrons left over for the molecular orbitals formed from the $p_z$ orbitals. For each of these units one $\sigma$ molecular orbital and one $\pi$ pair of molecular orbitals exist, and these are combined with corresponding orbitals of the two B—H units above and below the molecular plane, as shown in Fig. 3-8 for $B_6H_6^=$. The $B_3H_3^=$ unit, prepared for bonding, has no $\delta$ orbital, while the single $\delta$ orbital in $B_4H_4^=$ and the pair of $\delta$ orbitals in $B_5H_5^=$ are not of appropriate symmetry to bond to the BH units. Hence the

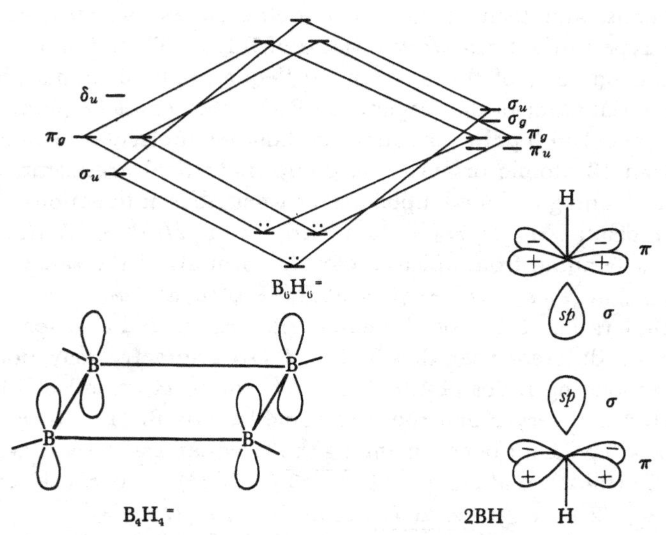


**Figure 3-8** *Simplified description of bonding in* $B_6H_6^=$. *Each B of planar* $B_4H_4^=$ *forms three bonds in the molecular plane. Energy levels of the* $\pi$ *system are shown above. Only the* $\sigma_u$ *and* $\pi_g$ *molecular orbitals of* $B_4H_4^=$ *are of appropriate symmetry to combine with linear combinations of corresponding symmetry from the two BH groups. Thus three electron pairs are required in addition to the four B—B bonds, making a total of seven pairs for boron-boron bonding in the octahedral* $B_6H_6^=$ *ion.*

bonding of two B—H units to $B_3H_3^=$, $B_4H_4^=$, requires, in each case, three electron pairs, as shown. Therefore $B_5H_5^=$, $B_6H_6^=$, and $B_7H_7^=$ may exist each with a $-2$ charge in its most stable state. Unfortunately, the electron-orbital ratios in this separation vary greatly, being 6/6, 8/8, 10/10 in the ring set and 6/9, 6/10, and 6/11 in the polar set for $N = 5, 6, 7$, respectively. Also the quantitative application of LCAO theory gives a totally symmetric orbital at a relatively much lower energy in the polar set for each species than has been found for the more complete LCAO calculation which does not make the ring-polar separation. In retrospect, $B_5H_5^=$ and $B_6H_6^=$ are correctly predicted by any of the approximations but $B_7H_7^=$ is correctly predicted[120] only by the full LCAO treatment.

### Polyhedral Fragments

Ring-polar separations are also applicable to the $N = 4$, 5, and 6 pyramids. There are $2(N-1)$ orbitals in the ring set and $N-2$ polar orbitals, and the full LCAO treatment or the ring-polar separation method both lead to the conclusions that close shells exist for $B_NH_N$ or $B_NH_N^{-4}$. Electron-orbital ratios are 6/6, 8/8, and 10/10 in the ring, and 2/6, 2/7, and 2/8 in the polar set for neutral $B_NH_N$ for $N = 4$, 5, and 6, respectively, and as expected these ratios are so different that the ring-polar factorization is a poor approximation. On the other hand, this separation is quite good for the $B_NH_N^{4-}$ series, for which the polar ratios are 6/6, 6/7, and 6/8 for $N = 4$, 5, and 6, respectively, and stability is predicted for $B_4H_4^{4-}$ ($C_{3v}$ subgroup of $T_d$), $B_5H_5^{4-}$ ($C_{4v}$), and $B_6H_6^{4-}$ ($C_{5v}$). Such highly charged species are not expected to be stable, but there are two general kinds of predictions to be made from these filled-orbital ions of $-4$ charge.

First, carborane analogues may exist, obtained formally by replacement of $B^-$ by C, in order to obtain neutral $C_4H_4$ (tetrahedrane), $C_4BH_5$, or $C_4B_2H_6$. Isoelectronic negative ions obtained by replacement of a fewer number of $B^-$ by C may be worth searching for. Orbitals around C are in fact more tightly held by the increased nuclear charge than those around B, and the compounds with large percentages of C atoms may tend to go over into less symmetrical species.

Second, one may add one or more $H^+$ to the edge-single bonds in order to produce bridge H atoms, thereby reducing the over-all charge. For example, addition of $3H^+$ to the ring single bonds of $B_4H_4^{4-}$ ($C_{3v}$) yields $B_4H_7^-$ ($C_{3v}$), the formula of which has been predicted earlier[173] and criticized[227] justifiably on the basis of unfavorable electron-orbital ratios in the ring and polar regions. However, rearrangement of bridge H atoms to yield $BH_2$ groups is at least a reasonably general property of negative ions,[175] and Onak,[215] who has recently prepared a salt of this ion, favors a formula in which just this sort of rearrangement occurs. Addition of $4H^+$ to the edge-single bonds in the $B_4$ ring of $B_5H_5^{4-}$ leads to the known $B_5H_9$ structure, and a similar addition of $4H^+$ to $B_6H_6^{4-}$ leads to $B_6H_{10}$.

Third, one may make both changes suggested in the two preceding paragraphs. Examples are $HCB_4H_6$ or $BeB_5H_{11}$. A $C_2B_4H_8$ may occur,[218] the formula of which could be obtained from the $B_6H_6^{4-}$ formula by replacement of two $B^-$ by C in the ring, plus the addition of two $H^+$ to the two remaining single B—B bonds in the ring to produce two hydrogen bridges. The $C_2B_4H_6Me_2$ structure described at the end of Chapter 1 is, however, based upon structural principles like those in $B_6H_{10}$.

Most of these descriptions of semilocalized molecular orbitals in polyhedral fragments have the comparatively simple $\sigma$, $\pi$ aspects present in the description of $B_5H_9$ given at the beginning of this chapter.

## $B_{11}H_{14}^-$ and Related Species[2,198]

We now combine the semilocalized molecular orbitals of the nearly complete polyhedron with the $\sigma$, $\pi$ bonding situation in the units bonding to the open part of the polyhedral fragment. Suppose, for example, that we remove one neutral BH unit from $B_{12}H_{12}^{=}$. This leaves $B_{11}H_{11}^{=}$ with the five orbitals pointing toward the now-vacant apex holding $6 - 2 = 4$ electrons. Now these five orbitals form one axially symmetric bonding molecular orbital and one degenerate pair requiring two more electrons. We could combine this set of orbitals with any sterically favorable group that would supply two more electrons in a $\sigma$, axially symmetric, orbital and a $\pi$ pair. For example, we could use $(:N—H)^{++}$ (with only four electrons) to make $B_{11}NH_{12}$, which might lose a proton to make $B_{11}NH_{11}^-$. Or we could use an $(H_3)^+$ group ($\sigma$ and $\pi$) in a triangle perpendicular to the molecular axis to give $B_{11}H_{14}^-$, which is possibly more stable than $B_{11}H_{13}^=$, which might require a direct B—B bond across the five-membered ring. A similar discussion of $B_{10}H_{10}^=$ yields the $B_9H_{12}^-$ ion, in which the nine BH units are in $C_{4v}$ symmetry, but combined by molecular orbitals to the triangle of $H_3$. However, a more promising structure for $B_9H_{12}^-$ is proposed in Chapter 2. A structure determination of $B_{11}H_{14}^-$ would be of considerable interest.

Modification of opposite vertices of $B_{12}H_{12}^=$ might occur to extend the sequence from $B_{11}H_{14}^-$ to a neutral hydride $B_{10}H_{16}$, in which the two $B_5$ pentagons are staggered in relative orientation. Unfortunately, not enough is known about boron rearrangements to suggest whether this particular geometrical structure for $B_{10}H_{16}$ is stable.

## 3–6 Three-Center and Resonance Descriptions of Polyhedral Species

### $B_4Cl_4(T_d)$

The four electron pairs in the framework molecular orbitals of $B_4Cl_4$ are in symmetries $a_1$ (two electrons) and $f_1$ (six electrons). Two independent studies[57,69] have shown that when these four orbitals are formed into equivalent orbitals made up from various sums and differences, there result four localized three-center bonds in the tetrahedral faces. For example, form the sum of

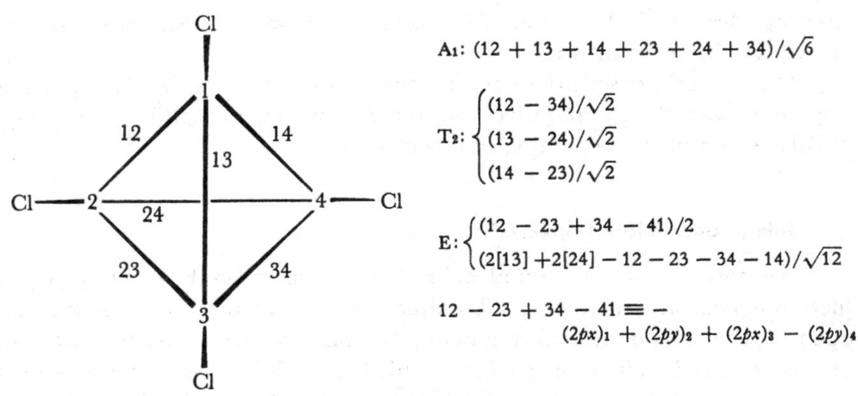


**Figure 3–9** *Molecular geometry and molecular orbitals in $B_4Cl_4$.*

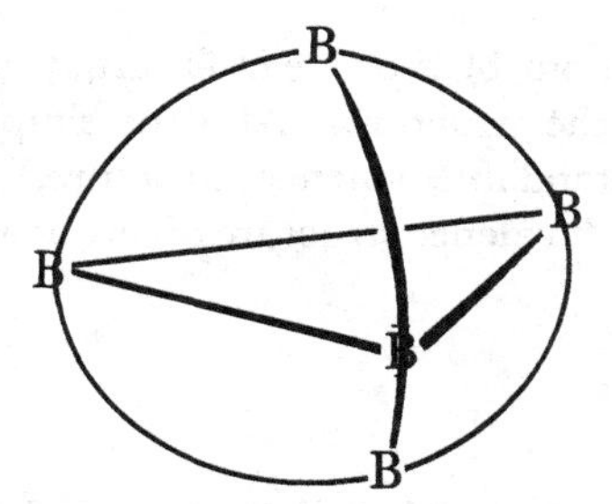


**Figure 3–10** *Framework bonds in $B_5H_5^=$.*

the $T_2$ orbitals and add to the $A_1$ orbital with use of proper normalization (Fig. 3–9). Each of these three-center bonds has one-fourth (square of coefficient) of the $a_1$ orbital. No resonance hybrid is required in this description.

$B_5H_5^{-2}(D_{3h})$

A similar transformation of the molecular orbitals[198] of $B_5H_5^=$ yields three single bonds in the equatorial triangular $B_3$ unit and three three-center bonds each joining an apex orbital through a $\pi$ orbital of an equatorial B to an orbital of the opposite apex (Fig. 3–10). Thus localized two-center and localized three-center bonds form a simple description of $B_5H_5^=$ and its isoelectronic analogues $CB_4H_5^-$ and $C_2B_3H_5$.

## 3–7 Borides

We now explore the molecular-orbital situation in the borides of the type $MB_x$ for $x \geq 2$, in all of which there are some aspects of localized concentration of electrons in the boron frameworks. The approach may be outlined as follows:

1. The boron framework is isolated into polyhedra which are connected to each other by localized bonds, occasionally with the use of multicentered orbitals.

2. The molecular orbitals of the individual polyhedra are investigated.

3. Electrons are transferred from the more electropositive element to the boron framework until the bonding orbitals are filled.

4. Excess valence electrons on the metal atoms are regarded as metallic in character and should lead to metallic optical and electric properties.

It is difficult to defend this procedure as a new one or, even less, as a general one applicable to all intermetallic compounds. In the borides themselves, this procedure was inferred in the group-theoretical arguments[69] relating to the $B_6$ octahedral arrangements, and was studied in detail[187] independently in $CaB_6$. A somewhat similar discussion[188] has been given previously for $B_{12}C_3$, in which vacant $p$ orbitals of the central C atom of the linear $C_3$ group are required to conjugate with the remainder of the structure. Also the simpler rhombohedral form of boron has an understandable valence structure,[47,191] but seems widely at variance with the apparent[188] valence structure of tetragonal boron.

### $MB_2$, $MB_6$, and $MB_4$

The compound most nearly approaching a closed-shell structure of the known $MB_2$ compounds is probably $MgB_2$. Similarly, $CaB_6$ is the closed-shell prototype of the $MB_6$ structures. The similarity of the valence orbitals of the intermediate structure[338] $MB_4$ to these two leads us to suggest, as we shall indicate below, that the closed-shell form is the $MgB_4$ structure in spite of the present limitations[138,253] of this phase to the rare earths,[241] Th, and U. When atoms with more than two valence electrons occur in $MB_2$ and $MB_6$ structures, the electrical and optical properties are much more marked.[295]

If M forms a 2+ ion the $B_2^{=}$ framework in $MB_2$ is completely isoelectronic with that of graphite, and partial transfer of these electrons back to $M^{++}$ would complete the description. It is relevant to $M^{++}B_6^{=}$ that there have been two independent discussions[69,187] of the internal orbitals of a $B_6^{=}$ octahedral arrangement with single bonds external to the octahedron. The seven internal bonding molecular orbitals of $B_6^{=}$ require seven pairs, and the external six orbitals prepared for bonding with similar orbitals from other octahedra require one electron in each, thus accounting for the 20 available electrons per $B_6^{=}$.

In the $MB_4$ structure there are two octahedral $B_6$ groups and four single B atoms in trigonal $sp^2$ bonding to other boron atoms. Hence we take $M_4B_{16}$ as our unit for discussion. Each $B_6$ octahedron requires 14 electrons for internal bonds and six more as its share of electron pairs in the six external bonds, thus requiring a total of 20 electrons per octahedron or a total of 40 electrons for two octahedra. The four additional B atoms are bonded in pairs, as in ethylene, and would require four per $B_2$ group to bind the pair and four more per $B_2$ group as its share of the electron pairs in the external orbitals for bonding to octahedra, or a total of 16 for all four B atoms of this kind. Thus a total of 56 electrons is required per unit cell, a total which would be provided nicely by the 16 B atoms and the $4M \rightarrow 4M^{++}$ ionization.

### $UB_{12}$ and $ZrB_{12}$

The geometrical coordination of boron atoms around a central Zr has been discussed previously.[18,240] However, we wish to emphasize that this structure can be regarded as an array of $B_{12}$ groups and metal atoms in the NaCl arrangement. The $B_{12}$ polyhedron is of symmetry $O_h$ and can be called a cube-octahedron (Fig. 3–11). Whereas earlier studies[69,187,188] had based boron hydride molecules or ions on the icosahedron, octahedron, or tetrahedron, or

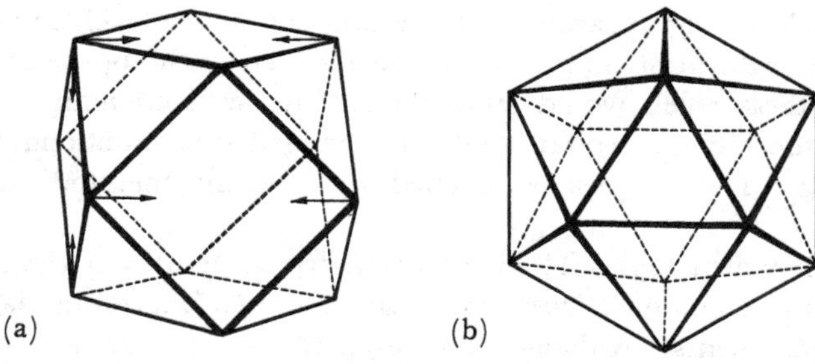


**Figure 3–11** *(a) The small distortions of the $B_{12}$ cube-octahedron which are required to convert it to the icosahedron (b) are shown by the arrows.*

on their fragments, this polyhedron as well as the recently discovered $B_8$ arrangement[135] in $B_8Cl_8$ and the $B_{10}$ arrangement[55,182] in $B_{10}H_{10}^{=}$ are relatively new in boron chemistry. Each of the B atoms in the $B_{12}$ group is bonded to a B atom of another $B_{12}$ group, and the Zr atoms lie in the interstices outside the $B_{12}$ groups.

If one considers the external bonds here as normal single covalent bonds the question of the bonding arrangement in the group arises. A simple LCAO molecular-orbital approach gives a reasonable answer to this question. It has been shown that the bonding orbitals will accommodate 13 pairs of electrons. Since the 12 borons have 36 electrons, of which 12 are needed for the external bonds, one more pair must be supplied by the metal atom. Therefore the group should be written as $B_{12}^{=}$ if all the bonding orbitals are filled with electrons. If the order of the magnitude of the resonance integrals[180] is estimated as $\epsilon > \zeta > \eta > \alpha > \beta > \gamma > \delta$, then the decreasing order of the bonding orbitals is

| | |
|---|---|
| $(p_z)\,f_{1u}\colon H_0$ | nonbonding |
| $(sp)\,f_{1u}\colon H_0 + 2\alpha - 2\gamma$ | |
| $(p_y)\,f_{1u}\colon H_0 + 2\zeta$ | |
| $(p_x)\,f_{2g}\colon H_0 + 2\epsilon$ | |
| $(p_y)\,e_g\colon H_0 + 2\zeta + 2\eta$ | |
| $(p_x)\,a_{2u}\colon H_0 + 4\epsilon$ | |
| $(sp)\,a_{1g}\colon H + 4\alpha + 2\beta + 4\gamma + \delta$ | lowest state |

The highest bonding orbital is a triplet level, as is the next higher orbital, which is nonbonding; therefore, it is exceedingly likely that the doubly negative ion is formed within this three-dimensional framework.

There are several sets of orbitals with the same symmetry. Hence a more thorough treatment should consider the shifting of their relative energies by the possible interactions between levels. There are two reasons why this refinement seems unnecessary in this approximation. The first is that it is apparent from inspection of a model of the polyhedron that all the resonance integrals other than the ones listed must be small, so that their inclusion would not shift the energy levels significantly. The second is that the levels in which the shifts would matter the most are the three $f_{1u}$ levels. To a first approximation the interactions of these levels would leave the middle level unchanged, and still bonding, and the highest orbital would be changed from nonbonding to slightly antibonding. The conclusions reached above would therefore not be altered.

The relation shown in Fig. 3–11 of the geometric structures of the cube-octahedron to the more compact icosahedron suggests two interesting aspects of these boron arrangements. Although both $B_{12}$ arrangements would a priori seem to be good candidates for $B_{12}H_{12}^{=}$ ions, it seems likely that the cube-octahedron would probably be unstable with respect to the small displacements that would lead it to adopt the icosahedral arrangement. On the other hand, the cube-octahedron arrangement is quite probably stabilized by the interactions of the highest filled orbitals of the $B_{12}$ groups with the $d^2sp^3$ orbitals of the Zr atom. The arrangement would thus lead to a closed shell for $M^{++}B_{12}^{=}$, and to a partial transfer of electrons back to $M^{++}$ from the boron orbitals. It might then be expected that $ZrB_{12}$ has two electrons per $ZrB_{12}$ in excess of the closed shells. The physical properties[254] of $ZrB_{12}$ are consistent with this assumption of excess electrons in $ZrB_{12}$.

### Boron

The simpler[191] rhombohedral form of boron has boron-boron bonds among atoms of icosahedra only. Half of the boron orbitals external to the icosahedra are involved in two-center bonds, and half are in three-center bonds, all between different icosahedra. Thus each $B_{12}$ is "prepared" for bonding by placement of one electron in each orbital to be used in a two-center bond, and by placement of one electron per 1.5 boron orbitals of the borons involved in the three-center bonds. Ten electrons are then required from each $B_{12}$ group for the external bonds to other icosahedra, thus leaving 26 electrons for the intraicosahedral bonds of each $B_{12}$. These 26 electrons just fill the bonding intraicosahedral molecular orbitals. This valence structure was recognized by Decker and Kasper.[47]

The complex rhombohedral form of boron[255] has just been solved by R. E. Hughes, and is discussed in the Concluding Remarks, p. 199.

Tetragonal boron[116,117] presents a difficult interpretive problem[180,188] in the framework of these assumptions. There are four $B_{12}$ icosahedra, and two additional boron atoms ($B_t$) per unit cell, furnishing a total of 150 electrons. Preparation of each $B_t$ to form four bonds requires eight electrons for $2B_t$, preparation of the $B_{12}$ for external bonds requires 48 electrons for $B_{12}$, and assignment of 26 electrons inside each $B_{12}$ requires 104 electrons, thus requiring a total of 160 electrons per unit cell, 10 more than the number available. Hoard et al.[117] found strong indications of approximately $\frac{3}{4}$ additional boron-atom impurity per unit cell. This excess can hardly be expected to account for the large discrepancy in the number of electrons, but might be involved if the actual discrepancy were much smaller.

In tetragonal boron the situation of the two unique boron atoms $B_t$ per unit cell merits further examination. The $B_t$—B distance, to a B of an icosahedral group, is very short, about 1.60 A, and the four bonds from $B_t$ are greatly flattened from a regular tetrahedral configuration such that the B—$B_t$—B angle is 135°. Let $B_t$ be initially unhybridized, and choose $p_z$ along the $c$ axis, and $p_x$ and $p_y$ in the $bc$ plane but oriented at 45° with respect to $b$ and $c$, respectively, so that $p_x$ and $p_y$ almost point toward the closest boron atoms of the four icosahedra bonded to $B_t$. At this short distance the overlap integral of $p_x$ on $B_t$ with a parallel $p$ orbital on the nearest B of the $B_{12}$ group is 0.30, greater than that for $\pi$—$\pi$ overlap in $C_2H_4$. Even the $p_z$ orbital of the B of $B_{12}$, which is perpendicular to the B—$B_t$ bond and therefore 23.5° away from the direction of $p_z$ of $B_t$, has an overlap integral very nearly equal to that for $\pi$—$\pi$ in $C_2H_4$. Therefore, these $p$ orbitals of $B_t$ must interact strongly with the $p$ orbitals on closest atoms of the four $B_{12}$ groups, and therefore with some of the quadruply degenerate $U_u$ levels[188] of the $B_{12}$ groups, which are the highest filled levels if we assign 26 electrons to each $B_{12}$ group. A simplified description of the $x,y$ interactions is obtained by replacing the icosahedral $U_u$ levels by a single $p$ orbital in the $xy$ plane (Fig. 3–12).

We now approximate a bonding scheme for tetragonal boron. Bonds to the $B_t$ at the center of the unit cell can be a five-center bond, involving the

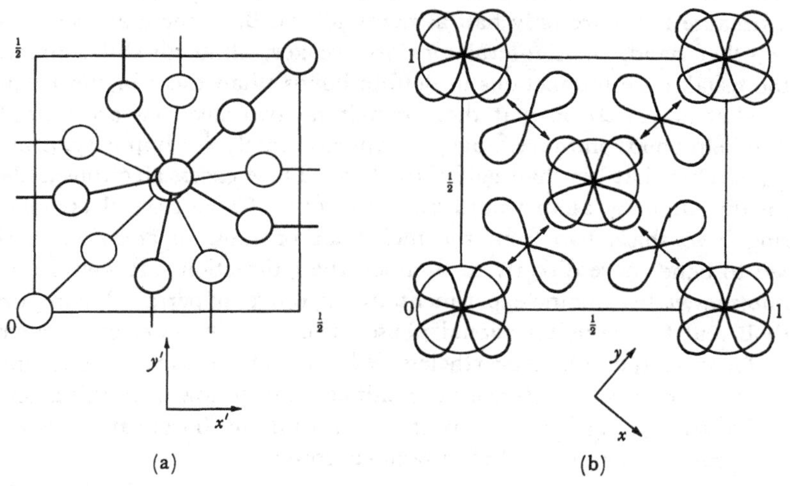


**Figure 3–12** *(a) One-quarter of the unit cell of tetragonal boron. (b) Full unit cell, in which $B_{12}$ has been replaced by a single $\pi$ orbital with two sp bonds pointing outward represented by arrows. The $\pi_z$ orbital of the $B_t$ at the origin and center are omitted, but $\pi_x$ and $\pi_y$ and the bonding 2s orbitals are shown.*

four $sp$ bonds interacting with the $2s$ orbital, and two three-center bonds, along $x$ involving an $sp$, $p_x$ on $B_t$, and $sp$, and, similarly, along $y$ involving an $sp$, $p_y$, and $sp$. The five-center bond does not interact with the highest filled levels inside the icosahedra, but the two three-center bonds interact so strongly that we propose that the electron pair of the appropriate icosahedral $U_u$ level fills each corresponding three-center bond: the interactions of these three-center bonds with icosahedral levels result in an infinite string of interactions along $x$ or along $y$, but parallel strings do not interact, nor do perpendicular strings, and the highest filled level could be considered as corresponding to bonding in pairs of these two localized molecular orbitals. At least this much bonding is required to ensure that the resulting bonding orbitals are at least as stable as the other $U_u$ icosahedral levels themselves. Thus two of the 26 electrons of an icosahedral level also fill a three-center bond about $B_t$ in these $xy$ interactions. Before we discuss the slightly weaker $\pi_z$ interactions of $B_t$ with the icosahedra let us recount the bonding electrons per unit cell:

| | | |
|---|---|---|
| $4B_{12}$ | internal bonding | 104 |
| $4B_{12}$ | external bonding to other $B_{12}$ | 40 |
| $2B_t$ | five-center bonds | 4 |
| | | 148 |

Of the 104 electrons in internal bonds, 96 are in bonds between B's inside icosahedra, and eight occupy the levels which interact strongly with $B_t$. We are left, at this stage, with an excess of two electrons. If the interactions of a different $U_u$ level with $p_z$ of $B_t$ were just as strong, we could require bonding of a similar type between a $p_z$ of B in $B_{12}$ to $p_z$ of $B_t$, but only along one direction $x$ or $y$, because there are only half as many $p_z$'s on $B_t$ as there are icosahedra. On the other hand, these interactions are weaker, although still comparable with the usual $\pi$—$\pi$ interactions in double bonds, than those in the $x,y$ directions, and it may well be that these remaining two electrons are indeed required to help bond, on the average, the approximately $\frac{3}{4}$ B impurity, probably along $p_z$ of somewhat less than half of the $B_t$'s. These excess electrons added to those of the impurity atom would give it a total of about six electrons, thus requiring it to share two from the molecular orbitals, in addition to those required to bond more directly to $B_t$ along the $z$ direction. It would thus be most informative to examine experimentally the ranges of permissible impurities of both B and C in this tetragonal phase, before a precise electron count is claimed for this structure. Nevertheless, reduction of the apparent discrepancy to a very small number of electrons per unit cell does follow from this examination of the very strong interactions of the two unique boron atoms with the four icosahedra in the unit cell of tetragonal boron.

### $B_{12}C_3$

Little can be added to the earlier discussion[188] of $B_{12}$ icosahedra and linear $C_3$ chains in $B_{12}C_3$.[42,339] If 26 electrons are required inside each icosahedron,

12 electrons for the 12 outward-pointing boron orbitals and six for the six orbitals pointing away from each $C_3$ group, only four electrons remain for bonding the central C atom to the outer two of each $C_3$ group. Single $\sigma$ bonds along the C—C—C axis would use these up, thus leaving the $\pi$ orbitals of the central C vacant for conjugation with the bonds from C to B. At least we may add that the eight $\pi$ electrons of the $B_3 \equiv C—C—C \equiv B_3$ unit can be formed into reasonable molecular orbitals which are doubly degenerate and increasing in energy, and, since they are based upon only one hybrid orbital from each B, they can be made orthogonal to the remaining orbitals of the structure. It may be that these orbitals are filled by varying amounts in different samples, since there is a rather large range,[116,254,288,289] from 4 to 26 at. %, of C in the structure.

### $AlB_{12}$, $AlB_{10}$, and B

A critique of the aluminum boride phases has recently been published[155] along with the identification of a new compound $AlB_{10}$. The general problems presented by these phases are rather formidable. It is easy to formulate a three-dimensional structure for $\alpha$—$AlB_{12}$ ($a$ = 10.16 A, $c$ = 14.28 A, $D_4^4$—$P4_12_12$) based upon 16 $AlB_{12}$ in a unit cell which is a superlattice of the cubic $UB_{12}$ or $ZrB_{12}$ structure. Unfortunately the density of the crystal requires 14.4 $AlB_{12}$ in the unit cell, and hence a defect structure is plausible. Similarly, it is easy to formulate an orthorhombic structure for $AlB_{10}$ ($a$ = 8.88 A, $b$ = 9.10 A, $c$ = 5.69 A, $C_{2v}^{12} - Bb2m$, $C_{2v}^{16} - Bbm2$ or $D_{2h}^{17} - Bbmm$), based upon four $AlB_{10}$ in $C_{2v}^{16} - Bbm2$. Unfortunately the density of the crystal requires 5.20 $AlB_{10}$ in the unit cell, and, although an excess structure could occur, it is surely not possible (except statistically) to have six Al in the unit cell. Both of these structures are based upon the principles outlined above, that boron polyhedra are joined to other boron polyhedra or atoms, and that the Al occur in the vacancies.

A survey of these problems, and of the relations between the other aluminum borides and boron, suggests that a principle of boron chemistry different from presently known principles may occur here. Kohn et al.[155] suggest the identity of the unit cells of monoclinic[94] $AlB_{12}$ ($a$ = 8.50 A, $b$ = 10.98 A, $c$ = 9.40 A, $\beta$ = 110°54′) and rhombohedral elemental boron[255] ($a$ = 10.2 A, $\alpha$ = 65°28′); Parthé and Norton[226] had earlier made "monoclinic $AlB_{12}$" from 99 per cent pure B. Similar close relations may be noted here between the unit cells of $\alpha$—$AlB_{12}$ ($a$ = 10.16 A, $c$ = 14.28 A) and a new form[304] of boron ($a$ = 10.12 A, $c$ = 14.14 A), and also between $\beta$—$AlB_{12}$[206] ($a/\sqrt{2}$ = 8.88 A, $c/2$ = 5.09 A) and tetragonal boron ($a$ = 8.73 A, $c$ = 5.03 A).

A simple hypothesis that correlates all these problems is that Al substitutes for B in these structure types. Whether it does so at random or in certain preferred positions is not clear in each instance, but the space-group indications are rather in favor of preferred positions if, indeed, there is any merit in these correlations. This hypothesis, moreover, provides a reasonable basis for understanding the "anomalous" densities of $AlB_{12}$ and $AlB_{10}$. In $AlB_{12}$ the 14.4 formula weights per unit cell corresponds to 187 atoms, only a few per cent less than the 192 atoms which would be required for 16 $(B, Al)_{12}$ groups, or eight $(B, Al)_{10}$ plus eight $(B, Al)_2$ groups, an assumption which would nicely account for the cleavage on (201). In $AlB_{10}$ the almost unreasonable number of 5.2 formula weights in space groups requiring multiples of four corresponds to 57 atoms, within a few per cent of 56, which could arise from four $(B, Al)_{14}$ groups, or four $(B, Al)_{12}$ plus four $(B, Al)_2$, or four $(B, Al)_{10}$ plus eight $(B, Al)_2$, etc. The partial or complete ordering of Al in these structures would make them similar to, but not identical to, the corresponding boron phases, and would stabilize the valence structures by removal of the degeneracy of the valence levels of the icosahedra or other polyhedra.

The covalent radius of B is[227] 0.88 A, or a bit smaller, and that of Al is 1.26 A. An Al—Al distance would probably greatly distort the structures, but the Al—B distance of 2.0 A is at least within the range of B—B distances found in boron hydrides and borides. Hence, limited substitution of B by Al would not be unreasonable from the point of view of interatomic distances. The apparent order introduced into the present controversies over the existence of higher Al, B phases, their relations to the boron structures, and the understanding of the unit cell and density measurements would seem to justify an approach to structure determinations based upon the hypothesis of limited partial substitution of Al for B, and may suggest directions in which to look for new types of phases in the Al, B system. As a final alternative we await a study of the possible relations between these structures and the newly established structure of complex rhombohedral boron (R. E. Hughes, private communication, 1963).

Finally, the simpler rhombohedral form[191] of boron corresponds to a cubic close packing of icosahedra, and hence it is a very natural suggestion that a corresponding structure may exist based upon hexagonal close packing. Such a structure,[180] suggested in Fig. 3–13, has not yet been discovered. This structure has lower symmetry than cubic, just as the rhombohedral form has lower than hexagonal symmetry, because the icosahedron is less symmetric than a sphere. This structure may be one of the unidentified phases noted recently by Hoard and Newkirk.[115] The bond-angle distortions, shown best by this particular projection, are quite comparable with those in simple rhombohedral boron.

Two modifications of the electron density undoubtedly occur in these idealized valence structures, but we feel that such changes would not modify the extent to which the orbitals are filled. First, the final electron-density distribution probably rearranges somewhat from these idealized valence structures, perhaps to approximate that described from bond-order arguments, such as that for tetragonal boron.[228] Also rearrangements of a "back coordination" type, such as that suggested[185] in more localized molecular orbitals, probably occur between the orbitals of M and the boron framework. The filled orbital frameworks can only be expected for the most saltlike of the intermetallic

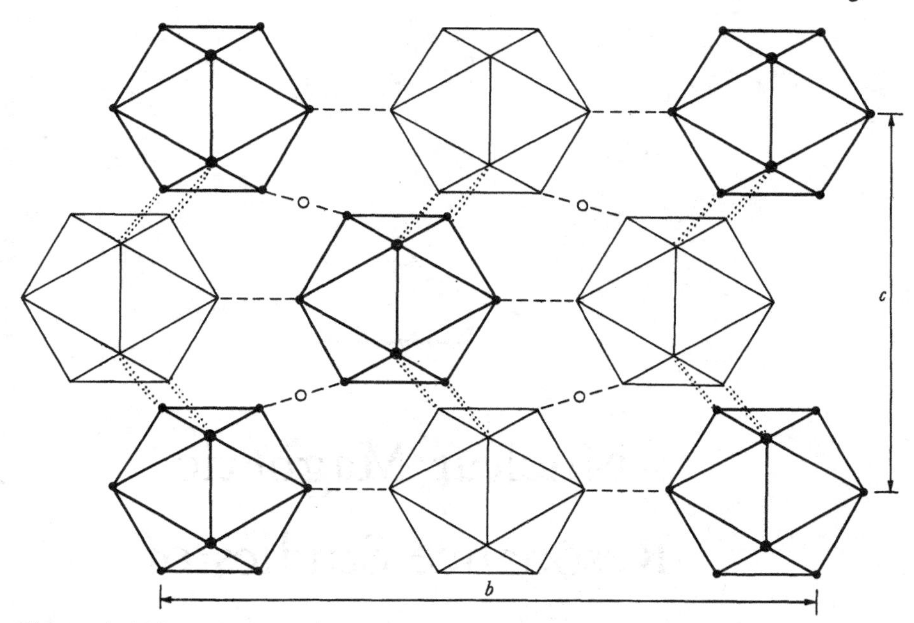


**Figure 3–13** *A conceivable structure for boron, related as hexagonal closest packing is related to simple cubic packing of icosahedra in the known simpler rhombohedral modification. The space group is $D_{2h}^{17}$-Cmcm with $a = 4.7$ A, $b = 9.3$ A, and $c = 7.8$ A. The five heavier-lined icosahedra are at $a = 0$, and the four lighter ones are centered on the plane $a = \frac{1}{2}$. Half of the boron atoms of each icosahedron are involved in three-center bonds which lie very nearly in the c glide planes at $b = \frac{1}{4}$ and $\frac{3}{4}$ relative to the centers of symmetry indicated by circles. Three-center bonds are indicated by dotted lines, and single bonds by dashed lines.*

compounds, and, indeed, may not be the most stable ones, but perhaps they do form some bridge in our understanding of these compounds in terms of the more usual rules of valence. It is at least of interest to see how far one can go in reducing the electron discrepancy in these unit cells, and it seems gratifying to us that the two known valence structures of boron are not as greatly inconsistent as was thought previously.

A further study is suggested of polyhedral approaches to valence structures, with the addition of topological rules somewhat like those formulated[49] for the boron hydrides, but generalized to other types of semilocalized molecular orbitals. Finally, the extension to structures like those of the lower borides[147] requires a much more detailed examination of the M—M interactions, and the extension to other classes of intermetallic compounds may involve placement of atoms inside polyhedra with due regard for orbital symmetries.

# 4

# Nuclear Magnetic Resonance Studies of Boron Hydrides and Related Compounds

The nuclear magnetic resonance (NMR) spectra, especially the $B^{11}$ spectra, are selected here as one of the most important physical methods for study of boron hydrides, ions, and substitution products. Structure determination by X-ray diffraction methods is the surest proof of a molecular formula and bonding in a new molecule, but this method invariably requires months of effort, and, in most cases, one to several years of work. Of the many methods, such as use of visible and ultraviolet spectra, infrared spectra, mass spectroscopy, etc., all of which are useful, the $B^{11}$ NMR spectrum stands out as one of the rapidly applicable methods from which both structural and molecular dynamical information may be obtained. More caution than is frequently applied, however, is required for reliable application of this method. Dynamical effects within molecules, exchange effects between different molecules, lack of understanding of the probable reasons for $B^{11}$ NMR chemical shifts, and coupling anomalies have all contributed to conclusions reached without sufficient test of their validity. On the other hand, these same effects make the information obtained by this method more interesting because they yield valuable chemical and dynamical information about these molecules when the structural information is known from other methods, or correctly interpreted from this method.

## 4–1 Compounds

### $BH_4^-$

The alkali borohydrides contain the tetrahedral $BH_4^-$ ion, as has been established by the early infrared work of Price[244] and powder-diffraction study by Soldate.[292] A confirmation of the tetrahedral structure was obtained by high-resolution NMR techniques (Fig. 4–1) by Ogg,[212] who also studied $B_2H_6$ by this method. A study of the second moment of the comparatively broad proton resonance in the solid alkali borohydrides[75] has led to the B—H distance of $1.25_5 \pm 0.02$ A, which is distinctly greater than one might extrapolate from the well-known distances of 1.092 A in $CH_4$ and 1.035 A in $NH_4^+$. This relative expansion of the hydrogen and boron orbitals is probably a major effect in promoting internal rearrangement, particularly in some of the negative hydride ions.

### $Al(BH_4)_3$

The trigonal, planar arrangement of 3B about a central Al has been established in an electron-diffraction study[11, 15, 286] which apparently disproved the correct hydrogen arrangement, later established from the infrared spectrum.[244] The remarkable nuclear resonance ($Al^{27}$, $B^{11}$, and H) spectra[213] (Fig. 4–2) suggest a time equivalence of the protons in each $BH_4$ group by a tunneling type of mechanism. In this process, collapse of the hydrogen fine structure of the $B^{11}$ resonance does not occur but rather all H's become equivalent, so that the quintet of relative intensities 1,4,6,4,1 is produced by interaction of four equivalent H's with the $B^{11}$ spectrum. The hydrogen resonance is broad. The temperature independence of these spectra and the noncollapse of the H fine structure argue strongly for this internal rearrangement, as opposed to some dissociation type of equilibrium.

### $B_3H_8^-$

The $B^{11}$ resonance[231] (Fig. 4–3a and b) of a solution of $NaB_3H_8$[126] in ether or $D_2O$ yields an apparent septet,[84, 231] with observed relative intensities of

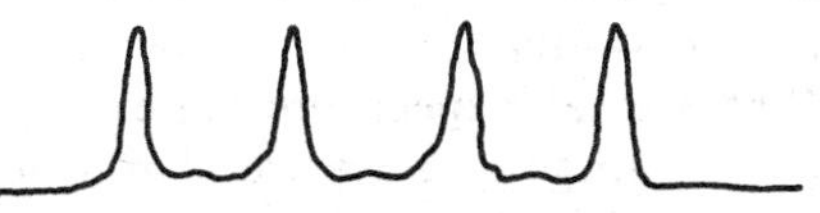

**Figure 4–1** *$BH_4^-$, $H^1$ NMR spectrum. The quartet is due to permitted transitions between $H^1$ levels, each of which is split into a quartet due to the four orientations, $m_s = -\frac{3}{2}, -\frac{1}{2}, \frac{1}{2}, \frac{3}{2}$ of the $B^{11}$ nucleus. The seven lines due to the $B^{10}$ orientation ($-3$ to $+3$) are of low intensity, and are not clearly seen here.*

1.0:3.5:6.3:7.8:6.3:3.5:1.0. The equality of spacings is very striking, but agreement is poor with the ratios 1:6:15:20:15:6:1 expected for a septet resulting from a group of six H atoms. However, if all B's are equivalent, and all B's see all eight H atoms, the relative intensities of the nonet are

0.1:0.9:3.2:6.4:8.0:6.4:3.2:0.9:0.1

The two very weak outer lines are predicted to be one-eighth of the intensity of the weakest members of the septet, and hence are presumably lost in the background. The proton resonance is broad, as is the case in $Al(BH_4)_3$. A static model in agreement with the observations seems impossible, and hence I have suggested[175] a model in which the electron pair bonding remains nearly invariant and internal exchange takes place. The correct structure is shown in Fig. 1–9 and is derivable from the predicted $B_3H_9$ structure by removal of either a bridge or terminal proton. Presumably this type of internal exchange is facilitated by negative charge transfer toward the boron and hydrogen by the Al in $Al(BH_4)_3$ and by the negative charge in the ions. The resulting expansion of the interatomic distances, similar to that observed in $BH_4^-$, probably makes this internal exchange much more likely in these compounds than in the hydrides themselves.

Other possible explanations of this effect, such as those based upon special relations of the coupling constants have not yet been ruled out, and hence further studies are desirable.

#### $B_3H_7$ Addition Compounds

A preliminary diffraction study[211] of $(CH_3)_3NB_3H_7$, and the preparation[128] and preliminary $B^{11}$ resonance study[231] of $(C_2H_5)_2O{\cdot}B_3H_7$ have been reported. The crystallographic evidence that the molecule has a threefold axis, and the somewhat rapid fall-off of intensities with increasing angle of scattering, together with the valence-theoretical observation that a single hydrogen on the threefold axis will be nonbonding to the $B_3$ molecular orbitals, have led to the suggestion[114] of a disordered crystal structure, in which there is a vacant orbital available for formation of a single bond to the lone pair of $NH_3$ or $N(CH_3)_3$ (Fig. 2–17b).

The $B^{11}$ resonance in $(C_2H_5)_2O{\cdot}B_3H_7$ is apparently a symmetrical sextet which, on the grounds of a similar argument to that given in Section 4–1 for $B_3H_8^-$, is very probably the more intense group of an octet with theoretical line ratios of 1/7:1:3:5:5:3:1:1/7. Again, the two outermost lines are presumably lost in the background. If so, we can expect internal exchange in $B_3H_7$– ether complexes, partly because of transfer of negative charge toward the $B_3H_7$. These etherates were first prepared by Edwards and co-workers.[128]

In $B_3H_7{\cdot}NH_3$ a crystal disorder similar to that found in $B_3H_7{\cdot}N(CH_3)_3$ has been observed above about $-16°C$. Nevertheless, C. E. Nordman and co-workers have obtained ordered single crystals of the low-temperature form,

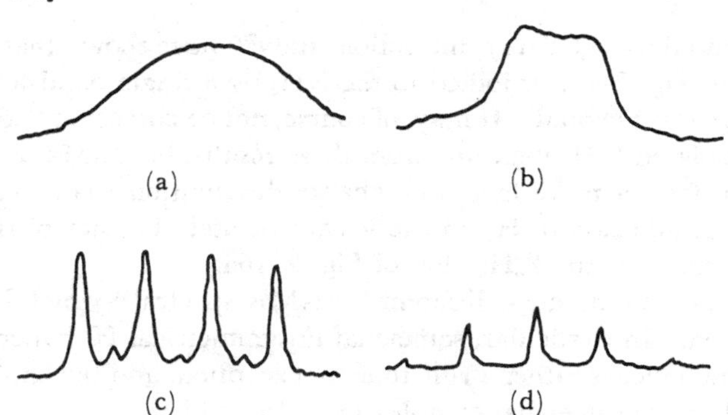


**Figure 4-2** $Al(BH_4)_3$. *(a) $H^1$ resonance, (b) $H^1$ resonance $B^{11}$ irradiated, (c) $H^1$ resonance $Al^{27}$ irradiated, (d) $B^{11}$ NMR spectrum. There is still some question of the purity of the sample on which these spectra were obtained. The minimum conclusion from these spectra is that there are $BH_4$ groups attached to Al.*

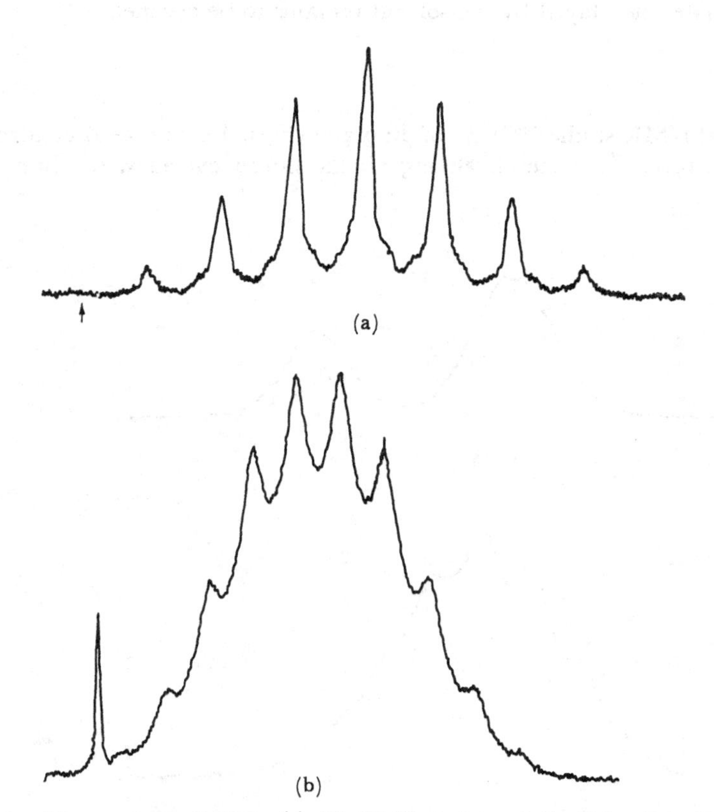


**Figure 4-3** $B_3H_8^-$. *(a) $B^{11}$ NMR spectrum of the $Na^+$ salt in $ND_3$. The apparent equivalence of all B and H atoms may be due to rapid intramolecular exchange of H atoms. (b) $H^1$ NMR spectrum of the $Na^+$ salt in $D_2O$. The sharp spike is residual $H_2O$.*

and in an excellent and detailed X-ray diffraction study[209] have shown that the $B_3H_7$ residue shown in Fig. 2-17b is joined to the $NH_3$ by a single bond to the boron atom having a vacant orbital. It may, of course, not be correct to deduce the structure of the isolated $B_3H_7$ molecule from these results, but the structure of Fig. 2-17b has, in fact, a more favorable charge distribution than that of Fig. 2-17a. Note that addition of $H^-$ to the vacant orbital of either of these possible $B_3H_7$ structures gives the $B_3H_8^-$ ion of Fig. 2-16a.

Various $B_3H_7L$ compounds show different $B^{11}$ NMR spectra[151] which have yet to be fully analyzed. In particular, somewhat unsymmetrical $H^1$ hyperfine structure in the $B^{11}$ resonance is rather a rule than an exception, and the analysis of these results may become somewhat complex (Fig. 4a and b).

#### $B_2H_7^-$

The $B^{11}$ NMR spectrum shows[328] that all H atoms are equivalent, and hence there is a dynamical process probably similar to the rotation tunneling in $Al(BH_4)_3$. The role played by the solvent remains to be clarified.[7]

#### $B_2H_6$

The several NMR studies[145,212,276] of diborane (Figs. 4-5 and 4-6) confirm the bridge structure. In general, among all the boron hydrides, the bridge

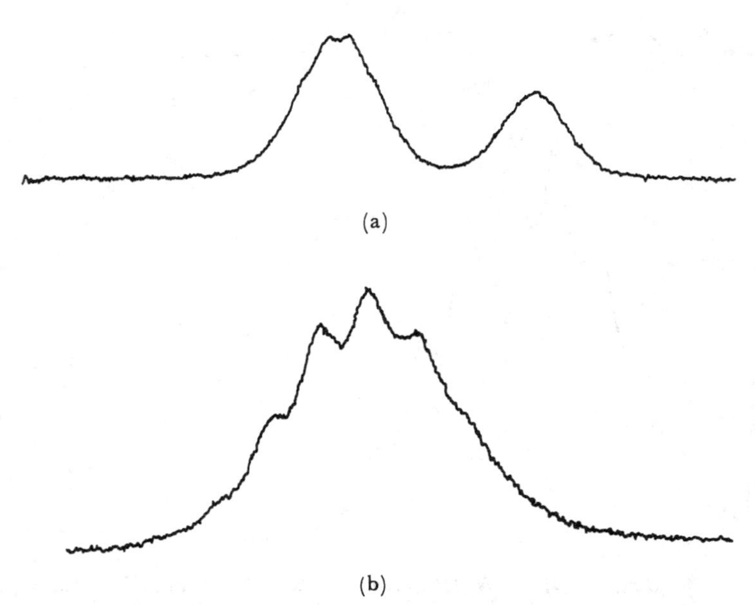


**Figure 4-4** *(a) $B_3H_7NH_3$. $B^{11}$ NMR spectrum in $Et_2O$. (b) $B_3H_7THF$. $B^{11}$ NMR spectrum in $Et_2O$.*

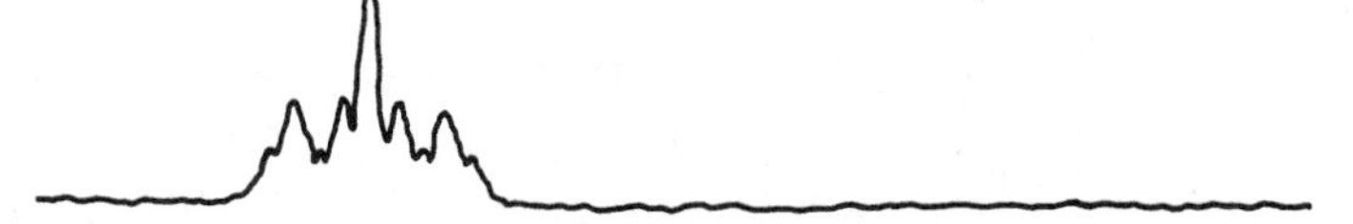

**Figure 4–5** *$B_2H_6$. $B^{11}$ NMR spectrum. The large triplet structure is due to the $BH_2$ groups, while the small triplet structure on each of the large components is due to the bridge H atoms.*

hydrogens tend to show H resonance at higher fields than the terminal hydrogens. This result is striking confirmation of the type of charge distribution expected on chemical grounds and predicted from molecular orbital theory.[95,146] The detail with which the NMR spectrum can be analyzed is shown by Schoolery,[276]

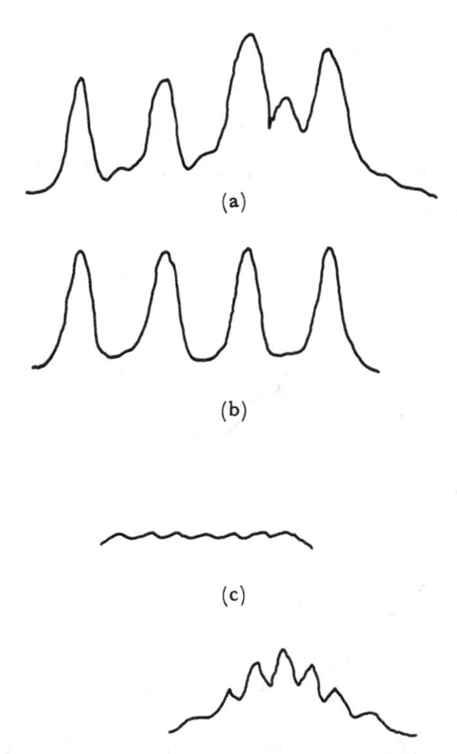


**Figure 4–6** *$B_2H_6$. (a) $H^1$ NMR spectrum. The large quartet is due to the $B^{11}$ spin of $\frac{3}{2}$. When this quartet (b) and the septet (c) due to the $B^{10}$ spin of 3 are subtracted out, there remains the $H^1$ spectrum of bridge H atoms which see mostly two $B^{11}$ nuclei with combined mutual orientations ranging from total spins (and statistical weights) of −3(1), −2(2), −1(3), 0(4), 1(3), 2(2), 3(1), as shown in (d).*

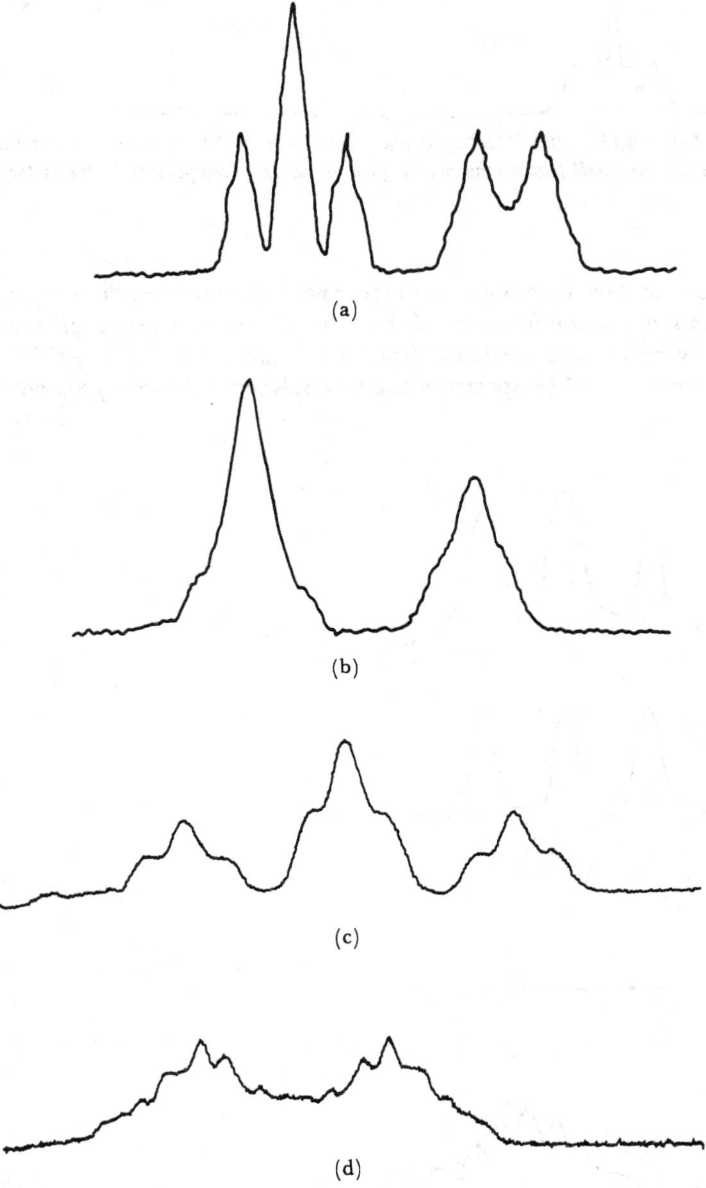


**Figure 4-7** *(a) $B^{11}$ NMR spectrum of $B_4H_{10}$.*[249] *The large triplet is due to the $BH_2$ groups, the smaller doublet to the B—H groups, while the smaller structure is associated for the most part with the bridge H atoms. (b) Tetraborane spectrum of 90 per cent deuterated sample.*[249] *$B^{11}$ NMR showing that hyperfine structure of the B—H doublet at higher field in $B_{10}H_{14}$ is due to a complicated $B^{11}$—$H^1$ interaction, not to $B^{10}$—$B^{11}$ coupling.*[332] *(c) $B_4H_{10}$. $B^{11}$ NMR spectrum of the triplet. (d) $B_4H_{10}$. $B^{11}$ NMR spectrum of the doublet.*

especially in the case of $B_2H_6$. In this compound and in the higher hydrides it has proved possible to simplify the observed NMR spectra very greatly by "nuclear stirring" techniques, in which a given nucleus (e.g., $B^{11}$) is induced to undergo many transitions during an observation. Under these circumstances, a neighboring nucleus sees only the average over these orientations, and the fine structure due to the nucleus being "stirred" collapses to a single line.

### $B_4H_{10}$

An early indication that the NMR spectrum of $B_4H_{10}$ was run was not followed up by any details, but a later study[332] indicates consistency of the $B^{11}$ resonance with the established structure. However, this latter study is incorrect with respect to the claim that $B^{10}$—$B^{11}$ coupling occurs, for this coupling does not occur when the $B^{11}$ NMR spectrum of $B_4D_{10}$ is examined[249] (Fig. 4–7a and b). Other $B^{11}$ NMR spectra of $B_4H_{10}$ are shown in Fig. 4–7c and d.

### $B_4H_8CO$

The $B^{11}$ NMR spectrum[263] of $B_4H_8CO$ is shown in Fig. 4–8. The structure of this compound has not yet been proved, but the spectrum suggests a $B_4H_{10}$-like B arrangement.

### $B_5H_9$

The use of $B^{11}$ "stirring" techniques[21] has not only confirmed[276] the tetragonal pyramidal structure including the ratio of bridge to terminal

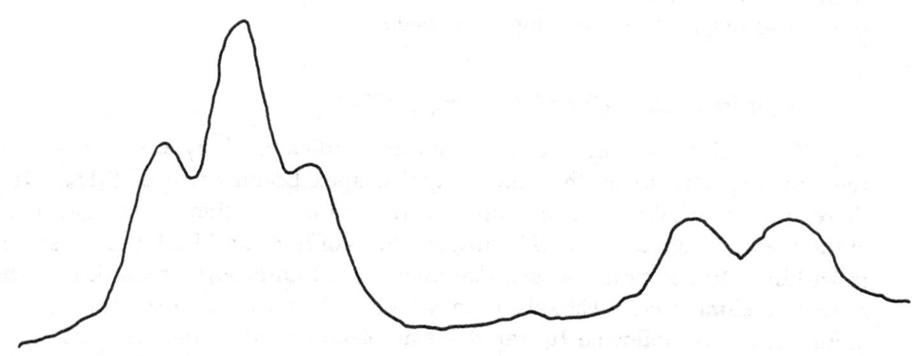

**Figure 4–8** · *$B_4H_8CO$. $B^{11}$ NMR spectrum.[263] The strong resemblance of this spectrum to that of $B_4H_{10}$ is suggestive that there are two $BH_2$ groups and two BH groups in $B_4H_8CO$. A structure topologically equivalent to 2113 $B_4H_9^-$ is proposed in which the B arrangement is like that in $B_4H_{10}$.*

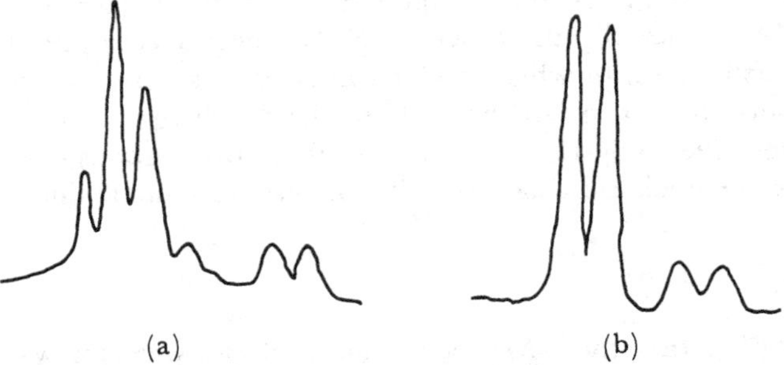


**Figure 4–9** *$B_5H_{11}$ (a) and $B_5H_9$ (b). $B^{11}$ NMR spectra. Some $B_5H_9$ impurity is present in the $B_5H_{11}$ sample. The $B_5H_9$ spectrum is consistent with the tetragonal pyramidal structure, with two $B^{11}$ resonances in the ratio of 4 to 1, each split by terminal H. The most interesting aspect of the $B_5H_{11}$ spectrum is that the second H atom attached to the apical B atom (high field doublet) shows no splitting,[266,333] and hence suggests that its bonding is much like a bridged bond involving the apex B and both of the B's of the $BH_2$ group in the molecule. This second H atom lies in the region of the icosahedral fragment, and is in roughly the region of the sixth B atom of $B_6H_{10}$.*

hydrogens, but has further indicated that the apex boron atom shows resonance at a higher field than the basal borons (Fig. 4–9). This result agrees with the simplified valence theory of this compound, which indicates that the apex boron is the more negative and is bonded more symmetrically, and is also indicative of the smaller contribution of temperature-independent paramagnetism and increased molecular diamagnetic contributions for the apex B as compared with the other B atoms of $B_5H_9$. Somewhat improved spectra obtained later[266] confirm these results. The hydrogen resonance[276] suggests that the apical hydrogen atom is slightly more negative than the basal terminal hydrogens, but not so negative as the bridge hydrogens.

### $B_5H_8Br$ (m.p., 32–34°) and $B_5H_8I$ (m.p., 53°)

The position of substitution is clearly indicated[267] by both H and $B^{11}$ resonance spectra to be the more negative apex boron atom of $B_5H_9$. It is, therefore, remarkable that the apparent rates of deuteration (in the gas phase) of the apex and basal terminal hydrogen atoms of $B_5H_9$ and $B_2D_6$ are the same[158] to within ±10 per cent. Koski, Kaufman, and Lauterbur[158] remark that they have not eliminated mechanisms in which only terminal apex (or base) H atoms exchange, followed by rapid intramolecular redistribution of H and D atoms among all terminal positions. Although they considered this mechanism unlikely and hence favored direct exchange, it may be important to reconsider this process. In fact, the rearrangement[179,214] of apex $MeB_5H_8$ to place the Me group on a different B atom may be similar to this D redistribution.

Mention of the terminal-hydrogen-substituted $B_2H_5Br$[45] seems desirable to complete this section, and perhaps to remind us that no bridge halogens have been found. But the existence of Cl bridges in $BeCl_2$, of $CH_3$ bridges in $Al_2(CH_3)_6$, and of N bridges[106] in $B_2H_5NH_2$ and $B_2H_5N(CH_3)_2$ suggest that these or similar possibilities may exist for substituted boron hydrides or boron hydride ions (cf. $EtNH_2B_8H_{11}NHEt$, below).

### $B_5H_{11}$

The H and $B^{11}$ resonances have been analyzed[266] to assign the most negative boron (type I, apex) of the valence theory to the boron resonance at highest field (Fig. 4–9). In the $B^{11}$-saturated hydrogen NMR spectrum, the peaks reading from low to high field are said to be of approximate relative intensities of 6:1:3:1. Schaeffer, Schoolery, and Jones[266] assign the six terminal H, except the apex H, to the large peak; then the terminal apex H to the next peak, the three bridge hydrogens to the next peak, and the unique hydrogen on the apex boron atom to the single peak at highest field. They emphasize the lack of certainty about this assignment, and further investigation is desirable by a combination of NMR, infrared, and isotopic-exchange techniques. In particular the relative peak weights may be 6:1:2:2, the bridge hydrogens may not all give peaks at the same place since two of the hydrogen bridges are probably unsymmetrical,[169] and the unique hydrogen may not be so greatly different from the other terminal hydrogens that it should show resonance at the highest field. It is exceedingly difficult to maintain the purity of $B_5H_{11}$ at room temperature, and it is not completely certain that the spectra[266,333] have been obtained on pure samples.

The position of the unique hydrogen almost in the basal plane of the four boron atoms has been reinvestigated twice[197] since the original structure study.[164] Among all the boron hydrides, this is the only extra hydrogen attached to a boron atom connected to more than two other boron atoms. The valency rules would be somewhat simpler if this hydrogen were a bridge hydrogen, and a satisfactory structure can be drawn in such a case,[114] although the hydrogen arrangement does not have the symmetry of the boron arrangement. However, a recent unpublished reinvestigation by M. G. Rossmann and myself of these X-ray diffraction data by the least-squares method, in which all atoms have been assigned different temperature factors, indicated no randomness in the hydrogen positions; and hence it is still very likely that this unique hydrogen lies in the apparent symmetry plane of the molecule and that the published structure is correct.

### $B_6H_{10}$

The $B^{11}$ spectrum[334] (Fig. 4–10) of $B_6H_{10}$ is remarkable. On the one hand it suggests B—H groups only (no $BH_2$ group) in the molecule, plus bridge H

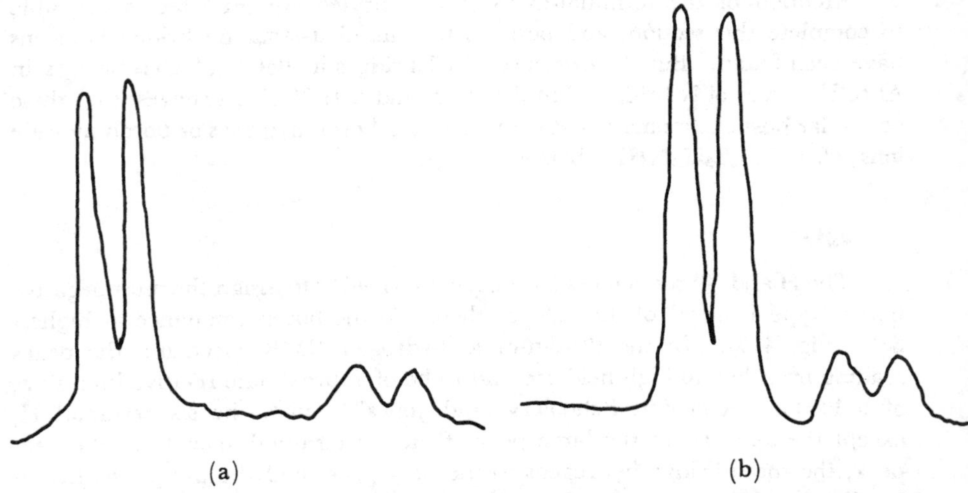


**Figure 4-10** *$B_6H_{10}$ (a) and $B_5H_9$ (b). $B^{11}$ NMR spectra. Although the spectrum of $B_6H_{10}$ is consistent*[334] *with no $BH_2$ groups in the molecule, the five basal B atoms of the pentagonal pyramid should not, in fact, show resonance at the same place unless some dynamical equivalence*[178,328] *is introduced by tautomerism, or unless some very unusual relations exist between the chemical shifts.*

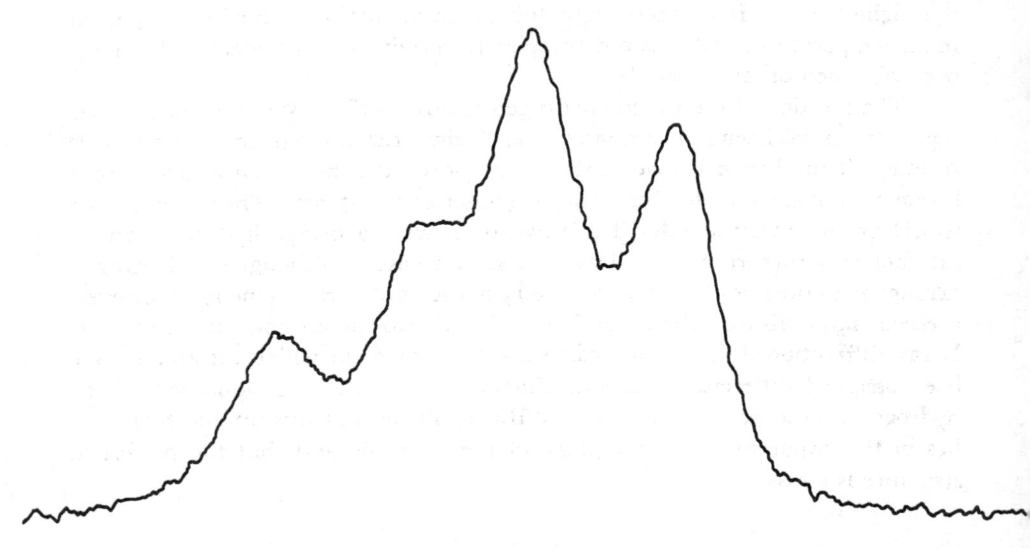

**Figure 4-11** *$B_8H_{13}^-$(?). $B^{11}$ NMR spectrum of $K^+$ salt. This hitherto unpublished spectrum is on a pure reaction product obtained*[151] *from reaction of excess K with $B_5H_9$ in tetrahydrofurane. Probable structures are based upon the compact $B_8$ arrangement discussed in Chapter 2, but H tautomerism of the $BH_2$ and B—H—B bonds is expected.*

atoms which are not so easily discernible in the higher hydrides. This result would be consistent with the ordered structure carefully established in the X-ray diffraction study,[71,114,250] except for the apparent exact equivalence of all five nonapical B atoms in the $B^{11}$ NMR spectrum. The ambiguity of the valence theory[50,172] is suggestive of a ready tautomerism,[176,178,328] but this point merits further study.

### $B_8H_{13}^-$(?)

For further study and comparison the $B^{11}$ NMR spectrum of a pure crystalline compound obtained[151] by reaction of excess K with $B_5H_9$ in tetrahydrofurane is given in Fig. 4–11. The chemical analysis is not completely certain. The two kinds of B atoms in a ratio of 3 to 5 suggest the 2513 structures shown in Fig. A–1, but H tautomerism around the periphery may be responsible for the apparent simplicity of the spectrum.

### $EtNH_2B_8H_{11}NHEt$

The only readily interpretable feature of this spectrum (Fig. 4–12) is that there are probably two B atoms in apex environments, thus showing resonance at high field. The structure is now known (Chapter 1).

### $B_9H_{13}NHEt_2$

The $B^{11}$ NMR spectrum (Fig. 4–13) is so poorly resolved that only the high-field doublet is a reliable feature. This doublet suggests that there are

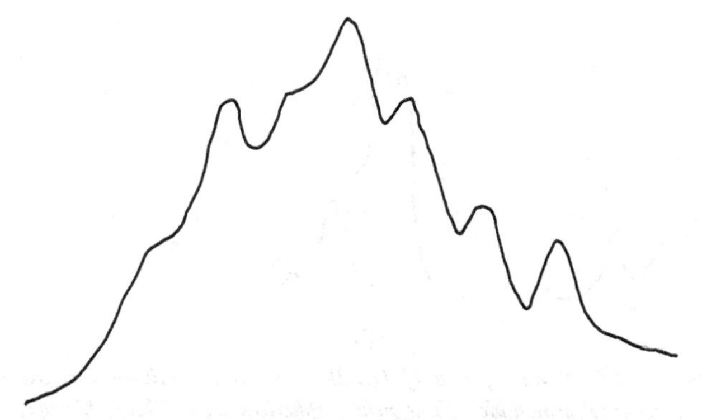

**Figure 4–12** *$EtNH_2B_8H_{11}NHEt$, $B^{11}$ NMR spectrum. The structure of this molecule has been determined by X-ray diffraction methods.[167] If the high-field doublet intensity had been interpreted as it was in the $B_9H_{13}L$ compounds, the correct conclusion that only two B atoms have apex environments would have been suggested, and the earlier suggestion[83] that the $B_9H_{12}L^-$ anion was responsible for this spectrum would have been questioned.*

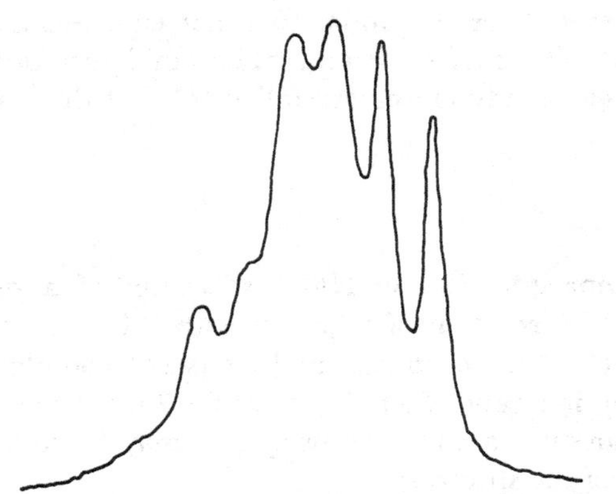

**Figure 4-13** *$B_9H_{13}NHEt_2$. $B^{11}$ NMR spectrum. The high-field doublet is associated with three B atoms, all now known[319,320] to be in apical environments.*

three B atoms in very nearly the same environment, originally suggested to be apex environments, and later confirmed by the X-ray crystallographic study.

### $B_9H_{12}^-$

The $B^{11}$ NMR spectra of two salts of this ion are shown in Fig. 4-14. There is some difficulty with decomposition of this ion under the conditions of these

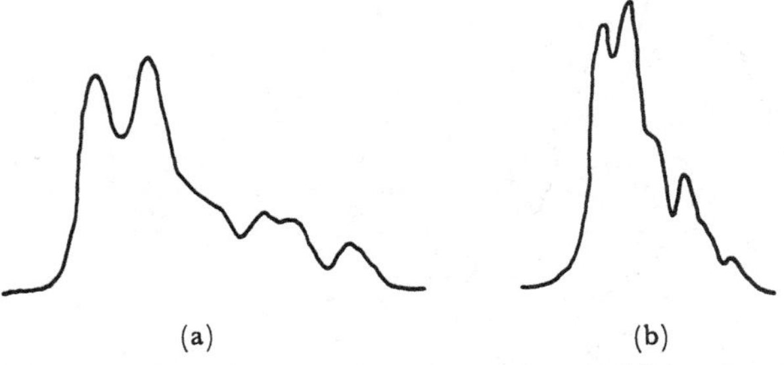


**Figure 4-14** *$B_9H_{12}^-$. $B^{11}$ NMR spectra of (a) the $Et_2NMe_2^+$ salt in $CH_2Cl_2$ and (b) the $Me_4N^+$ salt in dimethylformamide. The nearly spherical nature[318] of this ion, and the low symmetry of the H arrangement suggested by these spectra suggest that the ordered molecular structure suggested on the basis of the valence theory[179] may be a reasonable one. Chemical shifts in ppm relative to methyl borate are 25.5 and 35.2 for the large low-field doublet and 65.4 for the highest-field peak in (a); corresponding features of (b) are at shifts of 25.9, 36.3, and 65.4 ppm.*

experiments. Preliminary X-ray diffraction studies of these two salts suggest a threefold axis for minimum molecular symmetry. However, there is a transition in the solids at lower temperatures, there is an abnormally rapid fall-off of X-ray intensities with increasing angle of scattering, and hence the $B_9H_{12}^-$ ion may have some orientational disorder. It is therefore not possible to conclude from the X-ray results that the ion really does have a threefold axis. Although the NMR evidence suggests a molecule of low symmetry, the complexity of the spectrum may be due in part to some decomposition, and hence it seems unsafe to conclude that the molecule does not have a threefold axis on the basis of the NMR evidence. The two most plausible valence structures[179] have the same molecular geometry, and the same topology, 2621.

### $B_9H_{15}$

If, indeed, the $B_9H_{15}$ $B^{11}$ NMR spectrum (Fig. 4–15) can be analyzed on the assumption that it is a superimposed $B_4H_{10}$ spectrum and a $B_5H_{11}$ spectrum, both corrected for conversion of $BH_2$ to B—H groups where necessary, then the assignment is as follows. The highest field resonance is the apex of the $B_5H_{11}$ residue, and the second highest field resonance is due to the two singly bonded B atoms of the $B_4H_{10}$ residue. The lowest field bump seems to be too small to be anything but a component of the $BH_2$ group, and all other B atoms therefore contribute to the remainder of the low-field set of peaks. The long $B_4$—$B_5$ and $B_6$—$B_7$ distances (Fig. 1–6) tend to support this kind of analysis, which otherwise would not be attempted.

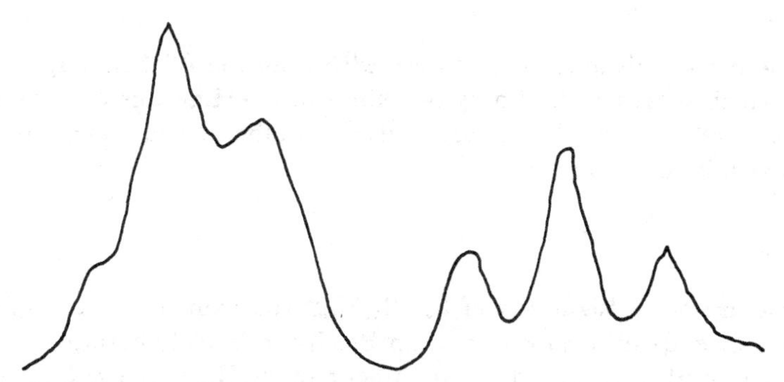

**Figure 4–15** *$B_9H_{15}$. $B^{11}$ NMR spectrum.[31] This spectrum is not greatly different from that obtained by addition of the $B_5H_{11}$ and $B_4H_{10}$ spectrum after corrections for conversion of the appropriate $BH_2$ groups to B—H groups. Chemical shifts in ppm from left to right are −1.6 (shoulder), −7 (highest), 0.2, 2.2, 3.3, and 4.3. The coupling constant J is 187 cps between the 2.2 and 3.3 peaks and 150 cps between the 3.3 and 4.3 peaks of the high-field group of three peaks. Reference sample is methyl borate.*

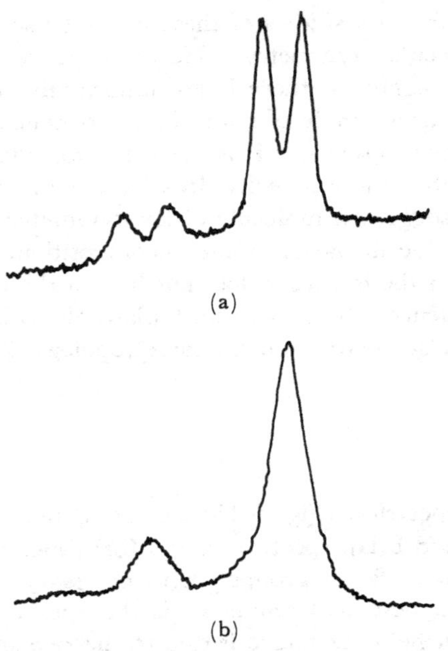


**Figure 4–16** *$B_{10}H_{10}^{=}$. $B^{11}$ NMR spectrum (a) of $K^+$ salt, (b) of $K^+$ salt in presence of about 1 M $FeCl_3$. The $Fe^{3+}$ has not yet produced a chemical reaction, but it does wash out the proton hyperfine structure without collapsing the $B^{11}$ separation, presumably by mostly an electron-spin-to-proton-spin magnetic interaction.[181] The separation of the two types of B atoms is 11.7 ppm.*

### $B_{10}H_{10}^{=}$

The structure of this ion was first derived[182] from the $B^{11}$ NMR spectrum (Fig. 4–16), which suggested that only two kinds of B—H groups were present in the ratio of 2 to 8. A crystal structure study[55,141] of the structure of $Cu_2B_{10}H_{10}$ has shown that this conclusion is correct.

### $B_{10}H_9OH^{=}$

The structure and assignment of the $B_{10}H_{10}^{=}$ spectrum make the assignment of apical or equatorial substitution on $B_{10}H_{10}^{=}$ relatively certain. When $B_{10}H_9OH^{=}$(?) is freshly prepared by treatment of $B_{20}H_{18}^{=}$ (see below) with $2OH^-$, the $B_{10}H_9OH^{=}$(?) is apex-substituted as shown by partial collapse of the small doublet and no effect on the large doublet (Fig. 4–17). A further reaction makes the large doublet unsymmetrical, and a rearrangement[141] has been proposed in which the OH substituent migrates without breaking the B—O bond. It would be worthwhile to test this suggestion by use of $O^{18}$ as a tracer.

**Figure 4–17** *$B_{10}H_9OH^=$(?). $B^{11}$ NMR spectrum of $K^+$ salt in $H_2O$, (a) immediately after preparation, (b) after standing at room temperature for several days. Apparent migration of the OH has occurred from the apex to substitute partially the equatorial positions.*

### $B_{10}H_{12}L_2$ and $B_{10}H_{14}^=$

The general resemblance of the $B_{10}H_{12}L_2$ and $B_{10}H_{14}^=$ spectra (Fig. 4–18) to that of $B_{10}H_{14}$ (see below) has led to an incorrect assignment,[221] which has since been corrected[205] by the data in Fig. 4–19. The reasons for this reassignment, while clear enough from the experiments, are not yet understood from the theory. The changes of the $B_{10}$ framework upon conversion from $B_{10}H_{14}$ to $B_{10}H_{12}L_2$ are large. For example, the range[197] of B—B distances from 1.71 to 2.01 A in $B_{10}H_{14}$ decreases to a range[248] of 1.74 to 1.88 A in $B_{10}H_{12}(NCCH_3)_2$. Perhaps when the molecular diamagnetism is analyzed it will be found to circulate mostly about $B_2$ and $B_4$ in $B_{10}H_{14}$, but about $B_1$ and $B_3$ in $B_{10}H_{12}L_2$. Clearly, the reasons for these large relative shifts are not yet understood. The spectrum of $B_{10}H_{14}^=$ is shown in Fig. 4–20.

### $B_{10}H_{12}I_2$

The assignment of substitution of I in $B_{10}H_{12}I_2$ to the 2,4 positions is based on a determination[261] of the intramolecular I · · · I distance by projection methods from X-ray diffraction data. The observed distance of 6.25 A com-

**Figure 4–18** *$B_{10}H_{12}(NHEt_2)_2$. $B^{11}$ NMR spectrum, for comparison with that of $B_{10}H_{12}(SEt_2)_2$ shown in Fig. 4–19.*

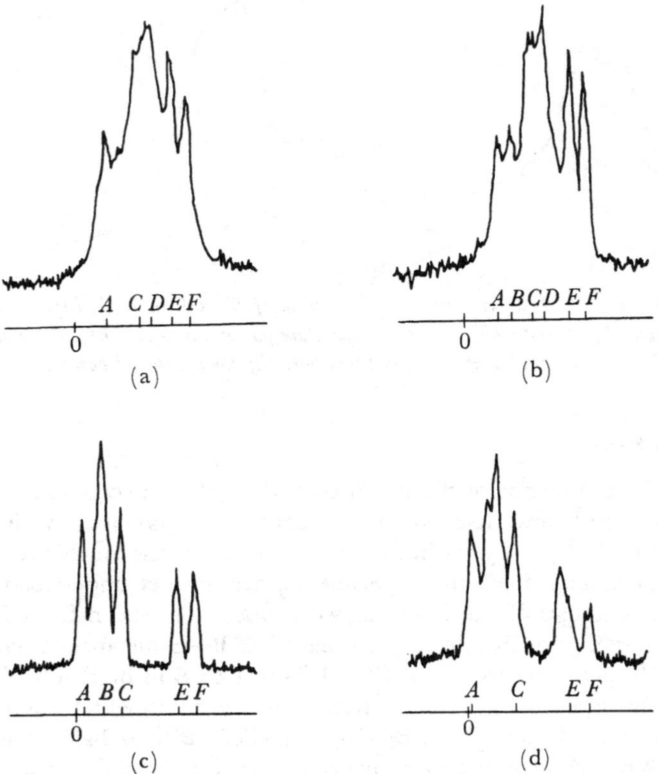


**Figure 4–19** *6,9-$B_{10}H_{12}(SEt_2)_2$. $B^{11}$ NMR spectrum*[205] *shown in (b). The 2-Br substitution product is shown in (a), thus identifying the* **low**-*field doublet with the 2,4 positions in $B_{10}H_{12}(SEt_2)_2$. It had previously been incorrectly assumed*[221] *that the 2,4 positions were at highest field in $B_{10}H_{12}L_2$ compounds, as they are in $B_{10}H_{14}$(c), in which substitution of Br at the 2 position partially collapses the high-field peak (d). Chemical shifts are 18.0, 24.4, 35.2, 41.5, 54.0, and 62.0 for A–F of (b); 17.1, 33.6, 39.9, 51.1, and 59.7 for (a); 3.6, 11.0, 22.8, 49.7, and 59.2 for (c); and 1.7, 23.4, 50.3, and 60.1 for A, C, E, and F of (d), all relative to trimethyl borate.*

**Figure 4–20** *$B^{11}$ NMR spectrum of $Na_2B_{10}H_{14}$. The assignment of $B^{11}$ resonances is like that in Fig. 4–19.*

pares favorably with that expected from the known molecular structure. If the B—H results of Kasper, Lucht, and Harker[142] are extrapolated to an expected B—I distance, the result is 5.92 A for the 2,4 substitution, and 7.28 A and 7.56 A for the other possibilities. If the further refinement of the $B_{10}H_{14}$ structure by Moore, Dickerson, and Lipscomb[197] is used, 2,4 substitution predicts a distance of 6.21 A, in somewhat better agreement with the observed value. The other possible positions lead to predictions of 7.88 A and 7.60 A, and hence are rejected as possibilities. In view of the importance of this structure determination in the following argument, the location in some future study of the boron framework in the X-ray diffraction study would be important as confirmatory evidence of the position of substitution.

The $B^{11}$ resonance study[267] of $B_{10}H_{12}I_2$ establishes that the I atoms are substituted on the B's giving $B^{11}$ peaks at highest field. It is important that in this and the following $B^{11}$ resonance studies of substituted boron hydrides, the positions of boron resonances are only somewhat shifted from their relative positions in the pure hydrides. This discovery has made the field of substituted boron hydrides easily studied by NMR techniques, and should lead to the identification of presently unassigned resonance peaks in the pure hydrides themselves and provide a basis for studies of internal exchange reactions. Thus we can expect more detailed analyses of the charge distributions in both the hydrides and their substitution products to appear in the near future.

### $B_{10}H_{14}$

Assignment of the high-field doublet (Fig. 4–21) to the 2,4 positions is based upon the I · · · I distance[261] in $B_{10}H_{12}I_2$, as discussed in the previous section. Assignment of the lowest-field region to the 1,3 and 6,9 set is based upon an analysis[267] of $B_{10}H_{14}$ and its substitution derivatives, and also on partial resolution[330] in the $B^{11}$ spectrum of $B_{10}D_{14}$. Here also, some further checks are desirable on this assignment.

Some evidence of a rather strong interaction propagated between B atoms 2 and 6 has been pointed out[331] in the $B^{11}$ resonance shift of one member of each pair when the other has been substituted.

### $B_{10}H_{13}^-$

The $B_{10}H_{14}$ molecule in 50 per cent ethanol-water behaves[89,299] as a strong monoprotic acid. The second ionization constant has a pK > 10, out of range of the studies made so far. The doubled peak at high field in the $B^{11}$ NMR spectrum (Fig. 4–22) shows that the initial stages of ionization do not involve the 2,4 hydrogen atoms. Infrared spectra of deuterium-hydrogen exchange of $B_{10}D_{14}$ indicate[66,99,283] that the four bridge hydrogens exchange first, followed by further exchange or internal rearrangement[99,283] involving the 5,7,8,10 terminal hydrogen positions adjacent to the bridge position. There is also some as yet

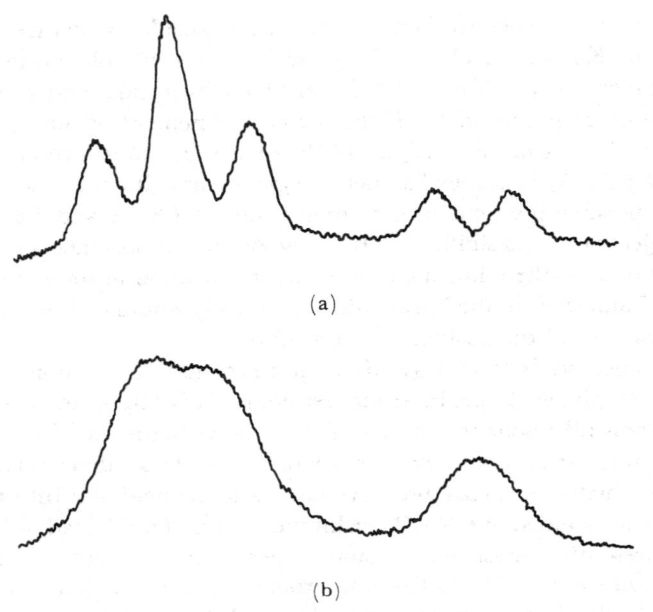


**Figure 4–21** *$B_{10}H_{14}$. $B^{11}$ NMR spectrum. In nonaqueous media (benzene) the $Fe^{3+}$ collapse of the $H^1$ hyperfine structure is not so clear, presumably because the acetonylacetate complex used here does not permit these interactions at such short distances. The high-field peak of the lower spectrum has been shown to be due to the 2,4 B atoms, and the low-field peak has been identified[267,330] with the 1,3 and 6,9 sets, which show some sign of resolution in $B_{10}D_{14}$. A structure study of a 6,9-$Me_2B_{10}H_{12}$ would be desirable to check this assignment.*

unpublished[89] infrared evidence suggesting a $BH_2$ group in $B_{10}H_{13}^-$. Possible structures for this ion are discussed in Chapter 5.

### $B_{10}H_{13}OR$

The alkoxylation[193] of $B_{10}H_{14}$ produces a derivative, $B_{10}H_{13}OR$, in which the point of substitution is probably 5,7,8,10 or possibly 6,9. The $B^{11}$ NMR spectrum (Fig. 4–23) shows that substitution is certainly not at the 2,4 position, and the probable mechanism involves nucleophilic attack by the solvent as a first step, so the 1,3 positions are also thought not to be involved. A structure study would settle this point.

This identification of the I substitution at the 2,4 positions in $B_{10}H_{12}I_2$ leads also, upon comparison with the charge distribution in $B_{10}H_{14}$, to the idea that the dominating mechanism of substitution is electrophilic attack of the halogen atom, as was predicted earlier.[171,172] Perhaps this is true in a majority of cases, but the existence of two different $B_{10}H_{13}I$'s leads one to the caution

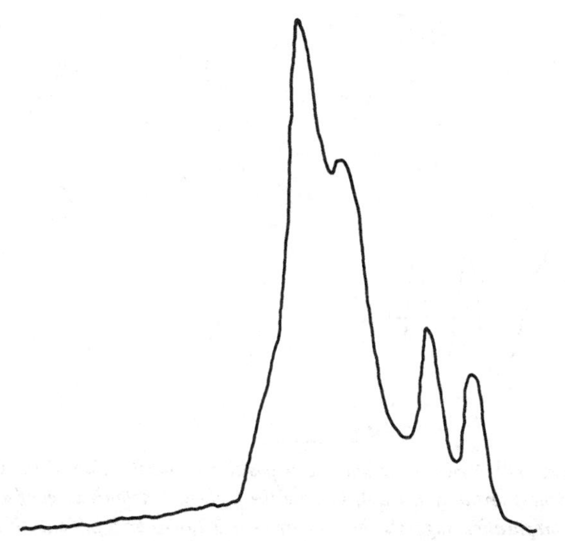

**Figure 4–22** *$B_{10}H_{13}^-$, $B^{11}$ NMR spectrum. The high-field doublet is probably due to the 2,4 B atoms, but the low-field group has not been analyzed. Chemical shifts in ppm relative to methyl borate are 16.8 (highest peak), 26.2, 47.3, and 57.1. Corresponding features in $B_{10}H_{14}$ are at 10.8 (highest peak), 22.7, 47.6, and 60.0 ppm (cf. Fig. 4–20).*

that other processes may be involved, perhaps a competing reaction by the same or a different mechanism.

### $B_{10}H_{13}I$ (m.p., 116°) and $B_{10}H_{13}Br$ (m.p., 105°)

The H and $B^{11}$ resonance spectra indicate[267] that substitution takes place in both of these compounds on the boron giving resonance at highest field. By the arguments given in Chapter 3, this is the boron atom of type 2,4.

### $B_{10}H_{13}I$ (m.p., 72°)

The H and $B^{11}$ resonance spectra show[267] clearly that the doublet at highest field is unaffected, and hence that substitution does not occur at the 2,4 position. The position of substitution is not known, and hence an investigation of this structure by X-ray diffraction methods is needed.

### $B_{10}H_{16}$

The $B^{11}$ NMR spectrum[85] (Fig. 4–24) is consistent with two $B_5H_8$ residues, like those in $B_5H_9$ molecules, joined at the apical B atom by a single bond. This result has been confirmed by the X-ray diffraction study.[85]

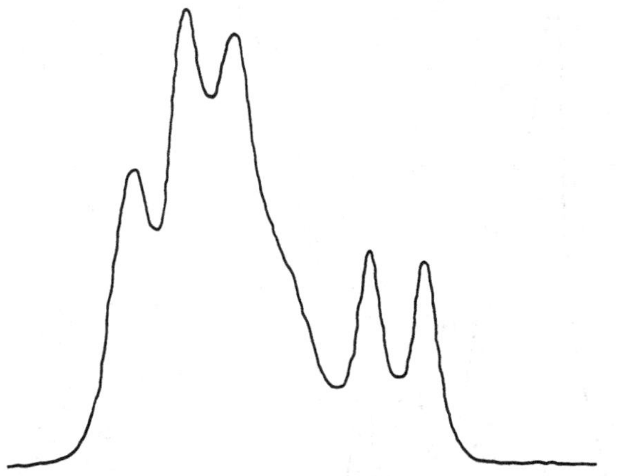

**Figure 4–23** *$B_{10}H_{13}OEt$. $B^{11}$ NMR spectrum. The position of substitution of the OEt group is not known but is almost certainly not 2,4, because the high-field doublet is unaffected. The nucleophilic method of preparation suggests that it may be a 5,7,8,10 or a 6,9 substituent.*

### $B_{12}H_{12}^{=}$

Only a doublet is found (Fig. 4–25),[233] consistent with the known[336] icosahedral structure, which was first proved by the X-ray diffraction study.[336] The single boron resonance is consistent with other structures, such as the cube-octahedral structure, or structures having internal tautomerism.

### $B_{18}H_{22}$

The structure, shown in Chapter 1, was established by the X-ray diffraction work,[291] but attempts to deduce it from this spectrum (Fig. 4–26) failed. However, it is now reasonable to view this spectrum as a decaborane-14 spectrum which has been mostly modified by loss of symmetry (Fig. 4–26). Thus the high-field triplet is probably two superimposed doublets, to be attributed to atoms 2 and 2′ (at highest field) and atoms 4 and 4′.

### Iso-$B_{18}H_{22}$

This spectrum (Fig. 4–27) of a byproduct[235] in the reaction from which $B_{18}H_{22}$ was formed suggests a close relation between these structures. Attempts

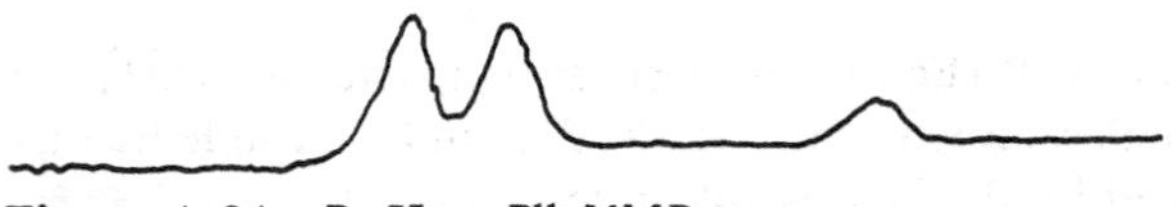

**Figure 4–24** *$B_{10}H_{16}$. $B^{11}$ NMR spectrum.*

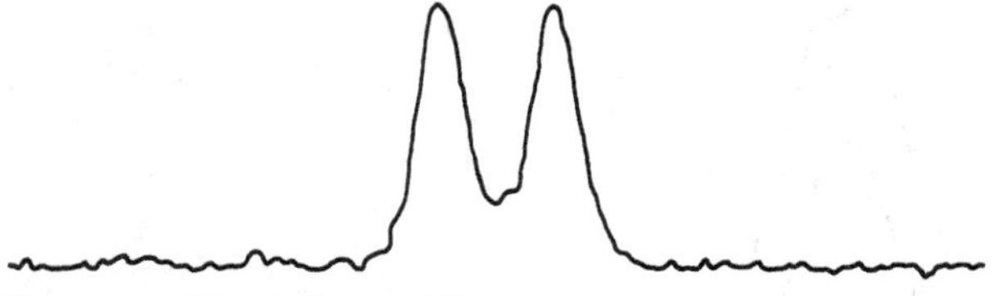

**Figure 4–25** $B_{12}H_{12}^{=}$. *$B^{11}$ NMR spectrum of $K^+$ salt in $H_2O$.*

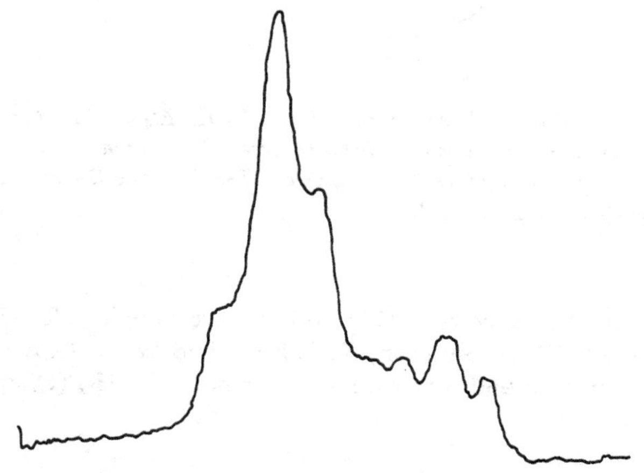

**Figure 4–26** $B_{18}H_{22}$. *$B^{11}$ NMR spectrum. This spectrum has not been completely analyzed. It does, however, seem safe to conclude that the high-field triplet is really two doublets, owing to the 2,2′ pair and the 4,4′ pair in order of increasing field (see Fig. 2–26 and the resonance discussion, Chapter 2). Chemical shifts (methyl borate = 0) are in ppm from left to right 0.0 (shoulder), 11.7 (highest peak), 22.3 (shelf), and for the high-field group, 43.0, 53.5, and 63.7.*

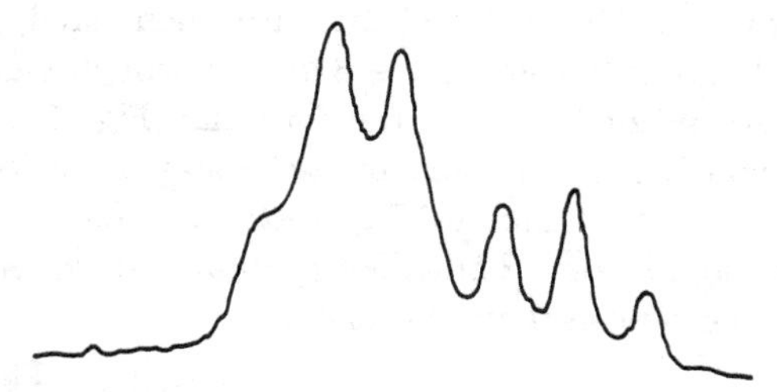

**Figure 4–27** *Iso-$B_{18}H_{22}$ · $B^{11}$ NMR spectrum. This product occurs in the solvolytic degradation of $2H_3O + B_{20}H_{18}^{=}$, which leads to $B_{18}H_{22}$. Its structure has been proved by X-ray diffraction methods. Chemical shifts are (methyl borate = 0) 10.1 (highest), 21.4 (next), and 38.9, 51.2, and 63.3 for the high-field group of three peaks.*

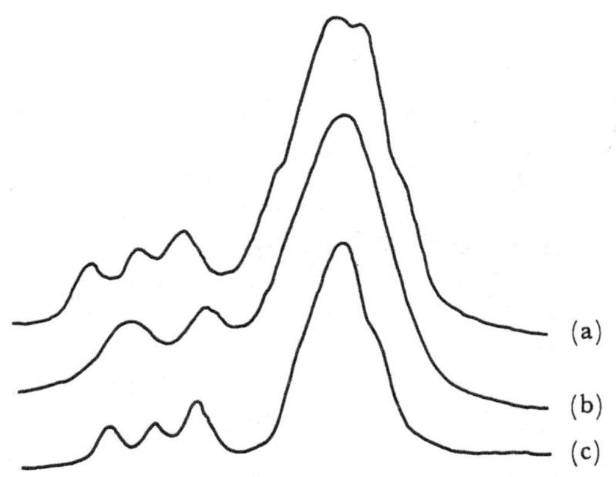


**Figure 4–28** *$B_{20}H_{18}^{=}$. The $B^{11}$ NMR spectra of $B_{20}H_{18}(HNEt_3)_2$. (a) The pure compound in perdeuteroacetone. (b) The compound in the presence of $Fe^{3+}$ in a methanol-acetone mixture. (c) The extensively deuterated sample in acetone. The free apex B—H remains undeuterated, contrary to the behavior in $B_{10}H_{10}^{=}$.*

to solve the structure from this spectrum (Fig. 4–27) have failed. However, an X-ray diffraction study by P. G. Simpson, K. Folting, and W. N. Lipscomb has established both the formula and structure as an isomer of $B_{18}H_{22}$ (Chapter 1).

### $B_{20}H_{18}^{=}$

We start from the preliminary crystallographic result[141] that the more stable ion, A, has a center of symmetry. The $B^{11}$ NMR spectrum of this ion (Fig. 4–28a) is a low-symmetry version of the $B_{10}H_{10}^{=}$ spectrum. Interpreted as such, this spectrum suggests that one apex atom of each $B_{10}$ unit is modified and that at least one equatorial atom is modified in the formation of $B_{20}H_{18}^{=}$. These results are supported by the collapse when $Fe^{3+}$ is added (Fig. 4–28b). Assuming that the valence structures inside the polyhedra are unchanged, we join the polyhedra with bridge H atoms[179] (use of B—B bonds will not give $B_{20}H_{18}^{=}$), and just two bridges are needed to give the formula (Fig. 5–21). Deuterium exchange in acid media occurs in some of the nonapex regions, probably exchanging about half of the H atoms (Fig. 4–28c). Here some further study is needed. Isomers should exist,[236] but the experimental evidence presently available is insufficient to establish their existence.

### $B_{20}H_{18}NO^{-3}$(?)

These nitroso derivatives[326] are discussed more fully in Chapter 5. The $B^{11}$ NMR spectrum of $B_{20}H_{18}NO^{3-}$ (Fig. 4–29) has not been completely

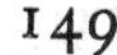

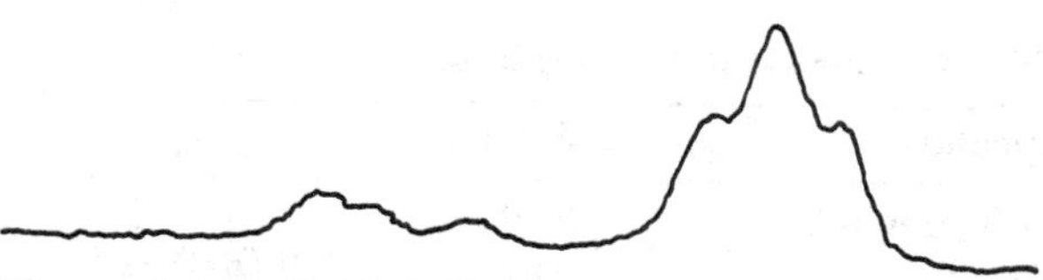

**Figure 4–29** *$B_{20}H_{18}NO^{3-}$(?). The $B^{11}$ NMR spectrum of the $HNEt_3^+$ salt. The number of H atoms is not quite certain, but this ion is assumed*[141] *to be related to $(B_{10}H_9^-)_2$, in which the units are joined by B—B, NO is substituted for one H, and one $H^+$ is added at an edge of a $B_{10}$ polyhedron. The high-field peak has been shown to be two distinct $B^{11}$ resonances at slightly different field, by addition of $Fe^{3+}$ to the aqueous solution.*

analyzed, but the low-field singlet is presumed to be associated with an apical-apical B—B bond[179] between $B_{10}$ polyhedra, and the low-field doublet with the opposite free B—H units at the unbonded apices. The large high-field peak is at least two distinct $B^{11}$ resonances of about equal intensity, as has been confirmed by examination of the $B^{11}$ NMR spectrum in the presence of $Fe^{3+}$. The position of substitution of the NO group on one of the equatorial positions and the point at which the extra $H^+$ is added are not known. The formula for this ion is partly established by a molecular weight study by X-ray diffraction methods.[141]

## 4–2 Coupling Constants and Chemical Shifts

Table 4–1 is a partial list[219,329] of coupling constants ($J$) and chemical shifts ($\delta$) in boron compounds. The progress in understanding chemical shifts even in diatomic molecules is so difficult that theoretical values are only recently available for the simplest first-row diatomic molecules. However, there is some hope that a suitable description can be developed within the framework of the approximations of molecular diamagnetism (clearly of importance in the boron hydrides in which molecular orbitals are most suitable), formal charge (clearly of some, but not of primary, importance) and temperature-independent paramagnetism (owing to unbalance of occupancy of atomic orbitals in the valence shell of boron). All three effects are in the direction to make B atoms in apex environments show $B^{11}$ resonance at high field. An apparent exception occurs in the 1,3 positions of $B_{10}H_{14}$, around which the diamagnetic circulation may be reduced by the long $B_5$—$B_{10}$ and $B_7$—$B_8$ distances, so that most of the diamagnetic circulation is around the already relatively negative and more symmetrically bonded $B_2$ and $B_4$ atoms. The bonding is, on the other hand, much more evenly distributed in $B_{10}H_{12}L_2$ compounds, and therefore presumably in $B_{10}H_{14}^-$, and hence the 1,3 atoms may be the essential centers of diamagnetic circulation. A detailed examination of the molecular orbitals may therefore be of some value in elucidating these

**Table 4–1** *$B^{11}$ NMR chemical shifts and spin coupling values*[219]

| Compound | $\delta \pm 0.5$ | J, cps |
|---|---|---|
| $B_5H_8I$ (apex boron) ($CS_2$ soln.) | 55.0 | |
| $B_5H_{11}$ (apex boron) | 53.5 | 170 (BH) ± 5 |
| $B_5H_9$ (apex boron) | 51.8 | 173 (BH) ± 5 |
| $B_6H_{10}$ (apex boron) | 51.2 | 182 (BH) ± 5 |
| $B_4H_{10}$ (BH) | 40.0 | 154 (BH) ± 5 |
| $NaBH_4$ (aq. soln.) | 38.7 | 82 ($BH_4$) ± 3 |
| $LiBH_4$ (ether soln.) | 38.2 | 75 ($BH_4$) ± 3 |
| $B_5H_8Br$ (apex boron) ($CS_2$ soln.) | 36.4 | |
| $B_{10}H_{14}$ (2,4 pos.) ($CS_2$ soln.) | 34.9 | 158 (BH) ± 5 |
| $(CH_3)_2HNBH_3$ (benzene soln.) | 15.1 | 91 ($BH_3$) ± 3 |
| $B_5H_9$ (base borons) | 12.5 | 160 (BH) ± 1 |
| $B_5H_8Br$ (base borons) ($CS_2$ soln.) | 12.5 | 161 (BH) ± 5 |
| $B_5H_8I$ (base borons) ($CS_2$ soln.) | 11.8 | 160 (BH) ± 5 |
| $C_5H_5NBH_3$ (pure liquids) | 11.5 | 96 ($BH_3$) ± 3 |
| $NaB(C_6H_5)_4$ (aq. soln.) | 8.2 | |
| $B_4H_{10}$ ($BH_2$) | 6.5 | 123 ($BH_2$) ± 3 |
| $BI_3$ (liquid salt) | 5.5 | |
| $B_5H_{11}$ (base B—H) | 2.3 (est.) | 133 (BH) (est.) |
| $NaBF_4$ (aq. soln.) | 2.3 | |
| $BF_3$·piperidine ($CS_2$ soln.) | 2.3 | |
| $NH_4BF_4$ (aq. soln.) | 1.8 | |
| $BF_3$·hexamethylenetetramine | 1.4 | |
| $BF_3 \cdot O(n\text{-}C_4H_9)_2$ | 0.0 | |
| $BF_3 \cdot O(C_2H_5)_2$ | 0.0 | |
| $HBF_4$ (50% aq. soln.) | −0.1 | |
| $BF_3 \cdot P(C_6H_5)_3$ ($CHCl_3$ soln.) | −0.4 | |
| $BF_3 \cdot S(C_2H_4\text{–}C_6H_5)_2$ ($CHCl_3$ soln.) | −0.5 | |
| $B_{10}H_{14}$ (5,7,8,10 pos.) ($CS_2$ soln.) | −0.5 (est.) | 141 (BH) (est.) |
| Tetraacetyl diborate ($CHCl_3$ soln.) | −1.1 ± 1 | |
| $NaBO_2$ ($B(OH)_4^-$) (aq. soln.) | −1.3 | |
| $NaB_5O_8$ (aq. soln.) (see text) | −1.3 ± 1 | |
| $B_5H_{11}$ (base $BH_2$) | −2.9 (est.) | 130 ($BH_2$) (est.) |
| $LiB(OCH_3)_4$ (methanol soln.) | −2.9 | |
| $NaBO_3$ (aq. soln.) | −5.5 | |
| $K_2B_4O_7$ (aq. soln.) | −7.5 ± 1 | |
| $HBCl_2 \cdot O(C_2H_5)_2$ | −7.9 | 152 (BH) ± 5 |
| $DBCl_2 \cdot O(C_2H_5)_2$ | −8.0 | ?(BD) |
| $Na_2B_4O_7$ (aq. soln.) | −8.9 | |
| $BF_3$ (gas) | −9.4 ± 1 | |
| $(NH_4)_2B_4O_7$ (aq. soln.) | −10.3 | |
| $BCl_3 \cdot O(C_2H_5)_2$ (ether soln.) | −10.5 | |
| $N(CH_2CH_2O)_3B$ (aq. soln.) | −10.7 | |
| $B_{10}H_{14}$ (1,3,6,9 pos.) ($CS_2$ soln.) | −12.4 (est.) | 138 (BH) (est.) |

Table 4–1 (*continued*)

| Compound | $\delta \pm 0.5$ | J, cps |
|---|---|---|
| $KB_5O_8$ (aq. soln.) | −13.0 | |
| Tri-*o*-chlorophenylborate (ether soln.) | −13.7 ± 2 | |
| $NaB_5O_8$ (aq. soln.) (see text) | −14.4 ± 1 | |
| $B_6H_{10}$ (base borons) | −15.0 | 160 (BH) ± 5 |
| Tri-*o*-cresylborate (ether soln.) | −15.0 ± 1 | |
| $B_2H_6$ (gas) | −16.6 | 128 ($BH_2$) ± 4 |
| Methylmetaborate (benzene soln.) | −17.3 | |
| *n*-Butyl metaborate (benzene soln.) | −17.5 | |
| $B(OCH_2CH{=}CH_2)_3$ | −17.5 | |
| $B(OCH_3)_3$ | −18.1 | |
| $B(OC_2H_5)_3$ | −18.1 | |
| $B(OH)_3$ (aq. soln.) | −18.8 ± 1 | |
| $B(OC_2H_5)_2Cl$ | −23.3 ± 1 | |
| $HB(OCH_3)_2$ | −26.1 | 141 (BH) ± 5 |
| $DB(OCH_3)_2$ | −26.7 | 24 (BD) ± 4 |
| $B(OH)_2(n\text{-}C_9H_{19})$ (ether soln.) | −29.3 ± 1 | |
| $B(OC_2H_5)Cl_2$ | −32.5 ± 1 | |
| $B_2H_2O_3$ | −33.6 ± 3 | 169 ± 5 (BH) |
| $BBr_3$ | −40.1 | |
| $BCl_3$ | −47.7 | |
| $B(C_2H_5)_3$ | −85 ± 1 | |

three effects in this crude approximation to the theory of chemical shifts. No other correlations are so clear, except for the general shift to low field associated with a change over to three-coordination and to bonding to electronegative atoms. Further detailed studies on the higher hydrides are desirable, but at this moment it would appear that good Hartree-Fock solutions of the molecular wave functions must be the starting point of any detailed study, except a semi-empirical one, and such studies are available[146] only for $B_2H_6$, for which there is reason[146] to suspect some of the values[337] of the molecular integrals.

**Table 4–2*** *$B^{11}$ NMR data on $B_{10}H_{10}^{=}$ derivatives*

| Compound | | | Point | cps | ppm |
|---|---|---|---|---|---|
| 1. Boric acid |  singlet | | | | |
| in | 0.53 *M*, pH ~ 4 | | | | 20 |
| water | satd. soln., pH ~ 9 | | | | 29 |
| 2. $K_2B_{10}H_{10}$ | 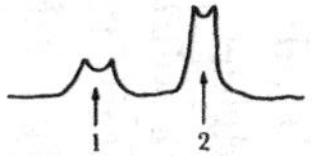 | | 1 | 610 | 40 |
| in water | from $B^{11}$ res. | from $H^1$ res. | 2 | 1050 | 69 |
| High-field coupling constant: | 129 ± 2 | 130 ± 2 | | | |
| Low-field coupling constant: | 140 ± 2 | 141 ± 2 | | | |
| (both in cps) | | | | | |
| 3. $(H_3O)_2B_{10}H_{10}$ in $H_2O$ | | | 1 | 613 | 41 |
| | | | 2 | 1033 | 68 |
| 4. $(NEt_3H)_2B_{10}H_{10}$ | | | 1 | 572 | 38 |
| 1 *M* in $H_2O$ | | | 2 | 1012 | 67 |
| 5. $K_2B_{10}H_{10}$ | 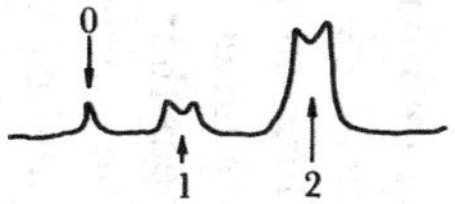 | | 6 | | 18 |
| | | | 1 | | 40 |
| | | | 2 | | 69 |
| 6. $Cu_2B_{10}H_{10}$ in acetone | 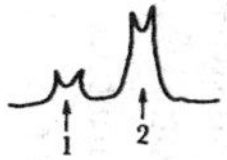 l.f. doublet coupling ≈ 126 cps | | 1 | 654 | 43 |
| | h.f. doublet coupling ≈ 100 cps | | 2 | 968 | 64 |
| 7. $K_2B_{10}H_9OH$(?) fresh, | 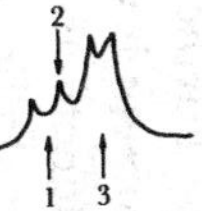 | | 1 | 495 | 33 |
| in $H_2O$ at pH 10 | | | 2 | 607 | 40 |
| | | | 3 | 1000 | 66 |
| 8. $K_2B_{10}H_9OH$(?) old, | 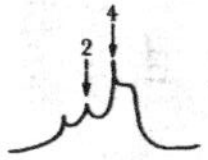 | | 2 | 610 | 40 |
| in $H_2O$ at pH 7 | | | 4 | 955 | 63 |

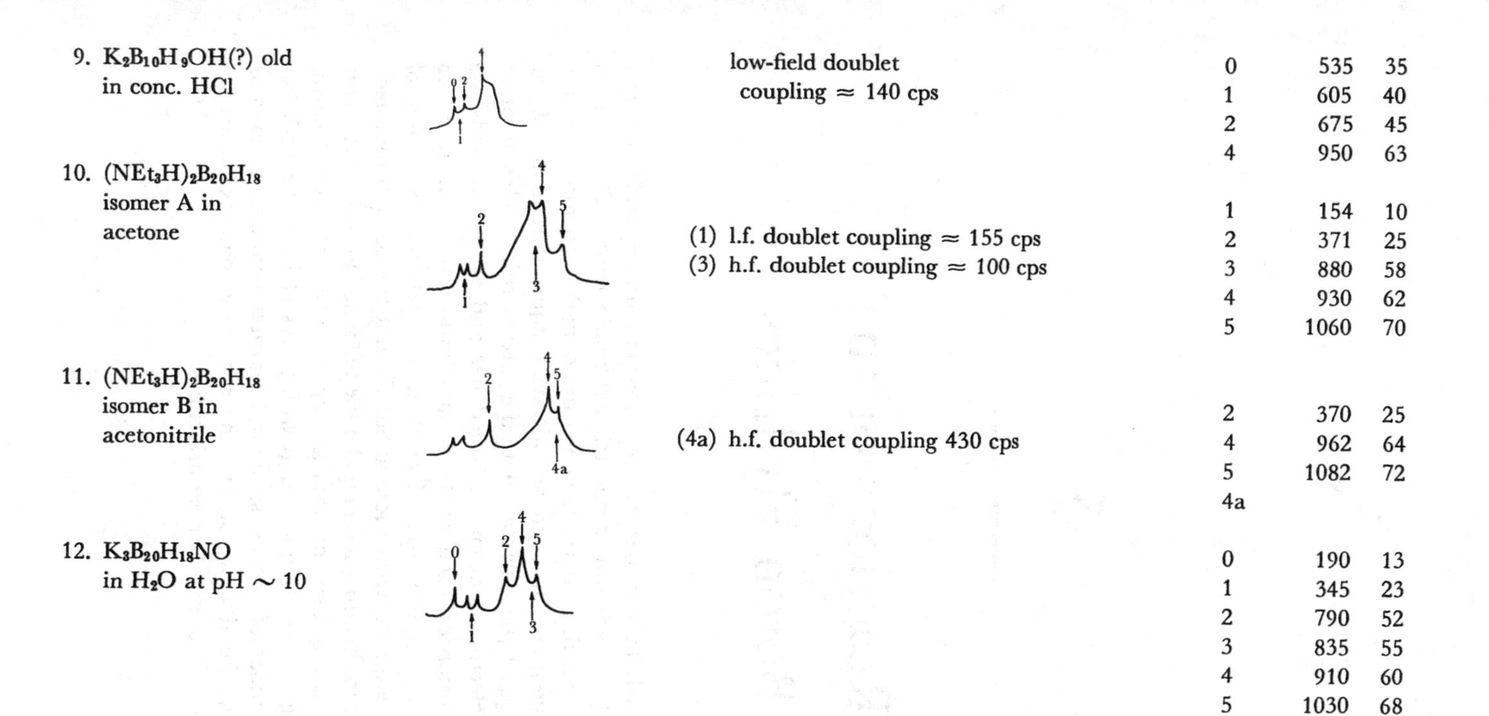

| Compound | Remarks | Peak | | |
|---|---|---|---|---|
| 9. $K_2B_{10}H_9OH$(?) old in conc. HCl | low-field doublet coupling ≈ 140 cps | 0 | 535 | 35 |
| | | 1 | 605 | 40 |
| | | 2 | 675 | 45 |
| | | 4 | 950 | 63 |
| 10. $(NEt_3H)_2B_{20}H_{18}$ isomer A in acetone | (1) l.f. doublet coupling ≈ 155 cps | 1 | 154 | 10 |
| | (3) h.f. doublet coupling ≈ 100 cps | 2 | 371 | 25 |
| | | 3 | 880 | 58 |
| | | 4 | 930 | 62 |
| | | 5 | 1060 | 70 |
| 11. $(NEt_3H)_2B_{20}H_{18}$ isomer B in acetonitrile | (4a) h.f. doublet coupling 430 cps | 2 | 370 | 25 |
| | | 4 | 962 | 64 |
| | | 5 | 1082 | 72 |
| | | 4a | | |
| 12. $K_3B_{20}H_{18}NO$ in $H_2O$ at pH ~ 10 | | 0 | 190 | 13 |
| | | 1 | 345 | 23 |
| | | 2 | 790 | 52 |
| | | 3 | 835 | 55 |
| | | 4 | 910 | 60 |
| | | 5 | 1030 | 68 |

* The $B^{11}$ NMR spectra were recorded relative to liquid $BBr_3$ at a frequency of 15.1 Mc.

# 5

# Reactions of the Boron Hydrides

While it may be too early to formulate the reactions of boron hydrides in terms of detailed steps, such an attempt is nevertheless made here. The detailed mechanistic studies to test these paths have yet to be made. Nevertheless, the theory itself is so suggestive, although not always unique, that many of the intermediates appear also to be consistent with the valency rules. This situation seems likely for the electron-deficient intermediates and for Lewis acid-base reactions. In other instances, such as B—H substitution, the analogies with carbon chemistry seem appropriate. However, the relative expansion of orbitals about B, as compared with those about C, and the comparative lack of exchange repulsions for bending forces make the tendency for rearrangement of either H positions, B positions, or both a very important feature of boron hydride chemistry. First, we outline the gross features of boron hydride reactions, then the reactions of certain structural features common to various members of the series. Finally, we outline the relatively complex derivative chemistry of decaborane in terms of these principles.

## 5–1 Summary of Reactions

**Polymerization**

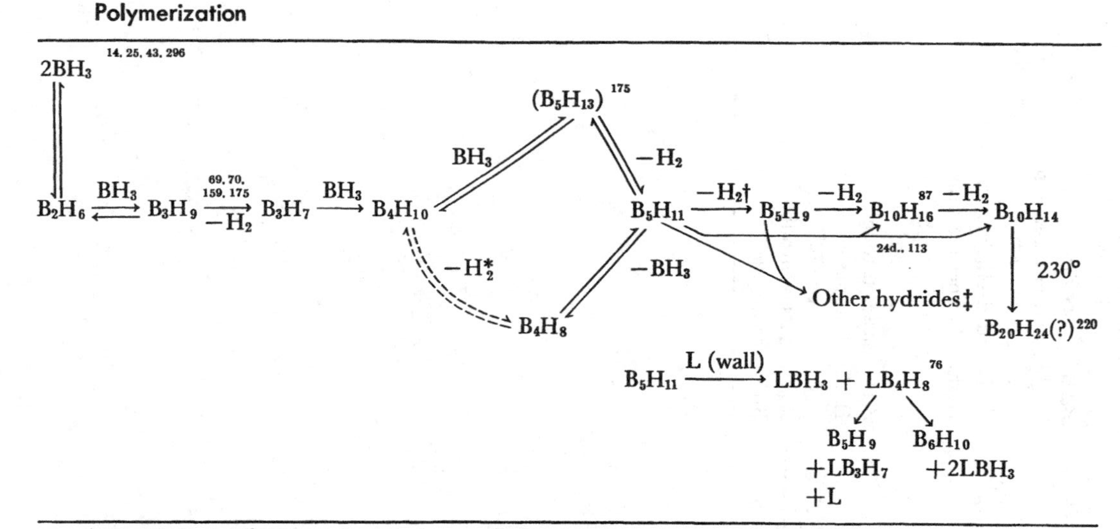


* Not if $T < 45°C$,[24b] and maybe not at higher temperatures; see also Ref. 26.

† This loss of $H_2$ appears to occur by the $B_4H_{10} + B_5H_{11} \rightarrow B_5H_9 + 2B_2H_6$ reaction.[262]

‡ It may be that the $B_6$, $B_7$, $B_8$ and $B_9$ hydrides are formed partly by L (wall) reactions, where L is an electron-pair donor such as $O^{=}$ of the glass wall.

**Separation**[86]

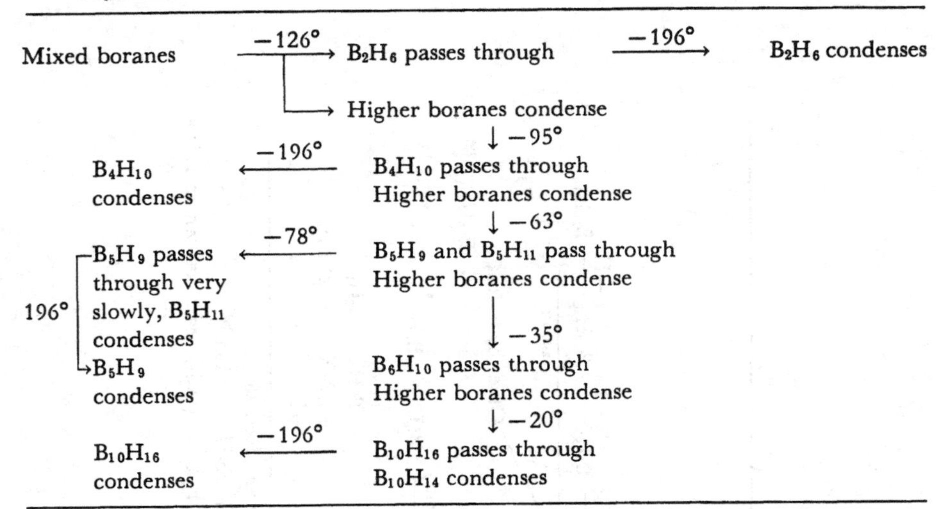

| **$B_2H_6$*** | References |
|---|---|
| $B_2H_6 + B_2D_6 \rightarrow B_2D_3H_3$, etc. | 156 |
| $B_2H_6 \xrightarrow{Br_2} B_2H_5Br$ (terminal) | 45, 299 |
| $2L + B_2H_6 \rightarrow 2BH_3L$ | 299 |
| $BMe_3 + B_2H_6 \rightarrow B_2H_2Me_4$ (1,2,3, or 4 R's, but not in bridges) | 272, 279 |
| $2EtB_2H_5 \rightarrow B_2H_6 + B_2H_4Et_2$(sym) | 293 |
| $B_2H_6 + 2NH_3 \rightarrow BH_4^- + NH_3BH_2NH_3^+$ | 150, 152, 225, 278 |
| $NH_3BH_2NH_3^+ \xrightarrow{NH_4Cl} H_3NBH_3$ | 132, 168, 284, 324 |
| $NMe_2H + B_2H_6 \rightarrow Me_2NB_2H_5 \xrightarrow{HCN} Me_2\overset{H}{N}BH_2NCBH_3$ | 32 |
| $Me_2NB_2H_5 \xrightarrow{B_2H_6} (Me_2NB_3H_8) \rightarrow Me_2NB_2H_5$ | |
| $B_2H_6 + 2NaH \rightarrow 2Na^+BH_4^-$ | 1 |
| $B_2H_6 + Na^+SCN^- \rightarrow H_3BSCN^-$ or $H_3BCNS^- + Na^+$ | 1 |
| $B_2H_6 + F^- \rightarrow H_3BF^-$ | 1 |
| $B_2H_6 + CN^- \rightarrow H_3BCNBH_3^-$ | 1 |
| $B_2H_6 + NH_4^+SCN^- \xrightarrow{0^\circ} NH_4^+BH_3SCN^- \xrightarrow{0^\circ} H_2 + H_3NBH_2SCN$ | 1 |
| $B_2H_6 + NH_4^+\ CN^- \xrightarrow[\text{temp.}]{\text{low}} NH_4^+ + H_3BCNBH_3^- \rightarrow H_2 + H_3NBH_2NCBH_3$ | 1 |
| or $H_3NMe^+$, or $H_2NMe_2^+$ cyanides, but | |
| $B_2H_6 + HNMe_3^+CN^- \rightarrow Me_3NBH_3$ | 1 |
| $B_2H_6 \xrightarrow{NaHg} (Na_2B_2H_6) \xrightarrow{B_2H_6} (Na_2B_4H_{12}) \rightarrow NaBH_4 + NaB_3H_8$ | 24e, 127 |
| $B_2H_6 \xrightarrow{NaHg} NaBH_3 \nearrow (Na_2B_2H_6)$; No $NaBH_4$; Not $B(BH_4)_3^=$ | |
| $B_2H_6 \xrightarrow{h\nu} \begin{pmatrix} BH_3 \\ B_2H_5 \end{pmatrix} \rightarrow B_5H_{11}, B_4H_{10}$ | 161, 162 |
| $B_2H_6 \xrightarrow{BH_4^-} B_2H_7^-$ (solvate) | 7, 24e, 28, 231 |

* Only typical reactions of $B_2H_6$ and reactions related to higher hydrides are included here.

## $B_4H_{10}$

| Reaction | References |
|---|---|
| $B_4H_{10}$ (intramolecular H exchange) (exchange of $BH_3$) (D exchanges more than 3 times faster than B) | 24b, 156, 157 |
| $B_4H_{10} + C_2H_4 \rightarrow C_2H_4B_4H_8$ [$B_4H_{10} \rightleftharpoons B_4H_8 + H_2$ does not participate at 45° (in 2 hr)] | 96, 97, 24b |
| $B_4H_{10} + 2NH_3 \rightarrow B_3H_8^- + H_3NBH_2NH_3^+$ ; (−78°, $Et_2O$, HCl) $\rightarrow B_3H_7Et_2O \xrightarrow[Et_2O,\,-78°]{NH_3} H_3NB_3H_7 + Et_2O$ | 150, 153, 154 |
| $B_4H_{10} \xrightarrow{CN^-} H_3BCNBH_3^-$ and $B_3H_7CN^-$ | 199 |
| $B_4H_{10} + NaH \rightarrow BH_3 + B_3H_8^- + Na^+$ | 128 |
| $B_4H_{10} + NaBH_4 \rightarrow B_3H_8^- + B_2H_6$ | 128 |
| $B_4H_{10} + HCl \rightarrow B_4H_9Cl$ (unstable) | 24a |
| $B_4H_{10} + NMe_3 \rightarrow Me_3NBH_3 + Me_3NB_3H_7 \xrightarrow[(C_6H_5)_3P]{NMe_3} Me_3NB_3H_7 +$ polymer; $Me_3NB_3H_7 \xrightarrow{(C_6H_5)_3P} [(C_6H_5)_3PBH_2]_2$ | 35, 81 |
| $2B_4H_{10} + 2Na(Hg) \xrightarrow{Et_2O} 2Na^+ + 2B_3H_8^- + \frac{1}{2} B_2H_6 + BH$ (polymer) | 24g |
| $NaB_3^{10}H_8 \xrightarrow{HCl} B_3^{10}H_9 \rightarrow \frac{3}{2} B_2^{10}H_6$ and $B_3^{10}H_7$; $B_3^{10}H_7 \xrightarrow{B_2^{11}H_6} B_3^{10}B^{11}H_{10}$ (selectively labeled) | 265 |

**$B_5H_9$**

| Reaction | Ref. |
|---|---|
| $B_5H_9 + B_2D_6 \rightarrow B_5H_8D$ (terminal only) $+ B_2D_5H$ (B does not exchange but $BD_3 + B_5H_9$ is the important step) | 24b, 156 |
| $B_5H_9 \xrightarrow[AlCl_3]{DCl} B_5H_8D$ (apex) | 64, 217 |
| $B_5H_9 \xrightarrow{Br_2I_2} B_5H_8X$ | 73, 267, 299 |
| $B_5H_9 \rightarrow B_5H_8R$ (apex) $\xrightarrow{\text{2,6 lutidine}} B_5H_8R$ (side) | 20, 73, 214 |
| $B_5H_9 \xrightarrow[-78°]{2NMe_3} B_5H_9{\cdot}2NMe_3(B_4H_7^-{\cdot}Me_3NBH_2NMe_2^+ ?)\ Me_3NBH_3 + Me_3NB_4H_6$ | 29, 129, 130 |
| $B_5H_9 + 4NH_3 \rightarrow B_5H_9{\cdot}4NH_3$ | 299 |
| $B_5H_9 + C_2H_4 \xrightarrow{AlCl_3} B_5H_8C_2H_5 + HCl$ | 73 |
| $B_5H_9 + C_2H_2 \rightarrow$ polymer, carbon chain with attached $B_5H_8$ groups | 151 |
| $B_5H_9 + 2LiH \rightarrow Li_2B_5H_{11}$ or $B_4H_7^-$, $BH_4^-$ or $B_7H_{12}^-$ | 151, 173, 269 |
| $B_5H_9 + Li \rightarrow Li_2B_5H_9\ (LiH + LiB_5H_8)(?)$ | 24e, 302 |
| $B_5H_9 + K(\text{excess}) \xrightarrow{THF} K^+B_8H_{13}^-(?)$ | 151 |
| $2B_5H_9 \xrightarrow[H_2]{\text{discharge}} B_{10}H_{16} + H_2$ (H forms $H_2 + B_5H_8$?) | 87 |
| $B_5H_9$ (exchange of terminal H) | 156 |
| $B_5H_9 \xrightarrow[Et_2O]{NaHg} Na_2B_5H_9 \xrightarrow{vac.} B_5H_9$ | 24e |

**$B_5H_{11}$**

| Reaction | Ref. |
|---|---|
| $B_5H_{11} + B_2D_6 \rightarrow$ exchange everywhere (both H and B at same rate within factor of 2) ($B_5H_{11} \rightarrow BH_3 + B_4H_8$, $D_{2d}$ ?) (intramolecular rearrangement of D) | 24b, 156 |
| $B_5H_{11} + L^* \rightarrow BH_3\ (\frac{1}{2}\ B_2H_6) + LB_4H_8$; $LB_4H_8 \rightarrow LB_3H_7 + B_5H_9 + L$; $LB_4H_8 \rightarrow 2L + B_6H_{10} + B_2H_6$; e.g., $B_5H_{11} + NMe_3 \rightarrow Me_3NBH_3 + Me_3NB_4H_8$ | 76 |
| $B_5H_{11} + L \rightarrow L + \frac{1}{2}\ B_6H_{10} + B_2H_6$ | 76 |
| $B_5H_{11} + 2L \rightarrow 2LBH_3 + (B_3H_5) \rightarrow \frac{1}{2}\ B_6H_{10}$ | 76 |
| $B_5H_{11}$ (intramolecular exchange) (exchange of $BH_3$) | 24b |
| $B_5H_{11} + 2CO \rightarrow H_3BCO + B_4H_8CO$ | 30 |
| $B_5H_{11} \xrightarrow[Et_2O]{NaHg} NaB_3H_8$ + polymers + other products | 24e |

* L is a Lewis base.

### $B_6H_{10}$

| Reaction | Ref. |
|---|---|
| $B_5H_{11} + 2L \rightarrow \frac{1}{2} B_6H_{10} + LBH_3$ (L = $NMe_3$) | 22, 23, 76 |

### $B_9H_{15}$*

| Reaction | Ref. |
|---|---|
| $B_5H_{11} \xrightarrow{N_4(CH_2)_6(s)} B_9H_{15}$ (13%) | 31 |

* For $B_9H_{12}^-$ and $B_9H_{13}L$, see $B_{10}H_{12}L_2$.

### $B_{10}H_{10}^{=}$

$B_{10}H_{10}^{=}$

$\rightarrow B_{10}X_{10}^{=}$* (and intermediates) — 149

$\rightarrow B_{10}H_9CO\phi^{=} \rightarrow B_{10}H_9OH^{=}$(?) — 149

$\quad B_{10}H_9CO\phi^{=} \xrightarrow{H_2O_2} B_{10}H_9OCO\phi^{=}$

$\rightarrow B_{10}H_9L^-$, e.g., L = $SMe_2$ — 149

$\rightarrow B_{10}H_8L_2$, e.g., L = $SMe_2$ — 149

$\rightarrow B_{20}H_{18}^{=}$ ($HNEt_3^+$, salt, m.p., 173–174°) — 141, 236

$\quad B_{20}H_{18}^{=} \rightarrow$ ($(H_3O^+)_2B_{20}H_{18}^{=}$ in EtOH concentrate, add $Et_2O$, $H_2O$) $\rightarrow B_{18}H_{22} \xrightarrow{OH^-} B_{18}H_{21}^-$; $B_{20}H_{18}^{=} \rightarrow$ Iso–$B_{18}H_{22}$ — 291

$\quad B_{20}H_{18}^{=} \rightarrow B_{10}H_9OH^{=}$(?) (apex) — 141

$\quad \downarrow$   } possibly $B_{20}H_{17}OH^{-4}$(?)

$\quad B_{10}H_9OH^{=}$(?) (equatorial) — 141

$\rightarrow B_{20}H_{18(?)}$ $NO^{3-}$   ($B_{20}H_{18}^{4-} \overset{H^+}{\rightleftharpoons} B_{20}H_{19}^{3-} \overset{H^+}{\rightleftharpoons} B_{20}H_{20}^{=}$ ?) — 141, 179, 326

$\quad B_{20}H_{18(?)}NO^{3-} \xrightarrow{H^+} B_{10}H_{10}^{=} + B(OH)_3$

$\rightarrow B_{20}H_{16(?)}(NO)_4^{=}$ — 141, 326

$\rightarrow B_{20}H_{14(?)}(NO)_6^{=}$ — 141, 326

$\xrightarrow{CuCl_2} Cu_2B_{10}H_{10}$ (covalent) $\xrightarrow{I^- \text{ or } BH_4^- \text{ or } B_{10}H_{10}^{=}}$ $Cu_2B_{10}H_xCl_y$ (free radical) — 141

$\rightarrow Cu_2B_{10}H_xCl_y$ (free radical) $\xrightarrow{Fe^{3+},\ Cu^{++}} Cu_2B_{10}H_{10}$ (covalent) — 141

* X = halogen.

**$B_{10}H_{12}L_2$**

| Reaction | Ref. |
|---|---|
| $B_{10}H_{12}L_2 \xrightarrow{L'} B_{10}H_{12}L'_2 + 2L$; $Me_2S < RCN < R_2NCN < RNC < (RO)_3P \sim R_3P \sim R_3N < P_Y < P\phi_3$ | 100, 102, 221 |
| $B_{10}H_{12}L_2\ (L = SMe_2) \xrightarrow[C_6H_6]{NMe_3} B_{10}H_{10}^{=} + 2HNMe_3 + 2L$ | 102 |
| $B_{10}H_{12}L_2\ (L = SMe_2) \xrightarrow{EtOH} B_9H_{13}L \xrightarrow{\phi_3PCH_2} B_9H_{12}^- \xrightarrow{CH_3OH} B_3H_8^-$; $B_9H_{13}L \xrightarrow{L'} B_9H_{13}L'$; $B_9H_{13}L \xrightarrow{EtNH_2} Et_2NH_2B_8H_{11}NHEt_2$ | 102 |
| $B_{10}H_{12}L_2 \xrightarrow{H^+ \mid H_2O} B_{10}H_{13}L^-$ | 102, 167 |
| $B_{10}H_{12}L_2 \xrightarrow{OH^-} B_9H_{12}^- \xrightarrow{EtOH} B_3H_8^-$ | 84 |
| $B_{10}H_{12}L_2 \longrightarrow B_{10}H_{12}L + L\ (L = SMe_2)$ | 148 |
| $B_{10}H_{12}L_2 \xrightarrow{HCCH} B_{10}C_2H_{12} + H_2 + 2L$ | 41, 315, 316 |

$B_{10}H_{14}$*

| Reaction | References |
|---|---|
| *a. Acid-base* | |
| $B_{10}H_{14} \xrightarrow[fast]{H_2O} B_{10}H_{13}^-$ (yellow) $\xrightarrow[slow]{} B_{10}H_{14}OH_2 \rightleftharpoons B_{10}H_{14}OH^- \rightarrow B_{10}H_{12}OH^-$ | 89, 258, 299 |
| $B_{10}H_{14} \xrightarrow[Et_2O]{NaH} NaB_{10}H_{13}\cdot nEt_2O$ (also $LiBH_4$ in place of NaH) | 24e, 125 |
| $B_{10}H_{14} \xrightarrow{\phi_3PCH_2} \phi_3PCH_3^+\cdot B_{10}H_{13}^-$ (also $Et_2NH$ in place of $\phi_3PCH_2$) | 98, 103 |
| $B_{10}H_{14} \xrightarrow{D^+} B_{10}H_{10}D_4$ (all $\mu$) | 64, 66, 99, 192, 283 |
| $B_{10}H_{14} \xrightarrow{D^+} B_{10}H_4D_{10}$ ($\mu$,5,7,8,10,6.9) | 99, 192, 283 |
| *b. Nucleophilic*[62] [see also (*d*)] | |
| $B_{10}H_{14} \xrightarrow{MeMgI} B_{10}H_{13}MgI \xrightarrow{RX} 6\text{-}RB_{10}H_{13}$ | 61, 77, 285 |
| $B_{10}H_{14} \xrightarrow[Et_2O]{L^-} B_{10}H_{12}L^-$, $L^- = CN^-, OCN^-, SCN^-$, etc. (6 subst. ?) | 1 |
| $NaB_{10}H_{13} \xrightarrow{\phi CH_2Br} \phi CH_2B_{10}H_{13}$ (6 subst. ?) | 222 |
| $B_{10}H_{14} \xrightarrow{LiMe} 6\text{-}MeB_{10}H_{13}$ + some $5\text{-}MeB_{10}H_{13}$ | 62 |
| *c. Electrophilic* | |
| $B_{10}H_{14} \xrightarrow{X=Br,I} 2,4\text{-}B_{10}H_{12}I_2 \rightarrow (B_{10}I_8)_x$ | 261, 299, 322 |
| $B_{10}H_{13}I$ (m.p. 100°) (5 subst. ?) and 2,5(?)-$B_{10}H_{12}I_2$ (m.p. 155°) | 111, 112, 267 |
| $B_{10}H_{14} \xrightarrow[CS_2]{RX,AlCl_3} B_{10}H_{13}R$ (2,4 preferred; 1,3 less for mono-, di-, polyalkyls) | 19, 20, 331 |
| $B_{10}H_{14} \xrightarrow{DCl,AlCl_3} B_{10}H_{10}D_4$ (1,2,3,4 equally preferred) | 63, 64, 66 |
| *d. Electron addition* (less specific solvation occurs, e.g., with acetone[231]) | |
| $B_{10}H_{14} \xrightarrow[NH_3(l)]{Na} B_{10}H_{14}^{=}$ (intermediate $B_{10}H_{14}^-$ ?) | 306, 307 |
| $B_{10}H_{14}^{=}$: $I_2$ or HCl → $B_{10}H_{14}$; $\xrightarrow{HCl}$ 50% EtOH → $B_{10}H_{15}^-$ | 65 |
| $B_{10}H_{14} \xrightarrow{L} 6,9\text{-}B_{10}H_{12}L_2$ (see above) $\rightarrow B_{10}H_{10}^{=}$ (see above) | 16, 36, 60, 72, 80, 100, 101, 102, 148, 201, 239, 247, 260 |

| Reaction | References |
|---|---|
| $B_{10}H_{14} \xrightarrow{L} 6\text{-}B_{10}H_{13}L^- \xrightarrow{HCl} B_{10}H_{14}L$ | 82, 102, 103 |
| $6\text{-}B_{10}H_{13}L^- \xrightarrow{HL'} B_{10}H_{12}LL'$ | 82 |
| $NaB_{10}H_{13}^- \xrightarrow[Et_2O]{L} B_{10}H_{13}L^- \rightarrow B_{10}H_{12}LL'$ | 82, 148 |
| $NaB_{10}H_{13}^- \xrightarrow[Et_2O]{I_2} B_{10}H_{13}OEt$ (6 subst. ?) | 193 |
| $B_{10}H_{14} \xrightarrow[H_2O]{NaCN} B_{10}H_{13}CN^-$; $B_{10}H_{14} \xrightarrow{Me_2S\ NaCN} B_{10}H_{12}CNSMe_2^-$ } (6,9 subst. ?) | 148 |
| $B_{10}H_{14} \xrightarrow{NH_3} B_{10}H_{13}^-\ NH_4^+ \xrightarrow{NH_3} [B_{10}H_{14}\cdot 2NH_3] \rightarrow B_{10}H_{14}\cdot 3NH_3 \rightarrow B_{10}H_{14}\cdot nNH_3$, $n=4,(5),6$; $B_{10}H_{14}\cdot nNH_3 \xrightarrow{95^\circ} B_{10}H_{14}(NH_2)_2 \xrightarrow{120^\circ} B_{10}H_{12}(NH_2)_2$ | 306 |
| *e. Build-up* | |
| $B_{10}H_{14} \xrightarrow{BH_4^-} B_{11}H_{14}^- \rightarrow B_{11}H_{13}^{=}$ | 2,198 |
| $B_{10}H_{14} \longrightarrow B_{12}H_{12}^{-2}$ | 233 |
| *f. Degradation* | |
| $B_{10}H_{12}L_2 \rightarrow B_9H_{13}L \rightarrow B_9H_{12}^-$ | 74, 83, 260 |
| *g. Dimers* | |
| $B_{10}H_{14} \rightarrow B_{20}H_{24}$(?) or/and $B_{20}H_{26}$ | 93, 220 |
| *h. H exchange* | |
| $B_{10}H_{14} \xrightarrow[gas]{B_2D_6} B_{10}H_{14-n}D_n$ (terminal H's exchange) | 24b, 144, 156 |
| *i. Internal rearrangement* | |
| $B_{10}H_{13}^- + B_2H_6 \rightarrow [B_{11}H_{16}^-$ or $B_{12}H_{19}^-] \rightarrow B_{10}H_{13}^-$ (B atoms scrambled) | 264 |

* $\mu$ = bridge, X = halogen, L = ligand, R = alkyl, Me = methyl, Et = ethyl, $\phi$ = phenyl.

| $B_{10}H_{16}$ | |
|---|---|
| $2B_5H_9 \rightarrow B_{10}H_{16} + H_2$ | 87 |
| $B_{10}H_{16} + HI \longrightarrow B_5H_9 + B_5H_8I$ (or HBr) | 85 |
| $B_{10}H_{16} + I_2 \rightarrow B_{10}H_{14} + 2HI$ | 85 |
| $B_{10}H_{16} + C_2H_4 \rightarrow B_{10}H_{14} + H_2$ + other products | 85 |
| $B_{10}H_{16} \xrightarrow[\text{in } CS_2]{AlCl_3} B_{10}H_{14}$ | 85 |
| $B_{10}H_{16} \xrightarrow{L} B_{10}H_{12}L_2$ and other derivatives | 85 |

| $B_{12}H_{12}^{=}$ | |
|---|---|
| $B_{12}H_{12}^{=} \rightarrow B_{12}X_{12}^{=}$ (and intermediates) | 149 |
| $\rightarrow B_{12}H_{11}NO_2^{=}$ | 149 |

| $B_{18}H_{22}$ | |
|---|---|
| $B_{18}H_{22} + OH^- \rightarrow B_{18}H_{21}^- + H_2O$ | 234 |

| $B_{20}H_{18}^{=}$ | |
|---|---|
| $B_{20}H_{18}^{=} + 2OH^- \rightarrow B_{10}H_9OH^{=}$(?) (apex) $\rightleftharpoons B_{10}H_9(OH)^{=}$(?) equatorial | 141 |
| $B_{20}H_{18}^{=} \xrightarrow{H^+} B_{18}H_{22}$; $\rightarrow$ Iso-$B_{18}H_{22}$ | 234 |

## 5–2 Structural Features and Reaction Types

The structural features that give rise to characteristic reactions are, in approximate order of increasing thermal stability:

1. The $BH_3$ group.
2. The doubly bridged $BH_2$ group.
3. The singly bridged $BH_2$ group.
4. The unsymmetrical B—H—B bond.
5. The symmetrical B—H—B bond.
6. The bridged B—H group.
7. The B—H group, unbridged.
8. The B—B bond.
9. The open three-center B—B—B bond.

10. The central three-center B—B—B bond.

11. The cooperative effect in a boron framework, which can sometimes lose or gain electrons, rearrange, or lose or gain atoms (usually with some rearrangement).

The types of reactions which boron compounds undergo may be classified in a manner similar to that used for carbon compounds, into five types (1) displacement or substitution reactions, (2) addition reactions, (3) elimination reactions, (4) rearrangement reactions, and (5) oxidation-reduction reactions. Examples are, respectively,

(1) $B_{10}H_{14} \xrightarrow{LiCH_3} B_{10}H_{13}CH_3$ (6 or 9) (nucleophilic)
$B_{10}H_{14} + I_2 \longrightarrow B_{10}H_{13}I + HI$ (electrophilic)
(2) $BH_3 + L \longrightarrow LBH_3$ where $L = NEt_3$ (nucleophilic)
$B_{10}H_{10}^{=} + 2Cu^I \rightarrow Cu_2B_{10}H_{10}$ (electrophilic)
(3) $B_5H_{11} \rightarrow B_5H_9 + H_2$
or $B_{10}H_{14} \rightarrow B_{10}H_{13}^- + H^+$
(4) apex-$MeB_5H_8 \rightarrow$ side-$MeB_5H_8$
(5) $B_{10}H_{14} + 2e \rightarrow B_{10}H_{14}^{=}$

The isolation of reactions (5) as extreme cases of reactions (2) is somewhat arbitrary, but probably useful. Naturally, most reactions observed in practice are combinations of these types, and indeed the examples given here are certainly not in every case the elementary steps in the mechanisms which almost always involve the solvent, walls, or other molecules. Nevertheless, when the elementary steps are elucidated we hopefully expect that they, too, will fall into a simple classification scheme like this one.

The fairly simple disproportionation $B_2H_6 \rightarrow 2BH_3$ is of type (3), while a cleavage reaction such as $B_{10}H_{16} + HI \rightarrow B_5H_8I + B_5H_9$ is a combination of (2) addition and (3) elimination. The apparently simple elimination $B_{10}H_{16} \rightleftharpoons B_{10}H_{14} + H_2$ involves a major rearrangement typical of the hazards in prediction of reactions of electron-deficient compounds.

## 5-3 The $BH_3$, $BH_4^-$, and $BH_5$ Units

The $BH_4^-$ ion can be thought of as a combination of $H^-$ and $BH_3$ in which all H atoms are symmetrically equivalent. The B—H distance of 1.255 ± 0.02 A argues for a boron radius[55] of 0.90 A for four coordinated boron. There appear to have been several[24e,28,37] independent discoveries of the $B_2H_7^-$ ion in diglyme (diethylene glycol dimethyl ether) or triglyme solutions. Formally, at least, the most reasonable structure[28,175] is two $BH_3$ units joined by a symmetrical or unsymmetrical (and possibly tautomerizing) H bridge. Hence its formation can be envisaged as $BH_3$ bonded to $BH_4^-$ in the same Lewis acid-base type of reaction as that described above for $BH_3$ and $H^-$. There are, however, strong indications of interesting complications in this simple idea that merit study. The possible dynamical equivalence of all protons in the hyperfine structure due to H in the $B^{11}$ NMR spectrum needs to be studied as a function of temperature, perhaps with double-resonance techniques. Also, the strong solvation[7] of $B_2H_7^-$ in $Et_2O$ solutions may indicate a further acidic behavior of $B_2H_7^-$ toward $Et_2O$ and consequent enlargement of the coordination sphere about B. Further enlargement of H-atom coordination about B occurs[137] in $BH_5$, isoelectronic with $CH_5^+$, which is no doubt a trigonal bipyramidal structure readily rearranging through the tetragonal pyramidal intermediate. Molecular orbitals of a trigonal pyramidal $H_5$ group give $\sigma'$ and $\sigma''$ symmetric and antisymmetric orbitals about the symmetry plane perpendicular to the threefold axis ($z$), and a $\pi$ pair of orbitals, all of which bond to the $2s$ ($\sigma'$), $2p$ ($\sigma''$), and $2p_y$, $2p_z$ ($\pi$) pair of the central B atom.

Although $BH_3$ itself has not been isolated, matrix methods of low-temperature spectroscopy may yield this unstable species. Several early kinetic studies[25,43] have clearly suggested the $B_2H_6 \rightleftharpoons 2BH_3$ steady state, and a recent estimate for $K_p$ is given[24f] by $\log_{10} K_p(\text{atm}) = 7.478 - 6205/T$. There are theoretical reasons[44] to believe that the ground electronic state of $BH_3$ is planar, and that the first excited state will distort toward the half-diborane geometry.

The typical reactions of $BH_3$ are associated with its electron deficiency. It adds to ligands (L = :$NR_3$) which have electron pairs, to $H^-$, to $BH_3$ itself, etc., in such a way as to increase the number of electron pairs. Even the H exchange reactions of some of the boron hydrides are apparently associated with this tendency. For example,[24b] exchange of $D_2$ with $B_4H_{10}$ does not occur at 45°C in 2 hours, but $B_2D_6$ does exchange D for H in $B_4H_{10}$ under comparable conditions. Hence, at least in the low-temperature polymerization of boron hydrides, the step

$$B_4H_{10} \rightleftharpoons B_4H_8 + H_2$$

is not a preferred one, and hence

$$B_4H_{10} + BH_3 \rightleftharpoons (B_5H_{13}) \rightleftharpoons B_5H_{11} + H_2$$

is therefore the suggested path. Thus $BH_3$ insertion, loss of $H_2$, and rearrangement of the boron framework are the characteristic reactions in the diborane pyrolysis. Inasmuch as the equilibrium

$$B_4H_{10} + \tfrac{1}{2} B_2H_6 \rightleftharpoons B_5H_{11} + H_2$$

is relatively rapid,[33] the $H_2$ probably then react quite readily with $B_5H_{11}$ or one of its decomposition products. Thus the reactions of $BH_3$ with itself to form $B_2H_6$ or with a Lewis base L to form $LBH_3$ (e.g., $R_3NBH_3$) are relatively simple, but even the H exchange of $BH_3$ probably involves fairly complex reactions, the simplest of which is the sometimes-overlooked H scrambling to be associated with $2BH_3 \rightleftharpoons B_2H_6$. In the higher hydrides, also, $BH_3$ may be in a steady state such as $B_4H_{10} \rightleftharpoons B_3H_7 + BH_3$, $B_5H_{11} \rightleftharpoons B_4H_8 + BH_3$, $B_5H_9 + BH_3 \rightleftharpoons B_6H_{12}$, $B_6H_{10} + BH_3 \rightleftharpoons B_7H_{13}$, $B_9H_{15} \rightleftharpoons B_8H_{12} + BH_3$, with the equilibrium far toward the side of the known hydride. The complete absence of two stable hydrides differing by a $BH_3$ unit has been noted,[114] but a violation of this observation may not be surprising.

## 5–4 Reactions of the Bridged $BH_2$ Group

### $BH_3$ Cleavage

Although $BF_3$, $BCl_3$, etc., exist, $BH_3$ does not, because there are no low-energy orbitals of the $H_3$ group which have the same symmetry as the vacant $\pi$ orbital of B. Hence $BH_3$ coordinates readily with a great number of electron donors such as $N(CH_3)_3$. Cleavage of $B_2H_6$ to $2BH_3$ (Fig. 5–1), followed by reaction of $BH_3$ with $B_2H_6$, has long been recognized as the initial reaction in the condensation of diborane to higher hydrides. Thus the existing evidence regarding $BH_3$ is from relatively indirect kinetic studies, but its coordination compounds are widely known.

The exchange of $BH_3$ as an entity in diborane has been established by study of isotopic exchange by Maybury and Koski.[190] There are indications[156] that $BH_3$ exchange also occurs as part of the mechanism of isotopic exchange in $B_5H_{11}$ and $B_4H_{10}$ as well. It is possible that $BH_3$ exchange is a general reaction[114] of all hydrides containing $BH_2$ groups.

Many of the reactions of $B_4H_{10}$ have been interpreted by Parry and Edwards[223,224] and co-workers as cleavage to $BH_3$ and $B_3H_7$, both coordinated to a Lewis base. For example, the reaction with $N(CH_3)_3$ is as shown in Fig. 5–2. This reaction can be followed by a further reaction of the $B_3H_7$ adduct with $N(CH_3)_3$. Other Lewis bases, except for $NH_3$, give similar reactions. The structures for $(CH_3)_3NB_3H_7$ are suggested on the basis of the very careful study by Nordman et al.[209,210] of $NH_3B_3H_7$.

A simple extrapolation suggests that $B_9H_{15}$ undergoes a similar cleavage

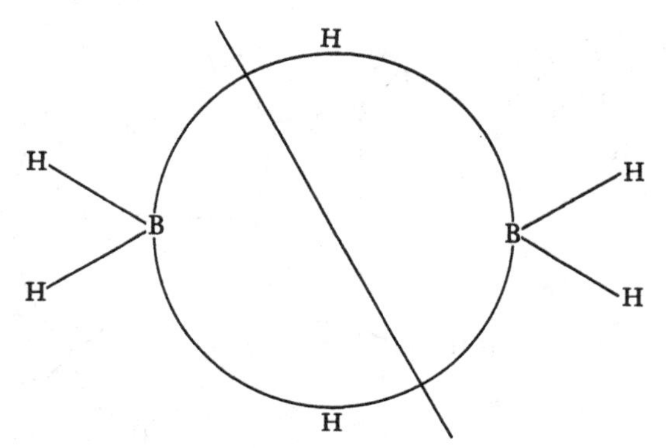


**Figure 5–1** *$BH_3$ cleavage in $B_2H_6$.*

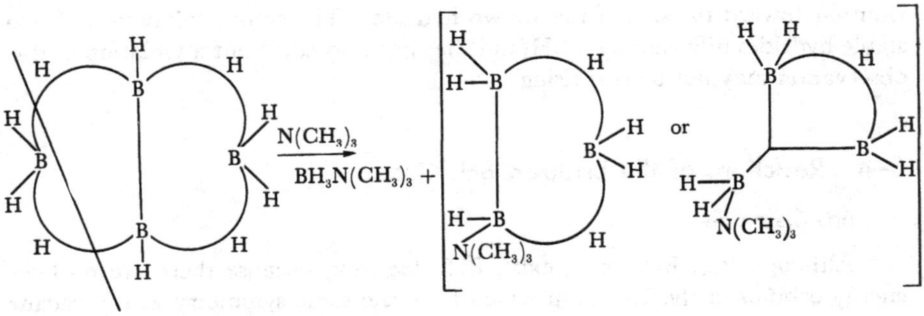

**Figure 5–2** *$BH_3$ cleavage in $B_4H_{10}$.*

(Fig. 5–3). Neither this reaction nor the reaction product are known, but the structure of $B_9H_{15}$ is known to be very similar to the $B_4H_{10}$ and $B_5H_{11}$ structures.

The $B_5H_{11}$ molecule reacts with CO to produce[30] $BH_3CO$ and $B_4H_8CO$ (see Fig. 5–4). Although it is possible to formulate $B_4H_8CO$ as a derivative of $B_4H_{10}$, further structural and chemical studies are desirable. In particular, there is a possibility that this reaction proceeds through an intermediate form of $B_5H_{11}$, which may have four bridge hydrogens.

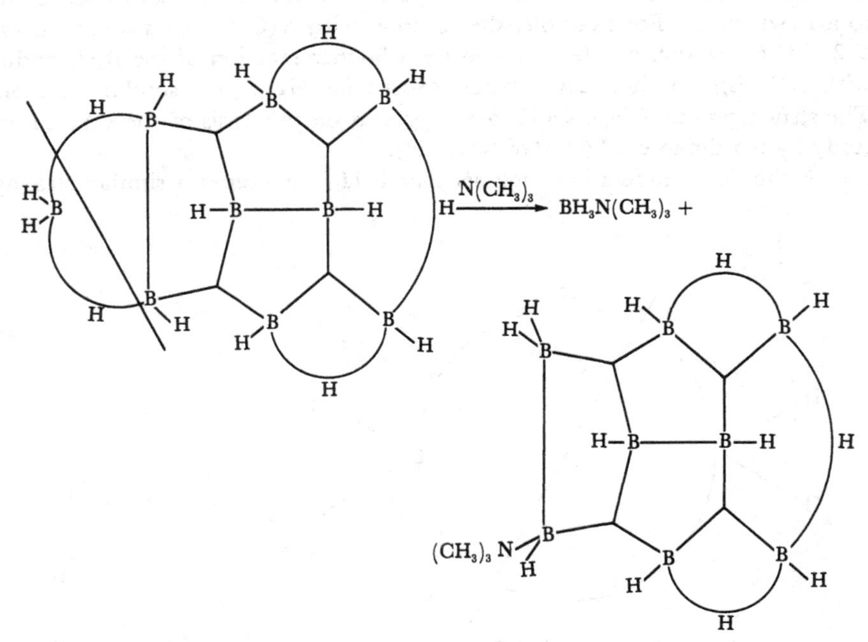


**Figure 5–3** *Probable course of $BH_3$ cleavage in $B_9H_{15}$.*

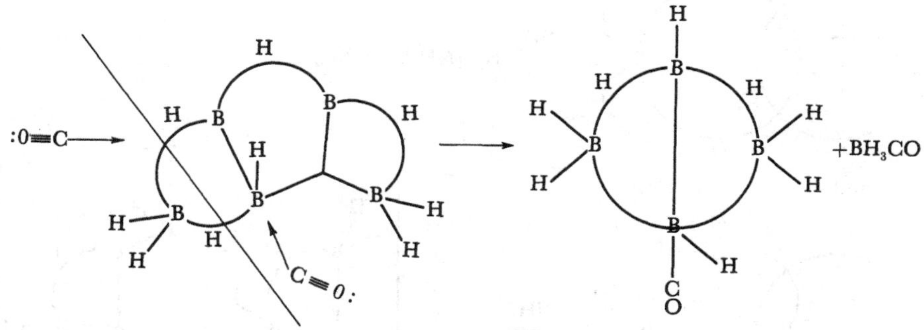

**Figure 5-4** *$BH_3$ cleavage in $B_5H_{11}$.*

The details of this cleavage of $BH_3$, and cleavage to coordinated $BH_2^+$, probably involve an intermediate[225] of the type shown in Fig. 5-5, formed by adding the electron pair of the ligand to one side of the hydrogen bridge. Addition of the next ligand L to the remaining three-center bond may then proceed either way to give symmetrical or unsymmetrical cleavage. Which process occurs seems to be more nearly correlated with steric effects than with base strengths of the ligands L. Only $NH_3$ of the simple $NH_mR_{3-m}$ series gives unsymmetrical cleavage; the substituted ammonias produce symmetrical cleavage.

### Cleavage Yielding $H_2BL_2^+$ Derivatives

Ammonia does not produce the $BH_3$ cleavage described above. In view of the earlier interpretation of the simpler ammoniates of boron hydrides as ammonium salts, the reformulation and proof[278] that these are $BH_2(NH_3)_2^+$ salts is of importance. The discovery[152] that $B_4H_{10}$ forms a diammoniate, not a tetraammoniate,[299] has led to the suggestion by Kodama[150] that $B_4H_{10}\cdot 2NH_3$ is also a salt containing the known $H_2B(NH_3)_2^+$ ion[278] and $B_3H_8^-$ ion.[126] Thus the reactions of $NH_3$ with $B_2H_6$ and $B_4H_{10}$ are as shown in Fig. 5-6. A simple

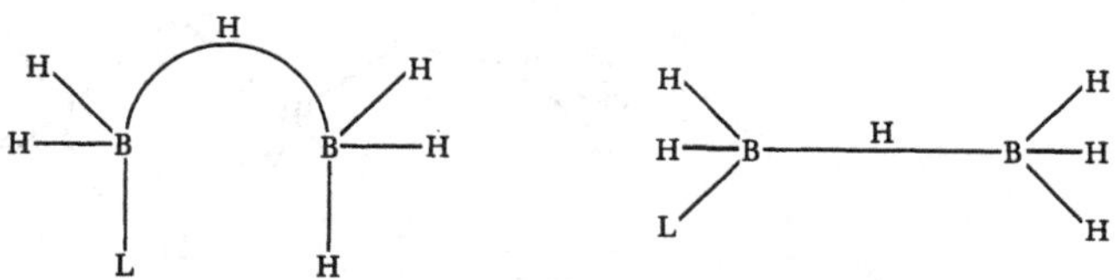

**Figure 5-5** *Presumed first step in cleavage reaction of $B_2H_6$ by a ligand L.*

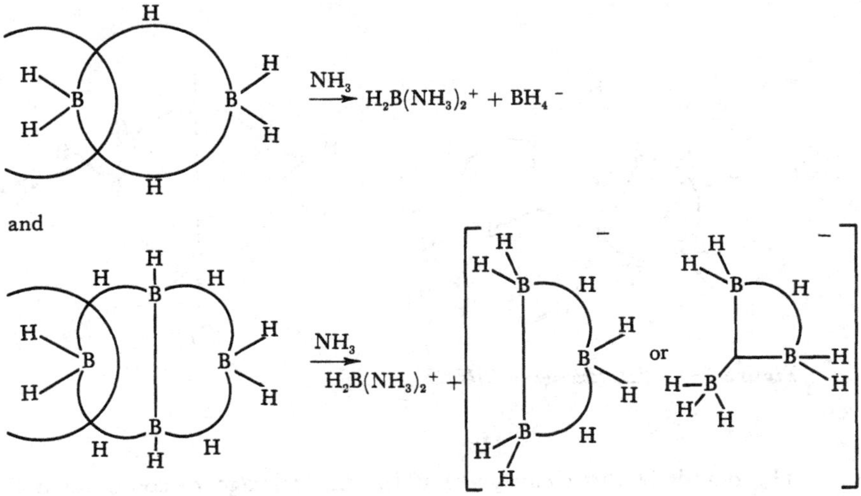

$\xrightarrow{NH_3} H_2B(NH_3)_2^+ + BH_4^-$

and

$\xrightarrow{NH_3} H_2B(NH_3)_2^+ +$ [ ... ]$^-$ or [ ... ]$^-$

**Figure 5-6** *$BH_2^+$ cleavage in $B_2H_6$ and $B_4H_{10}$.*

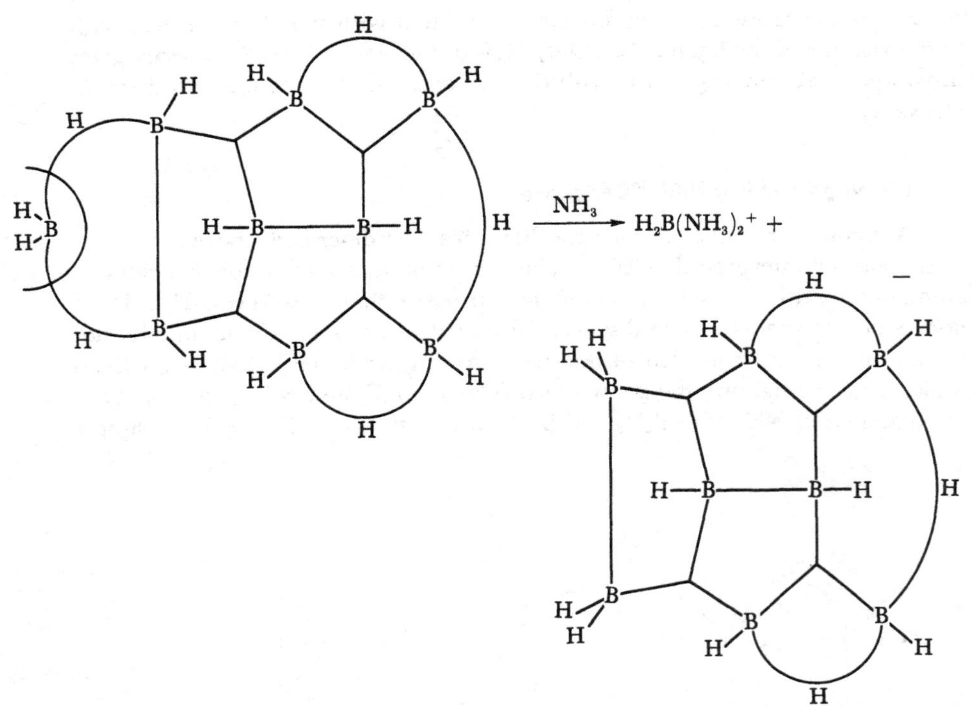


**Figure 5-7** *Extrapolation of $BH_2^+$ cleavage to $B_9H_{15}$.*

extrapolation[176] to $B_9H_{15}$ might be as shown in Fig. 5–7. This reaction has not been studied and the $B_8H_{13}^-$ ion has not been observed (however, cf. $EtNH_2B_8H_{11}NHEt$). However, it is among those suggested by the topology,[174] but the $B_8H_{14}$ obtained by addition of $H^+$ to the exposed single bond may have a steric problem. Addition of $H^+$ is suggested below as a possible reaction[174] of certain hydrides and ions.

A similar reaction of $NH_3$ with $B_5H_{11}$ occurs.[299] While the initial step is probably similar to that described for the other hydrides containing $BH_2$ groups, and may involve an intermediate form of $B_5H_{11}$, the ensuing reactions appear to complicate the details and hence further experimental data are desirable.

Details of the mechanism of formation of $BH_2(NH_3)_2^+$ from these compounds are not known as yet but it is expected[225] that the $NH_3$ adds at a hydrogen bridge region between the $BH_2$ in question and the bridge H.

Reaction products of $NH_3$ with hydrides as stable as $B_5H_9$ and $B_{10}H_{14}$ are known but have not been subjected to structural studies. It seems possible that they are based in part upon different principles than those for hydrides containing $BH_2$ groups, or at least they may contain hitherto-unknown ions. The possibility of a $HB(NH_3)_3^{++}$ ion may help in accounting for some of the higher ammoniates such as $B_5H_9{\cdot}4NH_3$, or some of those derived from $B_{10}H_{14}$, as described below.

## 5–5 Reactions of the BH Groups: Halides, Grignard Reagents, Alkyls

Substitution of the terminal H atom by a halogen atom has been observed several times. For example $B_{10}H_{12}I_2$,[299] $B_{10}H_{13}I$,[267,299] $B_5H_8Br$,[267] $B_5H_8I$,[267] and $B_2H_5Br$[45,299] have been isolated. The possibility of the substitution of halogen atoms at the most negative boron atoms was suggested[171] by analogy[183] with organic reactions, and indeed was one of the reasons for the detailed discussion of charge distribution in the valence theory.[69,123] Detailed charge distribution is a difficult aspect of valence theory, and hence only the major features can be expected. Nevertheless, the valence theory appears to be successful in choosing the most negative of the boron atoms in $B_5H_9$ and $B_{10}H_{14}$: It predicts successfully that halogen substitution takes place most readily at the apex boron atom in $B_5H_9$ and at the 2,4 type[171,261] of boron atoms in $B_{10}H_{14}$. The existence of two different $B_{10}H_{13}I$ molecules,[267] however, must lend caution to acceptance of this mechanism of substitution in all reactions, or at least should lead to further study. These halogen derivatives can be converted. The Grignard reagents[61,77,285] are, on the other hand, made by direct reaction of RMgX with $B_{10}H_{14}$.

Exchange of the more negative terminal H atoms also occurs in $B_5H_9$[64,217] and $B_{10}H_{14}$,[63,66] a process which requires $AlCl_3$ as a catalyst.

Replacement of a terminal H atom by an alkyl group, also in the presence of $AlCl_3$, is a well-established reaction.[19,20,62,73,331] In $B_5H_9$ this substitution occurs at the apex, and in decaborane it occurs with decreasing readiness at the 2,4 > 1,3 > 5,7,8,10 > 6,9 positions. This order is the same as that expected from a molecular orbital discussion[123,198] of charge distribution in the isolated molecule, and suggests that many of the aspects of this charge distribution are preserved in the intermediate transition state.

Similarly, reaction[151] of $B_5H_9$ with $C_2H_2$ in the presence of $AlCl_3$ gives $B_5H_8$ groups attached at the apical B atoms to a chain of C atoms.

## 5-6 Reactions of the BHB Bond

Both nuclear spin resonance results[146,212,276] and theoretical calculations[95,146,337] agree that the bridge hydrogens in diborane are more negative than the terminal hydrogens. If it is legitimate to extend these results to the higher boron hydrides it is indeed remarkable that the bridge hydrogens appear to be the more acidic. However, if a terminal hydrogen ion is lost, an electron pair remains adjacent to a valence situation that normally would require the electron pair for other interactions, and therefore quite probably a drastic change in the structure. But if a bridge hydrogen is lost, the structure of the ion requires far less drastic changes than if a terminal hydrogen ionizes off because the BHB bond is simply replaced by an ordinary B—B bond.[174]

That $B_{10}H_{14}$ has acidic hydrogens was shown in an unpublished preparation of $NaB_{10}H_{13}$ by Edwards and Hough. This result was established independently and in more detail by Guter and Schaeffer.[89] The important conclusion that the bridge hydrogens are involved appears to have been discovered by a number of groups: Edwards and Grillo, Atori and Kline, Taylor and Parry, Schaeffer and co-workers, and Hawthorne and Miller.[99] The task of an historian is made somewhat difficult by the fact that only one of these accounts appears in the published literature.

## 5-7 The B—B Bond

### The Bond between B Atoms with No Terminal H Atoms

In $B_{10}H_{16}$,[87] which is $B_5H_8$ dimer, there is a B—B bond between two B atoms, neither of which has a terminal H atom. The reactions of this new molecule have been but little investigated, but it has been established[85] that HI and HBr produce cleavage to yield $B_5H_9$ and $B_5H_8I$ or $B_5H_8Br$. This B—B bond is perhaps more nearly a single bond (whatever that is) than any other B—B interaction so far discovered.

### The Edge B—B Bond

A B—B bond on the edge of the boron fragment occurs in $B_6H_{10}$. Addition of $H^+$ might be expected, since the reverse reaction, loss of H from a B—H—B bond, is known to occur. If the edge B—B bond is adjacent to a bridge H atom, rearrangement may occur, as described below.

### The Internal B—B Bond

A B—B bond not on the edge of the boron framework has never been shown to take up $H^+$, but it may rearrange, as described below, if a bridge H atom is bound to one of the B atoms. However, a B—B bond appears to propagate[327] magnetic shift phenomena in $B_{10}H_{14}$. The internal single B—B bond in $B_4H_{10}$ has been said[332] to show $B^{10}$—$B^{11}$ coupling in the $B^{11}$ NMR spectrum, but the fact that this fine structure disappears when $B_4D_{10}$ is substituted for $B_4H_{10}$ disproves this particular interpretation of the hyperfine structure, which is due to a more complex coupling phenomenon.

## 5-8 Intramolecular Rearrangement of H

The possibility that hydrogen atoms can rearrange within boron-hydride-like molecules was first suggested[213] in the nuclear resonance study of $Al(BH_4)_3$. A similar process seems to me to be the simplest interpretation[175] of the relative intensities of the equally spaced proton hyperfine structure of the $B^{11}$ resonance lines of the $B_3H_8^-$. Less rapid intramolecular rearrangement of H atoms has been reported[190] to occur in $B_5H_{11}$ and $B_4H_{10}$ on the basis of isotopic-exchange experiments.

The mechanisms of intramolecular exchange is not certain, but a very suggestive transformation (Fig. 5-8) constitutes the only remaining ambiguity in the present valence theory,[49] and occurs in several pairs of possible structures and possible reaction intermediates. For example, this ambiguity[49] in $B_6H_{10}$ appears to favor the structure with no $BH_2$ groups (Fig. 5-9). On the other hand, the reverse reaction could possibly be favored in negative ions such as $B_{10}H_{13}^-$ because $BH_2$ groups connected to as many as three adjacent boron atoms are relatively negative.[175] This mechanism, or a similar one such as that described above, could easily account for the internal exchange which probably occurs in $B_3H_8^-$, and somewhat similar intermediates can be drawn for the other hydrides. This possibility of internal rearrangement, due probably to

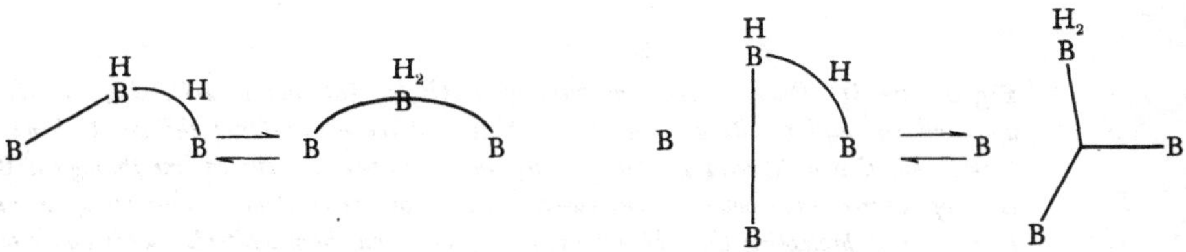

**Figure 5-8** *Tautomerism of H bridge adjacent to single bond.*

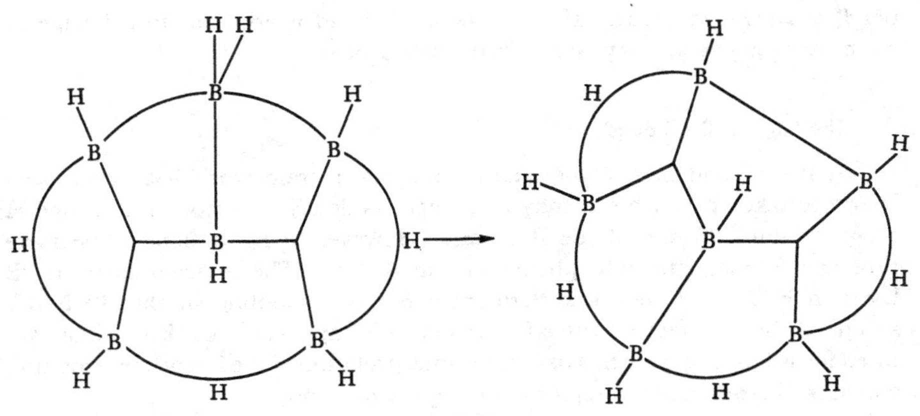


**Figure 5–9** *Probable rearrangement in $B_6H_{10}$.*

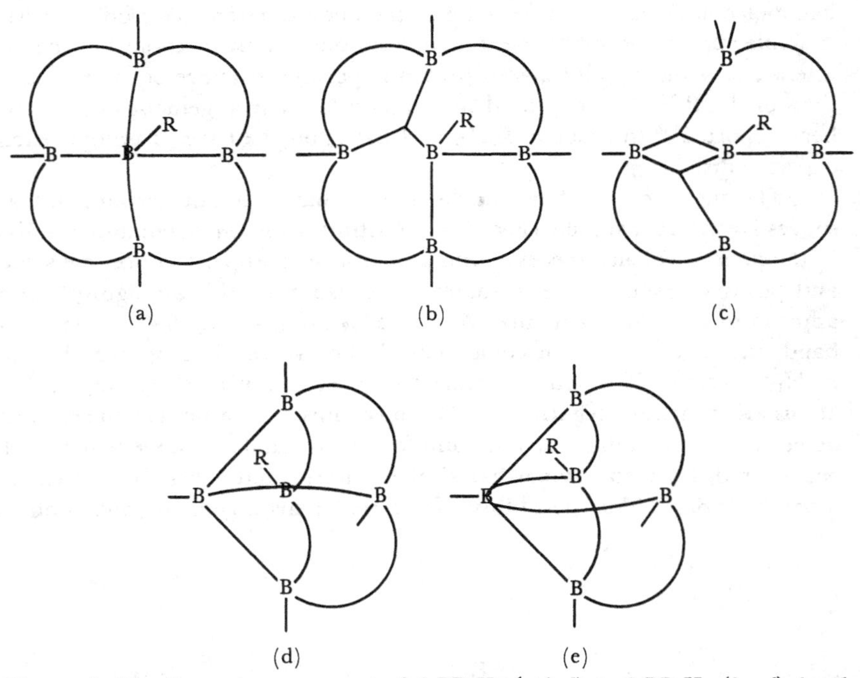


**Figure 5–10** *Proposed rearrangement of 1-$RB_5H_8$ (apical) to 2-$RB_5H_8$ (basal) based upon experimental results of Onak.*[214,216] *Note that we assume that the B—C bond is not broken, but that a different B atom becomes the new apex. A curved line through a B atom is an open three-center bond. Other curved lines are bridge H atoms. The role of an electron-pair donor in promoting these H-atom rearrangements has been omitted. This donor presumably would bond to an orbital freed by the conversion of a three-center bond to a two-center bond.*

the relatively expanded orbitals around boron, particularly in the negative boron hydride ions, adds to the usual difficulties of making predictions of structure and reactivity in this field of chemistry.

## 5–9 Rearrangement of $B_5H_8R$

Conversion[214] of 1-alkylpentaborane to 2-alkylpentaborane in 2,6-dimethylpyridine is said[85,224] to proceed, possibly, through an intramolecular symmetrical semicleavage, and hydrogen tautomerism followed by fast recombination (Fig. 5–10). Presumably, this process proceeds through an intermediate in which the electron-pair-donating property of the solvent is involved in an intermediate. Some further clarification of this mechanism by use of isotope tracer studies would be desirable.

The possibility has been mentioned[158] of internal rearrangement of terminal H atoms in $B_5H_9$ in the gas-phase exchange of D for H with the use of $B_2D_6$, but this rearrangement was considered less likely than the exchange of D in place of terminal H on $B_5H_9$ at both terminal positions at the same rate within $\pm 10$ per cent. It may be worthwhile to reinvestigate this question in the light of the $MeB_5H_8$ rearrangement.

The $B_{10}H_{16}$ rearrangement[85] to $B_{10}H_{14}$ upon loss of $H_2$ also can be described in terms of a series of steps which start with the $MeB_5H_8$ type of rearrangement.

## 5–10 Electron Addition to the Boron Framework

Of all the boron hydrides, only $B_{10}H_{14}$ appears to be capable of adding two additional electrons to give a satisfactory valence structure.[174] Again it may be noted that when an exposed single bond is adjacent to a bridge hydrogen, tautomers containing $BH_2$ groups may be observed.[175,176] If this situation occurs, as has been suggested,[175] in the $B_{10}H_{14}^{=}$ ion,[307] then $B_{10}H_{12}(CH_3CN)_2$ can be regarded[247] as a substitution product of $B_{10}H_{14}^{=}$. The linear ($CH_3CN$) groups require an additional pair of electrons in the boron framework.

## 5–11 Collapse of the Boron Framework

This reaction occurs when $B_{10}H_{12}L_2$ produces $B_{10}H_{10}^{=}$, as described below.

## 5–12 Degradation and Condensation of Boron Frameworks

These types of reactions occur in many of the other sections. Degradation occurs when, for example, a ligand L takes $BH_3$ or $BH_2^+$ away from a boron framework, and has been described above. Further degradation reactions of the $B_{10}$ framework are described below in the section on reactions of $B_{10}H_{14}$, which are considered separately.

A combination of this type of degradation and condensation apparently occurs in the several reactions[23] in which $B_5H_{11}$ is converted in high yields to $B_6H_{10}$, as well as to other hydrides, in the presence of electron-pair donors, such as $[(CH_3)_2N]_2BH$, $(CH_3)_3N$, $(CH_3)_2O$, or diglyme. The details of these reactions, especially the intermediates, are not understood.

Condensation of boron frameworks to a larger molecule occurs in the glow-discharge process for preparation of $B_{10}H_{16}$ from $B_5H_9$. One might guess that the reaction proceeds through the formation of $B_5H_8$ radicals. If so, it may be of some value to attempt the reaction in the glow discharge of these intermediates with other molecules, such as $C_2H_2$ or $C_2H_4$, as has been done by Keilin,[282] whose studies initiated the work on the smaller carboranes.[282]

Reaction of excess K with $B_5H_9$ in tetrahydrofurane has produced[151] a salt tentatively identified as $KB_8H_{13}$, for which the $B^{11}$ NMR spectrum is shown in Fig. 4–10. Proposed structures are shown in Fig. A–1, where the two 2513 structures have the same atomic arrangements, in which part of the structure is very similar to that of $B_5H_9$ and part is similar to that of $B_5H_{11}$. Proton tautomerism around the periphery of this negative ion may occur.

Other condensation of boron frameworks to produce higher hydrides occur in the section on $B_{10}H_{14}$ and its reactions, and in Section 5–13.

## 5–13 Loss of $H_2$ and Polymerization

Although $BH_3$ has not been observed experimentally, the kinetics of the initial stages of several reactions of diborane[25,43] render the steady-state equation,[14]

$$B_2H_6 \rightleftharpoons 2BH_3 \tag{1}$$

followed by the reaction

$$BH_3 + B_2H_6 \rightarrow \text{higher products} \tag{2}$$

as the generally accepted initial reactions. The discussion will be simplified if we consider first this and the later work relating to these initial reactions, after which we shall return to a consideration of the further reactions in the synthesis of the higher hydrides.

The initial stages of $B_2H_6$ polymerization,[25,43] of exchange[190] between $B_2H_6$ and $B_2D_6$, of exchange[281] between $B_2H_6$ and $B_2{}^{10}H_6$, and of reaction[325] of $B_2H_6$ with $C_2H_4$, all support this mechanism. Experimentally, all these reactions are $\frac{3}{2}$ order with respect to diborane concentration, as may be derived from the mechanism by writing Eq. (1) as a steady state, $K = (BH_3)^2/(B_2H_6)$, so that

$$\frac{d(\text{higher products})}{dt} = k(BH_3)(B_2H_6) = k\sqrt{K}\,(B_2H_6)^{3/2}$$

is the rate-determining step of the initial stage.

The $B_2H_6$—$C_2H_4$ reaction is of importance partly because it gives support to the idea that the initial higher product is $B_3H_9$ or $B_3H_7 + H_2$. The initial stages of the mechanism suggested[325] are

$$B_2H_6 \rightleftharpoons 2BH_3$$
$$B_2H_6 + BH_3 \rightleftharpoons B_3H_9$$
$$B_3H_9 + C_2H_4 \rightarrow B_2H_5Et + BH_3$$

The last two reactions are then repeated, starting with $B_2H_5Et$ in place of $B_2H_6$ to obtain $B_2H_4Et_2$, which reacts with $BH_3$ to give $B_3H_7Et_2$. At this stage an autocatalytic reaction appears,

$$B_3H_7Et_2 + B_2H_6 \rightarrow 2B_2H_5Et + BH_3$$

so that $B_2H_6$ and $C_2H_4$ compete for these later products. The autocatalytic effect increases considerably as the final product $BEt_3$ appears, and, if $2(B_2H_6) > (C_2H_4)$, the reaction becomes explosive.

The present literature[25,43,70,175,262] on the polymerization to higher hydrides around 100°C may be summarized in the sequence of reactions

$$B_2H_6 \rightleftharpoons 2BH_3 \qquad (1)$$
$$BH_3 + B_2H_6 \rightleftharpoons B_3H_9 \qquad (2)$$
$$B_3H_9 \rightleftharpoons B_3H_7 + H_2 \qquad (3)$$
$$B_3H_7 + B_2H_6 \rightleftharpoons B_4H_{10} + BH_3 \qquad (4)$$

or, combining (3) and (4),

$$B_3H_9 + B_2H_6 \rightarrow B_4H_{10} + BH_3 + H_2 \qquad (3')$$
$$B_4H_{10} + BH_3 \rightleftharpoons B_5H_{11} + H_2 \qquad (5)$$
$$B_4H_{10} + B_5H_{11} \rightarrow B_5H_9 + 2B_2H_6 \qquad (6)$$

Reaction (1) is apparently very rapid. Combination of (1) and (2) gives the reaction its $\frac{3}{2}$ power dependence on diborane concentration. Reaction (3) or (3′) presumably is rate-controlling. Combination of (4) and (5) is the recognizable reaction, $B_3H_7 + B_2H_6 \rightarrow B_5H_{11} + H_2$, which is questioned by Schaeffer[262] and others[175] as a single step and is currently being investigated. Reaction (5) is nearly an equilibrium, discovered by Burg and Schlesinger[33]; it has been invoked[262] to account for the sharp braking of the rate of $H_2$ formation as the $B_5H_{11}$ concentration rises. Reaction (6), which may not be a single step, is the slowest reaction in the above list. Reactions (5) and (6) account nicely for the later cleavage of $B_5H_{11}$ to $B_2H_6$ which is formed along with $B_5H_9$, apparently with the same activation energy and at an apparent rate about 2.5 times the rate of formation of $B_5H_9$.[262] Reaction (3′), which may be two steps with $B_5H_{13}$ as an intermediate, avoids formation of the electron-deficient intermediate, $B_3H_7$. The slow $H_2$ inhibition reaction, $B_4H_{10} + H_2 \rightarrow 2B_2H_6$, should also be included as a minor reaction. Nevertheless, hydrogen inhibition appears to merit further study in these reactions. Evidence[70] that $B_3H_9 \rightarrow B_3H_7 + H_2$ is the rate-controlling step in the initial stages of pyrolysis of $B_2H_6$ has been obtained from a comparison of the rates of pyrolysis of $B_2H_6$ and

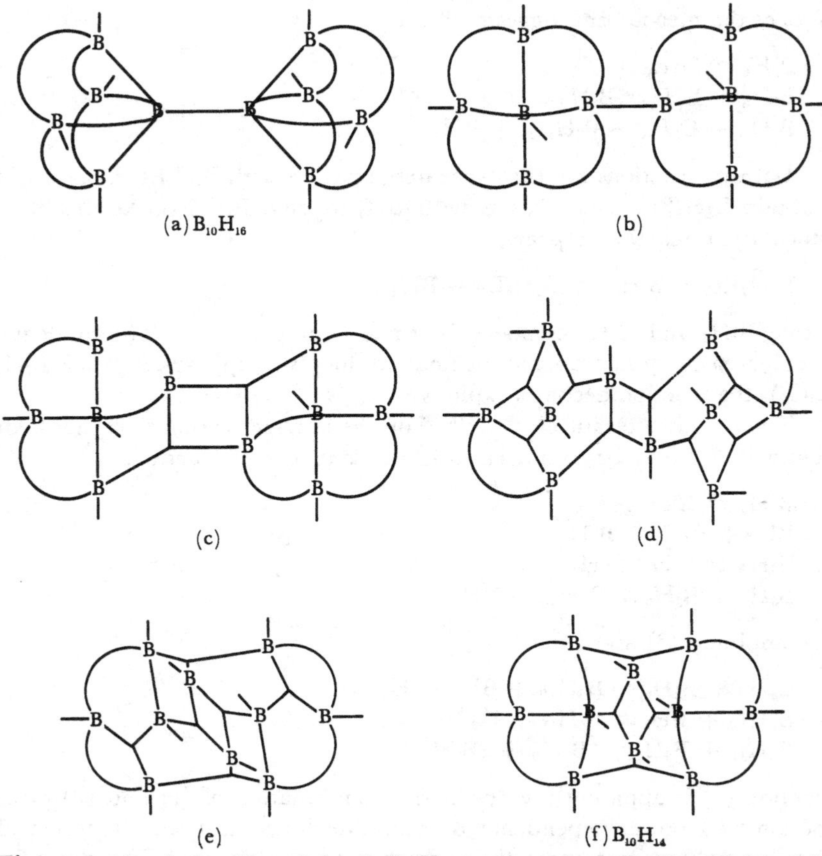


**Figure 5–11** *A plausible path for the $B_{10}H_{16}$ transformation to $B_{10}H_{14}$ with loss of $H_2$. After the apex-to-base rearrangements have occurred, a minimum of atomic motion is required. Again the role of the electron-pair donor has been omitted. In the discharge preparation of $B_{10}H_{14}$ a wall or surface may supply the electron-pair donor.*

$B_2D_6$. The latter compound is five times slower in its rate of decomposition, in spite of the fact that $BD_3$ is present in about twice the concentration of $BH_3$ under comparable conditions. This large isotope effect is consistent with the energy of D as compared with H in stretched B—D or B—H bonds, respectively, and further lends some support to the presence of $B_3H_9$ as the initial product formed from $B_2H_6$ and $BH_3$.

The importance of further isotope-exchange studies of the boron hydrides and simple reactions among them can hardly be overemphasized. It is conceivable[114] that all of the hydrides containing $BH_2$ groups, $B_2H_6$, $B_4H_{10}$, $B_5H_{11}$, and $B_9H_{15}$, undergo reversible dissociation into $BH_3$ and intermediates such as $BH_3$, $B_3H_7$, $B_4H_8$, and $B_8H_{12}$, which will add $BH_3$, $H_2$, or react with one another or other hydride molecules. It seems unlikely, on the other hand, that the more stable relatively highly condensed $B_nH_{n+4}$ hydrides undergo rapid associations with $BH_3$ to form a $B_nH_{n+6}$ hydride. The relation of these rather general speculations to the pyrolysis is that at higher temperatures, near 100°C, the reactions of $B_4H_{10}$ may be those of $B_3H_7$, and the reactions of $B_5H_{11}$ may be those of $B_4H_8$, which could add $H_2$ at high temperatures to give an unstable form of $B_4H_{10}$ or $B_3H_7 + BH_3$, as suggested by the reverse of reaction (5). Thus the mechanism of the pyrolysis can hardly be said to be understood clearly, but the stage of understanding of the structures of various possible reaction intermediates from the valence theory is sufficiently advanced that one can hope for some help from the structural concepts in the elucidation of the mechanisms and intermediates.

So far the important results of the isotopic-exchange experiments may be summarized as follows[156]: (1) Exchanges occur in which $BH_3$ remains an entity, as in the $B_2H_6$ self-exchange; the rate of deuterium exchange is three times the rate of boron exchange. (2) Exchanges occur when $BH_3$ exchanges with terminal hydrogen and not with bridge hydrogen, as in $B_5H_9$ and $B_{10}H_{14}$. (3) Intramolecular exchange appears to occur in $B_5H_{11}$ and in $B_4H_{10}$, as was noted in the $Al(BH_4)_3$ and in $B_3H_8^-$ and $B_3H_7$ addition compounds. Very little is known of the further reactions which lead to $B_{10}H_{14}$ and polymers, but a mass-spectroscopic study[24d,113] of the reaction at 78°C for 24 hours of isotopically enriched $B_2{}^{10}H_6$ and $B_5{}^{11}H_9$ indicates that half of the boron atoms in the resulting $B_{10}H_{14}$ come from $B_5H_9$. Also, exchange between $B_2H_6$ and $B_5H_9$ occurred, possibly involving a pair of B atoms at each step. The conversion[85] of $H_8B_5$—$B_5H_8$ smoothly at room temperature into $B_{10}H_{14}$ in the presence of $C_2H_4$, which takes up the $H_2$, and of $AlCl_3$ suggests that one route to $B_{10}H_{14}$ may involve loss of $H_2$ from one $B_5H_9$ molecule and a different $B_5H_x$ molecule, and then dehydrogenation and rearrangement (Fig. 5–11). At higher temperatures (210 to 250°) $B_{10}H_{14}$ forms a $B_{20}$ compound, possibly $B_{20}H_{24}$.[220]

## 5–14 Chemistry of $B_{10}H_{14}$

The complexity of decaborane chemistry is already becoming apparent, but what is known so far can only be the beginning. Again, we attempt some classifications of reactions, but lack of detailed studies of the mechanisms places many of the details at a level which may charitably be called suggestions for future research.

### Electrophilic Substitution

The order of charge distribution in $B_{10}H_{14}$ (Fig. 5–12) is, from the most recent molecular orbital work,[123,195,198]

$$\begin{array}{ccccccc} 6,9 & > & 5,7,8,10 & > & 1,3 & > & 2,4 \\ 0.33 & & 0.12 & & -0.04 & & -0.10 \end{array}$$

in units of the electronic charge. It is quite so that these charges refer to the molecule before formation of the four-center complex between, say, B—H and I—I when $B_{10}H_{13}I$ is formed. But to whatever extent this order of charges is preserved in formation of the complex, one might expect electrophilic substitution to occur first at 2,4 and then at 1,3 positions, provided steric factors do not interfere.

At least the substitution of two I atoms at 2,4 positions in[299] $B_{10}H_{12}I_2$ (m.p. 261°) has been established[261] with reasonable certainty in an X-ray diffraction study in which intramolecular I···I distance was found. This result depends upon predictions from the $B_{10}H_{14}$ geometry,[142] which has been slightly revised,[197] and hence upon H-atom locations in the $B_{10}H_{14}$ structure. It would still be of interest to locate the B atoms in the $B_{10}H_{12}I_2$ structure.

Electrophilic attack of $I^+$ (or I) in the $I_2$—$B_{10}H_{14}$ reaction is complex enough to yield two $B_{10}H_{12}I_2$'s and two $B_{10}H_{13}I$'s.[111,112]

$$\begin{array}{l} B_{10}H_{14} \longrightarrow B_{10}H_{13}I \text{ (m.p. } 100°) + B_{10}H_{13}I \text{ (m.p. } 119°) \\ \qquad\qquad\quad \downarrow \qquad\qquad\qquad\qquad\qquad \downarrow \\ \quad\;\; \longrightarrow B_{10}H_{12}I_2 \text{ (m.p. } 155°) + B_{10}H_{12}I_2 \text{ (m.p. } 261°) \end{array}$$

These reactions and X-ray results on $B_{10}H_{12}I_2$ (261°) therefore imply that the I on $B_{10}H_{12}I_2$ (119°) and one I in $B_{10}H_{12}I_2$ (155°) are in a 2 or 4 position, but the I in $B_{10}H_{13}I$ (100°) and the other I in $B_{10}H_{12}I_2$ (155°) are clearly elsewhere. The valence theory clearly suggests one of the 1,3 positions, but the $B^{11}$ NMR has been variously interpreted to include 5,7,8,10[267] or the 1,3 or 6,9 sets.[112] Obviously an X-ray diffraction study is needed.

Electrophilic alkylation by $CH_3Br$ in the presence of $AlCl_3$ in $CS_2$ yields[331] 2-$MeB_{10}H_{13}$, 2,4-$Me_2B_{10}H_{12}$, 1,2-$Me_2B_{10}H_{12}$, 1,2,3-$Me_3B_{10}H_{11}$, 1,2,4-$Me_3B_{10}H_{11}$, and 1,2,3,4-$Me_4B_{10}H_{10}$. In addition, a very small amount of 5-$MeB_{10}H_{13}$ was found. These results are consistent with electrophilic attack by a polarized $CH_3^{\delta+}$—$Br^{\delta-}$ structure, with $CH_3^+$ attacking $B_{10}H_{14}$ and $Br^-$ complexed with

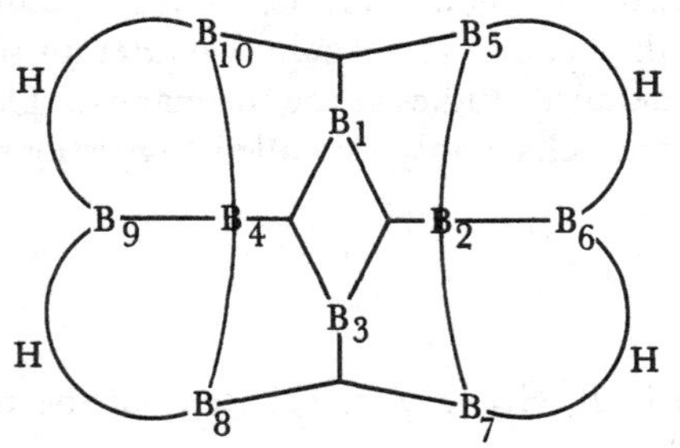


**Figure 5-12** *Atom numbering for* $B_{10}H_{14}$.

$AlCl_3$ in the transition state. The order (neg.) 2,4 < 1,3 < 5,7,8,10 < 6,9 (pos.) is indicated by these products and yields.

Electrophilic deuteration of $B_{10}H_{14}$ by DCl and $AlCl_3$ in $CS_2$ yields deuteration[103] at the 2,4 and 1,3 positions at so nearly the same rate that one might suspect a transition complex involving exchange between these two sets of positions. These results correct an earlier report of electrophilic attack elsewhere.[63]

### Nucleophilic Substitution

Lithium alkyls have been shown[61] to give principally 6-$MeB_{10}H_{13}$ or 6-$EtB_{10}H_{13}$. Small amounts of 5-$MeB_{10}H_{13}$, 6,5- or 6,8-$Me_2B_{10}H_{12}$, and 6,9-$Me_2B_{10}H_{12}$ are also formed. It is presumed that $Me^-$ or a polarized $Me^{\delta-}$—$Li^+$ attacks at positive regions of $B_{10}H_{14}$, and hence that the 6,9 positions are the most positive, with the 5,7,8,10 positions next. The intermediate complex $B_{10}H_{14}Me^-$ or $B_{10}H_{14}Et^-$ is topologically equivalent to $B_{10}H_{15}^-$ recently reported[65] and earlier predicted.[248] The valence structures[179] suggest H-atom tautomerism in $B_{10}H_{15}^-$.

Grignard reagent MeMgI also appears to produce[61] $MeB_{10}H_{13}$ by a direct $Me^{\delta-}$—$Mg^{\delta+}$—I attack of $Me^{\delta-}$ at a 6,9 position in much the same way as MeLi or EtLi does. However this is a minor reaction. The major reaction product of MeMgI with $B_{10}H_{14}$ is a "Grignard reagent" itself,[77,285] $B_{10}H_{13}MgI$, which gives reactions[61] more characteristic of $[B_{10}H_{13}]^{\delta-}Mg^{\delta+}I$ than of $B_{10}H_{14}$. Thus $B_{10}H_{13}MgI$ yields (1) about equal amounts of 5-$MeB_{10}H_{13}$ and 6-$MeB_{10}H_{13}$, when treated with $Me_2SO_4$[61]; (2) 5-$EtB_{10}H_{13}$, when treated with triethyloxonium fluoroborate or with $Et_2SO_4$; or (3) 6-benzyl $B_{10}H_{13}$, when treated with benzyl chloride. Rearrangement of H atoms in the $B_{10}H_{13}^-$ ion is discussed below.

### Addition Reactions

The $B_{10}H_{14}\cdot Py_2$ complex[36,72] is undoubtedly incorrectly formulated. The *bis*-2-Br-pyridine complex has been shown to be $B_{10}H_{12}(2\text{-}BrPy)_2$ in an X-ray diffraction study,[133] and to have molecular geometry similar to that[248] of $B_{10}H_{12}(NCCH_3)_2$. The tilt of the aromatic rings away from a plane containing the molecular twofold axis is substantial, and may lead to some revision of the assumptions[80] regarding the electronic transition. If steric problems in the open face of decaborane did not arise, it seems possible that a $B_{10}H_{14}L_2$, or a topologically equivalent $B_{10}H_{16}^{=}$, could exist,[179] but it probably loses $H_2$ readily to form $B_{10}H_{12}L_2$ compounds (discussed more fully below).

No solvate of $B_{10}H_{14}$ has yet had a full structure determination but there is evidence of a hydration complex,[89,90] and $B^{11}$ NMR evidence[231] of shift to higher field of the 5,7,8,10 B atoms in an acetone complex. The possibility that the $B_{10}H_{14}$ can absorb one or two electrons is discussed below.

### Elimination of $H^+$

A stronger Lewis base, such as $NH_3$, $NR_3$, or $OH^-$ forms initially $B_{10}H_{13}^-$ from $B_{10}H_{14}$. In fact, $B_{10}H_{14}$ is a strong monoprotic acid,[89] but the second dissociation constant has not been found, even though a $Na_2B_{10}H_{12}$ salt has been described.[306] A number of research workers (see Ref. 176) have shown that a bridge H atom ionizes off during this dissociation. However, the reaction is quite probably more complex than a simple ionization in that rearrangement of H atoms is presumed[176] to occur (Fig. 5-13) for reasons discussed below in connection with $B_{10}H_{14}^{=}$.

### Electron Addition

The expectation from the valence theory[174] and the discovery[307] that salts of the form $Na_2B_{10}H_{14}$ were independent and nearly simultaneous studies. No structure determination has been made on $B_{10}H_{14}^{=}$, but its correct structure can probably reliably be inferred from the $B_{10}H_{12}(NCCH_3)_2$ X-ray study.[248] Although at first sight $B_{10}H_{12}L_2$, where L = $NCCH_3$ or another electron-pair donor, appears to be a substituted decaborane, an electron count indicates that the boron framework of $B_{10}H_{12}L_2$ has two more electrons than $B_{10}H_{14}$, as may be seen from the formal equivalence of $H^-$ and L, both of which donate an electron pair. Also, somewhat surprisingly, the H-atom arrangement has been shown with certainty to be different from that in $B_{10}H_{14}$. It is therefore inferred that $B_{10}H_{14}^{=}$ has H bridges in the 5,10 and 7,8 positions and $BH_2$ groups at the 6,9 positions, again in order to accommodate the negative charge in the regions of the $BH_2$ groups (Fig. 5-14).

### H-atom Rearrangement

The usual evidence for H-atom rearrangement is some equivalence or higher-than-expected symmetry in the $B^{11}$ NMR spectrum, such as that noted

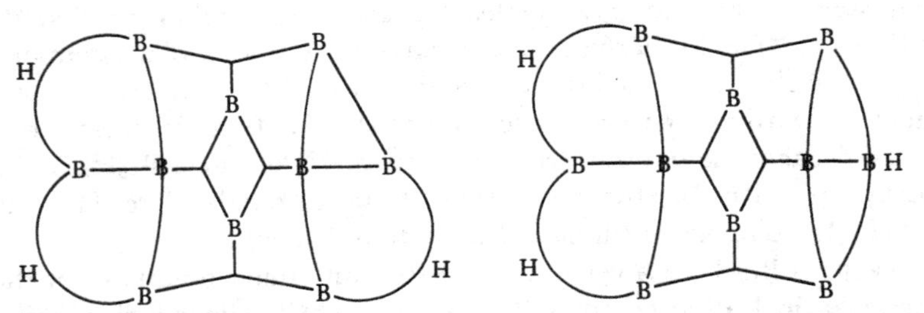


**Figure 5-13** *The $B_{10}H_{13}^-$ structure. The presumably less stable form on the left results from removal of a bridge proton in accord with exchange experiments in acid solutions. The probably more stable form on the right has one $BH_2$ group, which accommodates the negative charge.*

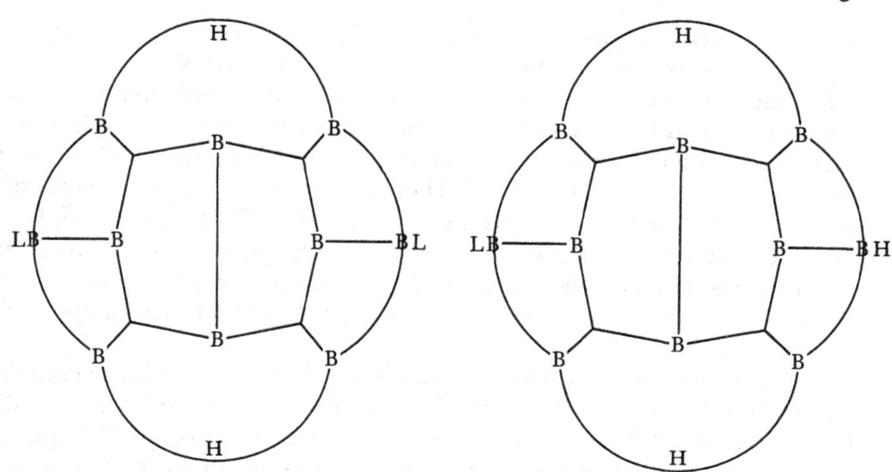


**Figure 5–14** *The $B_{10}H_{12}L_2$ structure ($L = NCCH_3$). The presumed $B_{10}H_{13}L^-$ structure is shown at the right, and the structure of $B_{10}H_{14}^=$ is presumed to be analogous, with $H^-$ replacing L formally.*

above for $Al(BH_4)_3$, $B_2H_7^-$, $B_3H_8^-$, or $B_6H_{10}$. Unfortunately, the migrations of D atoms as a function of temperature have been rarely studied, and most intramolecular H-atom migrations are only tentative conclusions based upon $B^{11}$ spectra at a single temperature, without even the full exploration of possible strong-coupling models by double resonance experiments. Only the relatively slow intramolecular exchanges of H atoms in $B_5H_{11}$ and $B_4H_{10}$,[24b] and migration in dioxane of bridge D atoms to the 5,7,8,10, and the 6,9 positions in $B_{10}H_{14}$[192,283] have been shown with reasonable certainty.

Nevertheless, the orbitals around B are considerably expanded as compared with those in C compounds. The electron-deficient bonding decreases exchange repulsions between interacting electron pairs. Also, the valence theory provides many examples of alternative H arrangements for a given B arrangement in some species. Hence, there are ample theoretical reasons for expecting intramolecular exchange, either spontaneous or solvent-promoted, in many of the hydrides, ions, and derivatives. The rearrangements which probably occur for $B_{10}H_{13}^-$ and $B_{10}H_{14}^=$ to form $BH_2$ groups from bridge H atoms adjacent to a single bond have just been referred to. The conjectured rearrangements for $B_{10}H_{14}$ and $B_{10}H_{14}^=$ are shown in Figs. 5–15 and 5–16, in which the more stable structures are shown on the left. Other H-atom rearrangements in derivatives of $B_{10}H_{14}$, and some B-atom rearrangements, are discussed below.

### Nucleophilic Attack by Ligands, L = $HNEt_2$, etc.

The most probable first step is loss of a proton from $B_{10}H_{14}$ to produce $B_{10}H_{13}^-$ and $H_2NEt_2^+$, but the next product isolated is[74,82,102] $B_{10}H_{13}NHEt_2^-$.

$H_2NEt_2^+$. Hence addition of $NHEt_2$ to $B_{10}H_{13}^-$ very likely occurs, accompanied by H-atom rearrangement to give the $B_{10}H_{13}L^-$ structure shown in Fig. 5–14. Substitution of the ligand at 6,9 positions is inferred not only because these are the most susceptible positions for electrophilic attack, but because it is suspected that the B—N bond formed here is not subject to rearrangement when a further reaction[260] to form the known[248] $B_{10}H_{12}L_2$ structure occurs. The details of loss of $H_2$ are not known when $B_{10}H_{13}L^-$ reacts with $HL^+$ to give $B_{10}H_{12}L_2$, and hence D tracer studies would be useful. In addition, it is not known whether the intermediate here is the $B_{10}H_{13}L_2^-$ structure, analogous to a possible $B_{10}H_{15}^{-3}$ of 1634 topology, or whether $H_2$ is lost prior to the addition of the second ligand.

Ligand displacement reactions are known[100,221] for $B_{10}H_{12}L_2$ compounds. The order is $Me_2S < MeCN < Et_2NCN < HCONMe_2 = AcNMe_2 < Et_3N = C_5H_5N = \phi_3P$. A $B_{10}H_{12}L_3$ transition state would be topologically equivalent to a $B_{10}H_{15}^{3-}$, but the details of this exchange reaction are not known as yet.

#### Nucleophilic Attack by $NH_3$

The action of $NH_3(g)$ at 0° on $B_{10}H_{14}$ yields[306] $B_{10}H_{14}\cdot nNH_3$ for $n = 6$, 4, and 3. There is also some evidence[306] for $n = 5$, 2, and 1. The $B_{10}H_{14}\cdot 3NH_3$

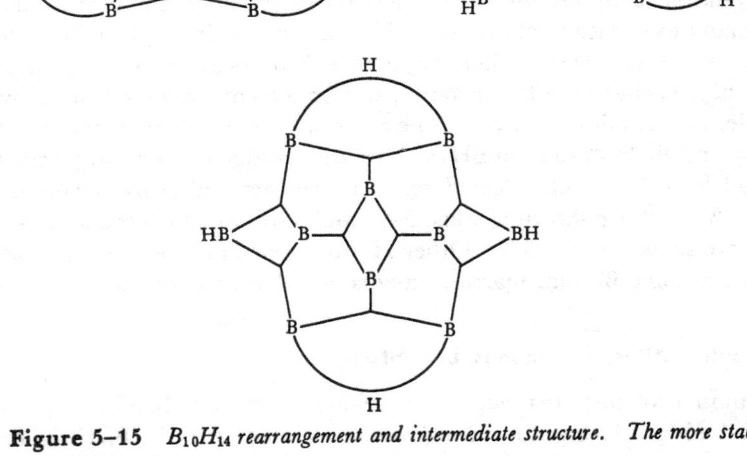

**Figure 5–15** *$B_{10}H_{14}$ rearrangement and intermediate structure. The more stable structure is on the left.*

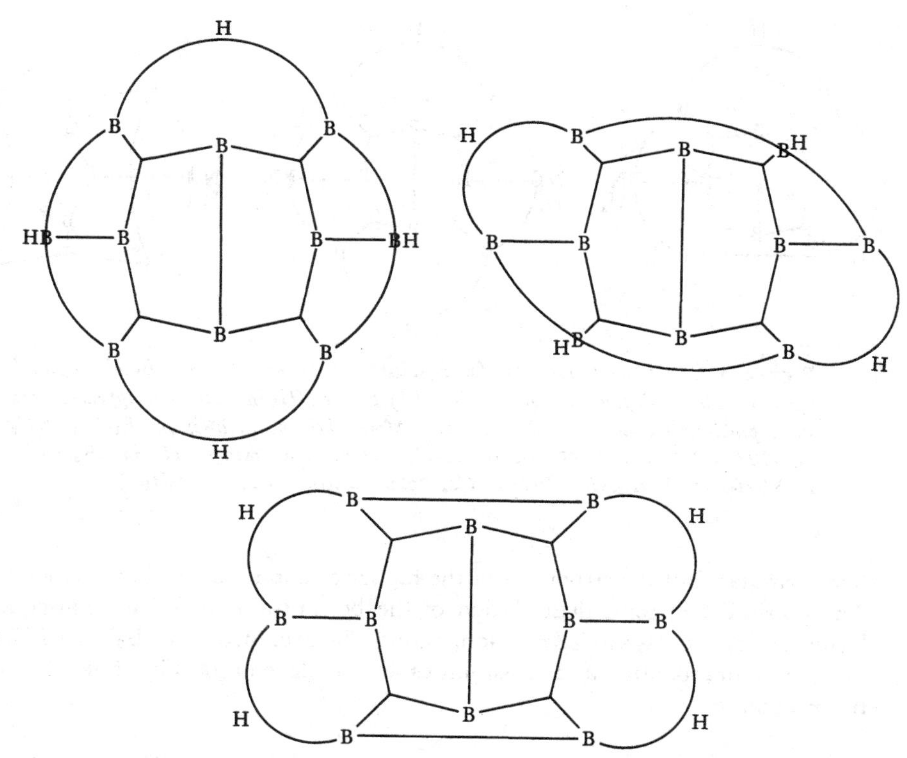


**Figure 5–16** *$B_{10}H_{14}^{=}$ rearrangement and intermediate structure. The presumably correct structure is on the left.*

compound is especially stable, salt-like, and yields[306] at 95° a compound which is probably $B_{10}H_{18}N_2$, and at 120° a compound which is probably $B_{10}H_{12}(NH_2)_2$. At 80° a 1:1 molar ratio of $B_{10}H_{14}$ and $NH_3$ yields[306] $B_{10}H_{12}NH$. Recovery of $B_{10}H_{14}$ from solutions in $NH_3(l)$ is possible only if they are freshly prepared.[306] Stock[299] mentions reversible formation of the hexammoniate. Reactions of $Me_2NH$ with $B_{10}H_{14}$ yield compounds[74] $B_{10}H_{14}{\cdot}mMe_2NH$ for $m = 1$, 2, and 3, with some evidence for $m = 4$ and 5. The reason for the red color of the $B_{10}H_{14}$·quinoline complex is unknown, but may be a charge-transfer spectrum.

It needs hardly to be stated that a series of $B^{11}$ NMR and X-ray diffraction structure determinations will be useful in providing some certainty with respect to these compounds. The suggestions of valence theory are somewhat complicated by the possibilities that bridging $NH_2$ groups and terminal $NH_3$ groups may occur as alternative formulations. Nevertheless, some reasonable guesses can be made. Very probably the 1—1 compounds are $B_{10}H_{13}^-$ salts with $NH_4^+$ or a substituted $NH_4^+$ as the positive ion. The 1—2 compounds may be formulated as $LH^+{\cdot}B_{10}H_{13}L^-$ salts, where L = $NH_3$ or $HNEt_2$. The stable, salt-like $B_{10}H_{14}L_3$ could be $B_{10}H_{13}NH_2^={\cdot}2NH_4^+$ or $B_{10}H_{12}NH_3^={\cdot}2NH_4^+$ (Fig. 5–17,

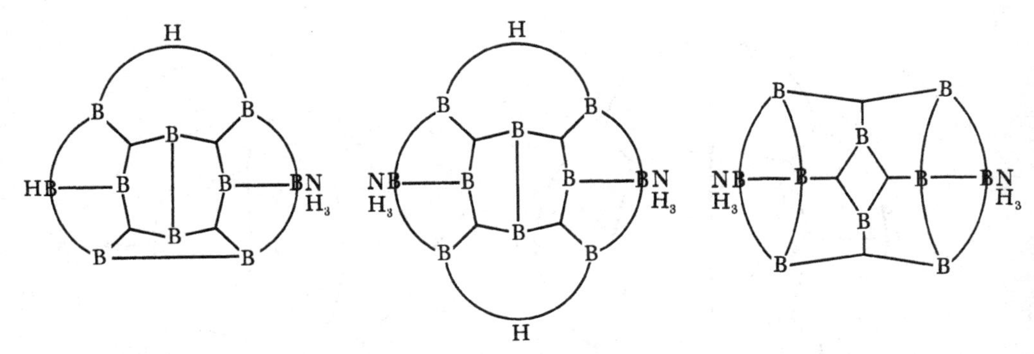


**Figure 5-17** *Possible formulas for $B_{10}H_{14}\cdot 3NH_3$, $B_{10}H_{18}N_2$, and $B_{10}H_{16}N_2$ all based upon $B_{10}$ units. Replacement of a bridge H by bridge $NH_2$ in a $B_{10}$ unit appears to produce steric problems due to H · · · H contacts. More likely possibilities for $B_{10}H_{14}\cdot 3NH_3$ are $H_3NBH_2NH_3{}^+B_9H_{13}NH_2{}^-$ (where $NH_2$ replaces a bridge H in $B_9H_{14}{}^-$), or $H_3NBH_2NH_3{}^+B_9H_{12}NH_3{}^-$ (where $NH_3$ replaces a terminal $H^-$ in $B_9H_{13}{}^=$).*

first formula), but the structures of the higher ammoniates are less predictable. The possibility of some degradation of the boron framework, and consequent formation of the $H_3NBH_2NH_3{}^+$ ion, cannot be excluded, e.g., $B_9H_{12}{}^-LBH_2L^+$ for the diammoniate, and possibilities in the legend of Fig. 5-17 for the triammoniate.

### Solvolytic Degradation of $B_{10}H_{12}L_2$

Ethyl alcohol degrades[83,102] $B_{10}H_{12}L_2$ to $B_9H_{13}L$, where L = $SMe_2$, for example. The degradation of $B_{10}H_{13}L^-$ by $H^+$ in aqueous solution also yields $B_9H_{13}L$. This degradation product, $B_9H_{13}NHMe_2$, was first observed several years ago[74,260] but was not then completely characterized. It has recently been subjected to a full X-ray study[319,320] in which the H-atom arrangement (Fig. 5-18) is shown to be consistent with the valence theory of boron hydrides. The existence of a topologically equivalent $B_9H_{14}{}^-$ is inferred[319,320] but has not yet been reported.* The detailed processes by which $B_9H_{13}L$ is formed undoubtedly involve a large number of steps, since the B atom removed appears as $B(OH)_3$ in the acid aqueous hydrolysis. However, the first step may be attack by an electron pair at a three-center bond (Fig. 5-18).

The reaction of $B_9H_{13}NH_2Et$, originally thought[83,102] to yield $EtNH_3{}^+\cdot B_9H_{12}NH_2Et^-$ as a prototype of the $B_9H_{12}L^-$ anion, has been shown[167] actually to yield $EtNH_2B_8H_{11}NHEt$, a molecular compound (Fig. 5-19). The possibility that a $B_9H_{12}L^-$ anion might exist in the case that L = pyridine has some chemical support in the reverse formation of the original $B_9H_{13}L$ compound.[83]

The $B_9H_{12}{}^-$ ion is produced[83] by proton abstraction from $B_9H_{13}SMe_2$ by

* However, see the Concluding Remarks, p. 198.

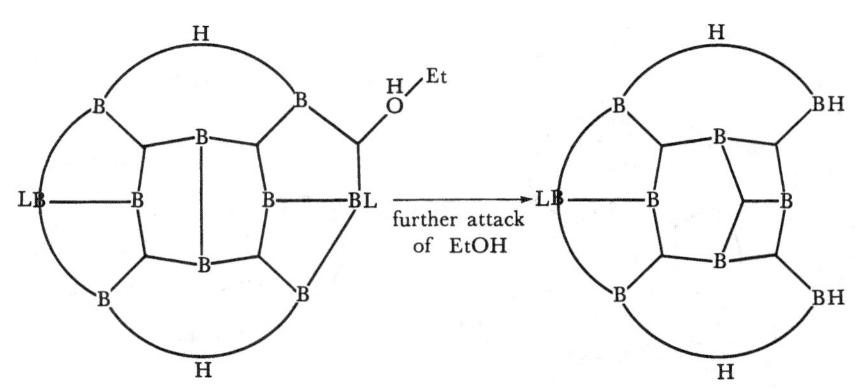


**Figure 5–18** *Possible initial mode of attack of EtOH to begin the formation of $B(OEt)_3$ and $B_9H_{13}L$ from $B_{10}H_{12}L_2$.*

the reagent $\phi_3PCH_2$ in a reaction which also yields $\phi_3PCH_3^+$ and free $SMe_2$. Preliminary X-ray diffraction data[318] on single crystals of the $NMe_4^+$ and $NEt_2Me_2^+$ salts of $B_9H_{12}^-$ indicate that the $B_9H_{12}^-$ ion is definitely monomeric, probably quite nearly spherical in shape and very likely orientationally disordered in these crystal structures. The valence theory indicates a unique preferred geometrical structure[179] based upon three-center bonds, but this structure has not been proved.

The $B_9H_x(P\phi_3)_2$ compound[199] is very probably $B_9H_{11}(P\phi_3)_2$, which is of the $B_9H_{11}L_2$ type. Of the two structures consistent with the valence theory[320] (Fig. 5–20), one has two symmetry equivalent P atoms, whereas the other does not. Because of the relatively heavy P atom the X-ray study would be facilitated, except for H-atom location, but a study of the $F^{19}$ NMR spectrum might settle the question of the structure of this compound.

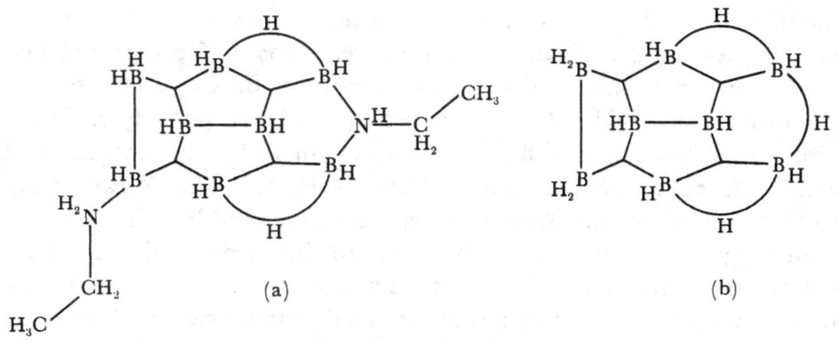

(a) (b)

**Figure 5–19** *Relation of a valence structure of the known $EtNH_2B_8H_{11}NHEt$ to an unknown $B_8H_{13}^-$.*

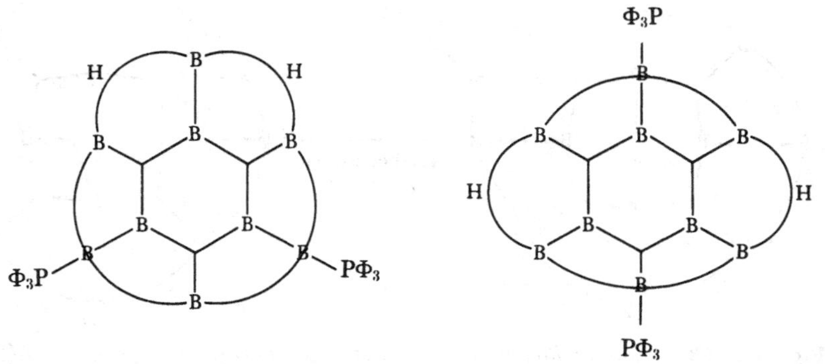

**Figure 5–20** *Proposed structures for* $B_9H_{11}(P\Phi_3)_2$.

Further degradation of $B_9H_{12}^-$ in $CH_3OH$ yields[84] $B_3H_8^-$ through intermediates which are as yet unidentified.

### Synthesis of $B_{11}H_{14}^-$

The preparation and properties[2] of $B_{11}H_{14}^-$ from $B_{10}H_{14}$ and $BH_4^-$, and from $R_2OBH_3$ and $B_{10}H_{13}^-$ confirm one of the more interesting predictions[198] of the valence theory. Also, the $B_{11}H_{13}^=$ ion, expected from the valence theory, has now been prepared,[2] and the equilibrium $B_{11}H_{14}^- = B_{11}H_{13}^= + H^+$ has been studied. The $B^{11}$ NMR spectrum shows no appreciable localization of the three extra H atoms over those in the terminal BH units of $B_{11}H_{14}^-$, in agreement with the nonlocalized valence description.[198]

### The $B_{12}H_{12}^=$ Ion

A 4 per cent yield of $B_{12}H_{12}^=$ was obtained[233] from $2\text{-}IB_{10}H_{13}$ upon successive treatments with $NEt_3$ and acetone. However, a new method,[149] as yet unrevealed,[200] gives larger yields. The only reactions so far reported are $H^+$-catalyzed four-center substitution reactions to form $B_{12}X_{12}^=$ and intermediates (X is halogen) and $B_{12}H_{11}NO_2^=$. Interesting products, possibly $B_{24}H_{22}^=$, may be formed upon treatment of $B_{12}H_{12}^=$ with $Fe^{3+}$ or $Ce^{4+}$ solutions, inasmuch as the analogous reactions are known[141] for $B_{10}H_{10}^=$. The X-ray study has proved[336] the icosahedral structure, earlier predicted.[188] The $B^{11}$ NMR spectrum shows[233] only one kind of B atom, but this evidence did not prove the icosahedral structure, since it is also consistent with the cube-octahedral structure, some intermediate structure, or rapid tautomeric equilibrium among these structures.[141] The extraordinary hydrolytic stability of $B_{12}H_{12}^=$ makes it a candidate for tumor therapy by neutron irradiation of the $B^{10}$-enriched material, and as possible elements in polymers stable to very high temperatures.

### $B_{10}H_{10}^{=}$ and its Reactions

**Preparation and Structure.** Formation of $B_{10}H_{10}^{=}$ from $B_{10}H_{12}L_2$ occurs upon further reaction[101,182] of excess L, to form salts $2HL^+ \cdot B_{10}H_{10}^{=}$ in which the positive ion may be readily exchanged. The indication of two kinds of B atoms in the $B^{11}$ NMR spectrum in the intensity ratio of 2 to 8 suggested[128] the polyhedral structure and mechanism of formation[67,179] shown in Fig. 5–21. A study[67] of $B^{11}$ resonance of a selectively deuterated $B_{10}H_{12}L_2$ followed through the formation of $B_{10}H_{10}^{=}$ is consistent with this structure, which has been completely established by a careful X-ray diffraction study.[55,141] An earlier report[306] of the preparation of $Na_2B_{10}H_{10}$ by heating of $Na_2B_{10}H_{14}$ to 190° merits a further investigation by X-ray powder diffraction photography in order to compare this material with a well-established $B_{10}H_{10}^{=}$ salt. Molecular orbitals in the polyhedron are delocalized,[182] and the excess of diamagnetism[68] over that for $K^+$, atomic B, and atomic H is $-40 \times 10^{-6}$ cm$^3$/mole, an exceedingly large value which is consistent with the great stability of this ion. In most of its reactions the bonding in the $B_{10}$ unit is essentially undisturbed.

**Four-Center Substitution Reactions of $B_{10}H_{10}^{=}$.** The positive ion may also be $H_3O^+$, and this hydrated acid form may be prepared by ion exchange. The proton catalyzes the reaction of nucleophiles, and gives a large variety of substituted $B_{10}H_9X^{=}$, $B_{10}H_9L^-$, and more highly substituted derivatives. Among those reported[149] are $B_{10}Cl_{10}^{=}$, $B_{10}H_9Br^{=}$, $B_{10}H_6I_4^{=}$, $B_{10}I_{10}^{=}$, $B_{10}H_9COC_6H_5^{=}$, $B_{10}H_9OCOC_6H_5^{=}$, $B_{10}H_9OH^{=}$(?), $B_{10}Cl_9OH^{=}$, $B_{10}H_9OCHO^{=}$, $B_{10}H_9NMe_2H^-$, $B_{10}H_8(SMe)_2^{=}$, $B_{10}H_3Br_7^{=}$, $B_{10}H_8(SMe_2)_2$, and $B_{10}H_9SMe_2^-$. Substitution at the tetragonal-pyramidal apex is indicated where the $B^{11}$ NMR has been reported, in agreement with the molecular orbital prediction that this atom is the more negative of the two kinds of atoms in $B_{10}H_{10}^{=}$.

**Oxidation of $B_{10}H_{10}^{=}$.** Either $Fe^{3+}$ or $Ce^{4+}$ produces[141] $B_{20}H_{18}^{=}$ by taking up the electrons in the reaction $2B_{10}H_{10}^{=} = B_{20}H_{18}^{=} + 2H^+$, in aqueous solu-

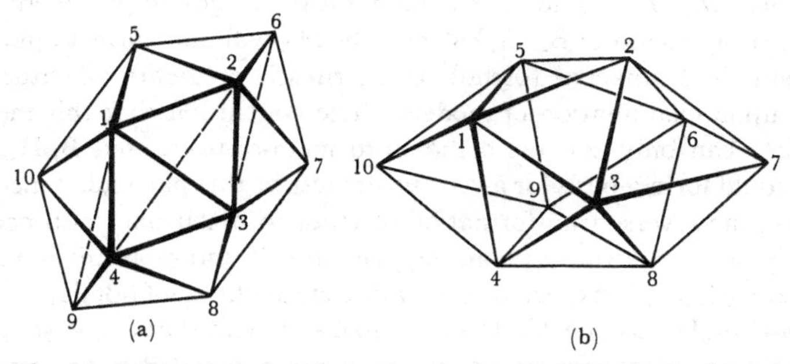


**Figure 5–21** *Collapse of the $B_{10}$ framework of $B_{10}H_{12}L_2$ (a) when $B_{10}H_{10}^{=}$ (b) is formed. New contacts are formed between atoms 6 and 9, between 6 and 8, and between 5 and 9 as indicated by the dashed lines in (a). Either the 7,10 pair or the 5,8 pair of boron atoms become apices of the $B_{10}H_{10}^{=}$ polyhedron.*

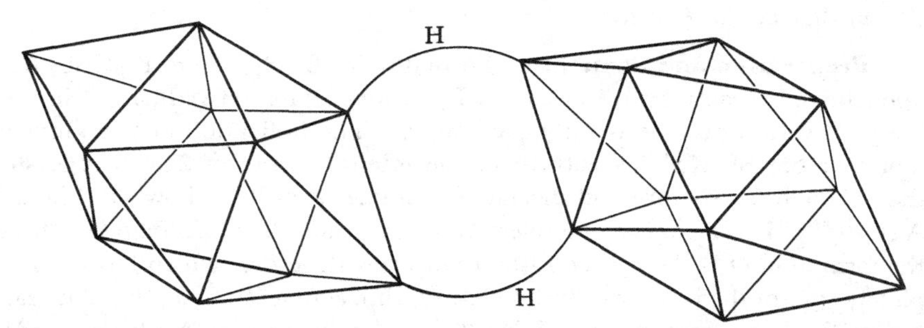


**Figure 5-22** *Proposed structure for $B_{20}H_{18}^{=}$. The $B^{11}$ NMR spectra suggest that an apex B and an equatorial B are involved in each polyhedron.*

tion. It is known[141] that the $B_{20}H_{18}^{=}$ ion is centrosymmetric, and the NMR spectrum[141] suggests that one apex and one equatorial position of each $B_{10}$ unit are involved in the formation of $B_{20}H_{18}^{=}$. The structure most nearly consistent with the known valence principles[179] is shown in Fig. 5-22.

**Polyhedral Rearrangement of $B_{10}H_9OH^{=}$(?).** This structure is also consistent with the $B_{10}H_9OH^{=}$(?) obtained[141] in essentially 100 per cent yield when $B_{20}H_{18}^{=}$ is treated with $OH^-$ in aqueous solution. The $B_{10}H_9OH^{=}$(?) is initially apex-substituted, as indicated by the modification of the small, low-field $B^{11}$ resonance of $B_{10}H_{10}^{=}$, but after a time the very nicely symmetrical, large, high-field doublet becomes quite unsymmetrical. A mechanism for this reaction has been suggested[141] so that the B—OH bond is not broken, but in which the B atoms change their coordination by slight movement of 0.5 A or less (Fig. 5-23). It must be emphasized that this mechanism has not been tested, although it could be tested by D or Me substitution. Also it has not yet been established that this $B_{10}H_9OH^{=}$(?) is the same as that reported by a very different method of preparation.[149] However, the lesser exchange repulsions and consequent delocalization of bonding makes rearrangements of this kind likely.

*$B_{12}H_{10}X_2^{=}$ and $C_2B_{10}H_{12}$.* The cube-octahedral intermediate proposed[141,124] as possible in rearrangements of $B_{12}H_{10}X_2^{=}$ or $C_2B_{10}H_{12}$ will not convert opposed (para) substituents into adjacent (ortho) or intermediate (meta) substituents, as may be seen upon examination of models. The conclusion that this mechanism (Fig. 5-24) can only convert ortho- into meta-disubstituted $B_{12}H_{12}^{=}$ is surprising, and could form a basis for a very severe test of this particular mechanism. Of course, the reverse transformation of meta to ortho may also occur, but it has already been argued from bonding principles[120] and molecular-orbital energies[121] that *m*-$C_2B_{10}H_{12}$ is expected to be more stable than *o*-$C_2B_{10}H_{12}$. The stability of the *p*-$C_2B_{10}H_{12}$ is expected to be comparable with that of *m*-$C_2B_{10}H_{12}$. As negative charge of $-1$ charge or, better, $-2$ charge is added to these three isomers of $C_2B_{10}H_{12}$, the meta and para are expected to become comparable in stability to the ortho ion. While a reversal in the ortho↔meta transforma-

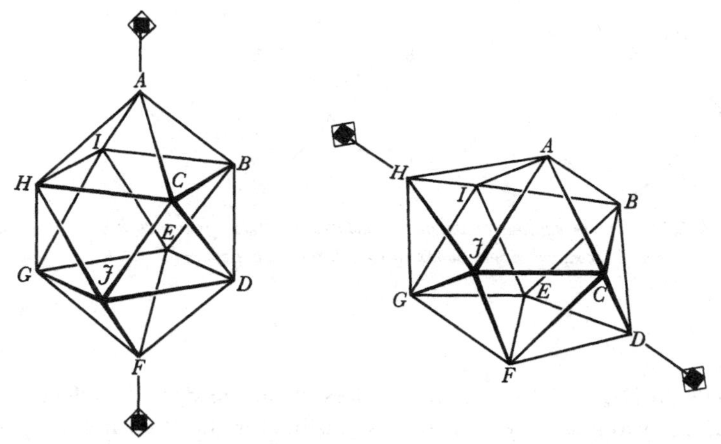


**Figure 5–23** *A plausible explanation of rearrangement of apex $B_{10}H_9X^=$ to equatorial $B_{10}H_9X^=$, in which the B—X bond remains unbroken and in which no atom moves more than 0.5 A. A four-coordinated apex atom may move toward five-coordination in order to start the rearrangement.*

tion is not predicted for neutral $C_2B_{10}H_{12}$ isomers, the equilibrium situation may become accessible in the corresponding negative ions. Labeling of three sites, e.g., in $C_2B_{10}H_{11}X$, gives other possibilities of ortho to meta transformations, including the possibility of racemization of an optical isomer, provided that the driving force is present.

*$B_{10}H_{13}^- + B_2H_6$.* One more rearrangement is of interest here. When $B_{10}H_{13}^-$ is exposed to labeled $B_2{}^{11}H_6$, it is found[264] that all B atoms of the $B_{10}H_{13}^-$ are rearranged. A reasonable intermediate is a 3814 $B_{12}H_{19}^-$ (related to a

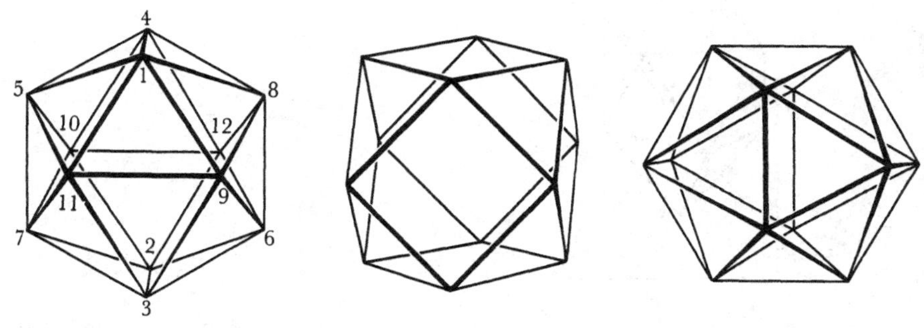


**Figure 5–24** *Rearrangement of the icosahedral structure through cube-octahedral intermediate to another icosahedral structure. A para-disubstituted compound will always remain para, but ortho-to-meta and meta-to-ortho transformations may take place.*

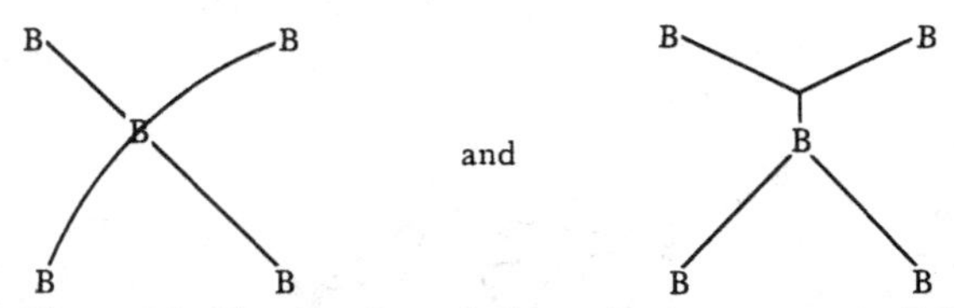


**Figure 5-25** *Nearly equivalent valence structures in $B_5H_9$ framework orbitals, showing possible positions of localized single bonds in complexes with electrophiles.*

plausible 4804 $B_{12}H_{20}$ of $C_4$ symmetry by loss of one bridge $H^+$. Rearrangements within $B_3$ triangles may then be responsible for the B scrambling, but a more highly coordinated inversion may also occur through a $B_{12}$ arrangement consisting of a cube with four more B atoms in a plane over four of the cube faces.

**Covalent bonds to $B_{10}H_{10}^{=}$.** The $Cu_2B_{10}H_{10}$ salt, insoluble in water and soluble in acetonitrile, has Cu—B bonds in the range[55,141] 2.0 to 2.3 A. These bonds are closest to the B—B bonds toward the apices of the tetragonal pyramidal structure in the crystal (Fig. 5-23). Each $B_{10}H_{10}^{=}$ is bonded to four Cu and each Cu to two $B_{10}H_{10}^{=}$ ions. It is clear that sufficient stability has been achieved in the boron framework that covalent Cu—B bonds are formed. In fact, the bonding at the two apices suggests that the $Cu^I$ makes a three-center bond with the edge B—B bond, and that the bonding arrangements (Fig. 5-25)

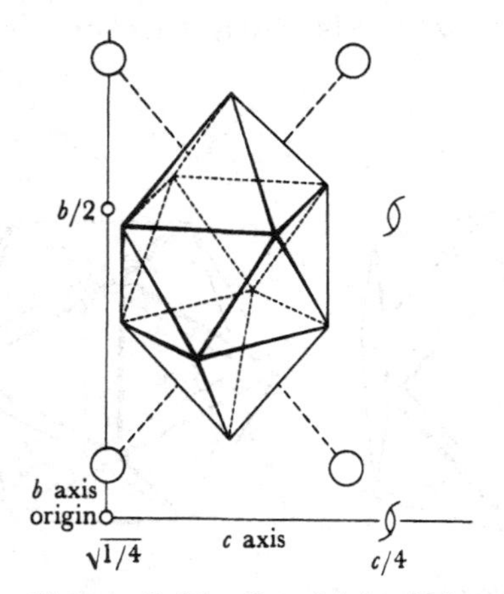


**Figure 5-26** *One-eighth of the unit cell of $Cu_2B_{10}H_{10}$, showing Cu-polyhedral-edge interactions as dashed lines. Relations of atomic positions to centers of symmetry (small circles) and twofold screw axes of Pcab are indicated.*

are equally good choices. The bonding in the crystal indicates that one apex uses one type and that the other apex uses the other type of covalent bonding (Cu—B = 2.14 to 2.33 A). The nature of the soluble complex formed when $Cu_2B_{10}H_{10}$ dissolves in acetonitrile is unknown. Covalent complexes similar to these may be formed in strongly ($H^+$) acidic solutions of $B_{10}H_{10}^=$. Indeed, the strong boron hydride–like odor which arises from these acidic solutions may indicate volatile complexes or decomposition products worthy of further study.

Presumably, covalent bonding also occurs in the ether-soluble free radical[141] having two Cu per $B_{10}H_{10}^=$ formed from the reaction of $CuCl_2$ with $K_2B_{10}H_{10}$. A clear electron spin resonance signal at $g = 2.018$ indicates either that the free electron is mostly in the boron framework or that the orbital contribution of $Cu^{II}$ is very strongly quenched by the covalent bonding to B. The present guess, that the radical is $Cu^{II}B_{10}H_{10}Cu^{I}Cl$, needs further study before acceptance.

**Oxazahydroborate Ions.** Alkylammonium salts of three ions, assigned formulas $B_{14}H_{12}NO^{-2}$, $B_{10}H_6(NO)_2^-$, and $B_{10}H_4(NO)_3^-$ or some multiple, with the numbers of H atoms uncertain, have been prepared[326] by reaction of $NO_2$ with aqueous $B_{10}H_{10}^=$ salt solutions. Of these, the $B_{14}H_{12}NO^{-2}$ ion is in sharp disagreement with the valence theory, considering the relatively mild conditions of preparation, and in fact has since been shown[141] to be $B_{20}H_xNO^{3-}$ where the value of $x = 18$ has been assigned on the assumption that two $B_{10}H_9^=$ units

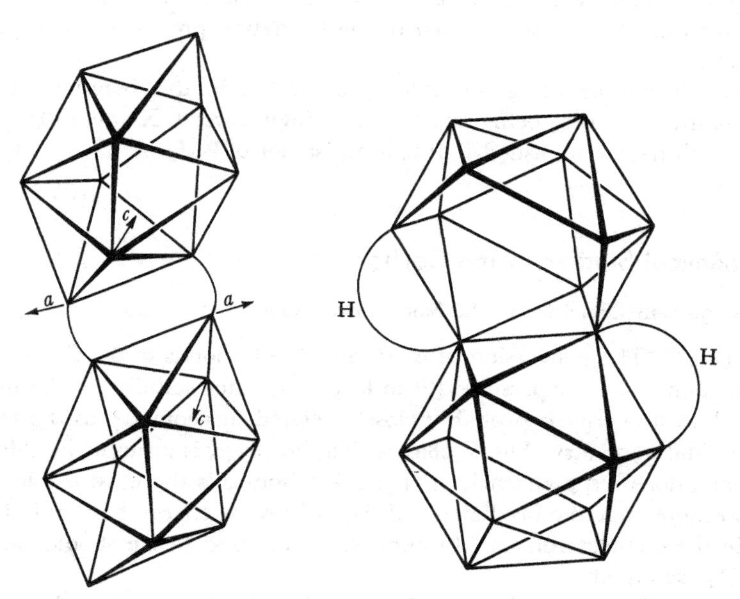


**Figure 5–27** *Least-motion hypothesis for the $B_{20}H_{18}^=$ to $B_{18}H_{22}$ transformation. The attack of O ($H_3^+O$) occurs probably near an apical region of the $B_{10}$ unit, possibly withdrawing the B atoms in the region labeled a, with resultant collapse of the $B_{20}$ unit to a $B_{18}$ unit.*

are joined by a single B—B bond to give $B_{20}H_{18}^{4-}$, to which a proton is added on a single bond, and on which one H is replaced by NO. On the same basis, the $B_{10}H_6(NO)_2^-$ ion is possibly $B_{20}H_{16}(NO)_4^{=}$, and the $B_{10}H_4(NO)_3^-$ ion is possibly $B_{20}H_{14}(NO)_6^{=}$. It needs hardly to be emphasized that further studies of these structures are desirable.

### $B_{18}H_{22}$

This compound has been prepared[234] by partial hydrolysis of a diethyl ether solution of the acid of $B_{20}H_{18}^{=}$. The X-ray molecular weight[234,291] of 216 ± 1 places the boron number at 18 but leaves the hydrogen number at either 20 or 22. However, the location of all H atoms in the X-ray study[291] removes this doubt. The molecular structure of this centrosymmetric hydride is shown in Chapter 1. The only chemical reaction so far reported is its behavior as a strong monoprotic acid, thus suggestive of the $B_{18}H_{21}^-$ ion as a stable species in aqueous solution. Indeed, the outer regions of this molecule resemble $B_{10}H_{14}$ strongly in geometrical structure, valence structures and charge distribution. Hence, one may easily conclude that a bridge $H^+$ is lost, and that rearrangement of the neighboring bridge H to a $BH_2$ group may occur, as seems probable in the case of $B_{10}H_{14}$. The 5,6 B atoms, which are relatively inaccessible sterically, are the most positive, and hence electron-pair donors would have to interact with other atoms. For this reason the molecule is very stable toward most Lewis bases. Part of the formation process is conjectured in Fig. 5-27.

Another crystalline boron hydride, also formed in this reaction but in smaller quantity, is also being studied by single-crystal X-ray diffraction methods, which have established that it is an isomer of $B_{18}H_{22}$.

## 5-15 Chemical Studies of the Carboranes

### Rearrangement Chemistry of the Small Carboranes[124]

*$B_5H_5^{=}(D_{3h})$.* The conversion of B—B and C—C bonds to B—C bonds in an isomerization process is presumed[179] to lead to greater stability. The intermediate of lowest energy is probably closely related in geometry to the tetragonal pyramidal geometry (Fig. 5-28a) which, however, is unstable in $C_2B_3H_5$. LCAO calculations[121] show clearly that the 4-5 isomer is the most stable. In addition we note that the predicted $C_2B_5H_5^{=}$ of lowest energy has orbital degeneracy in the one-electron LCAO approximation, and hence should not be stable in $D_{3h}$ symmetry.

*$B_6H_6^{=}(O_h)$.* The trans form of $C_2B_4H_6$, like the trans form of $C_2B_3H_5$, arises from reactions of boron hydrides with acetylene.[282] However, the cis form is also present, which is presumably the initial product inasmuch as the

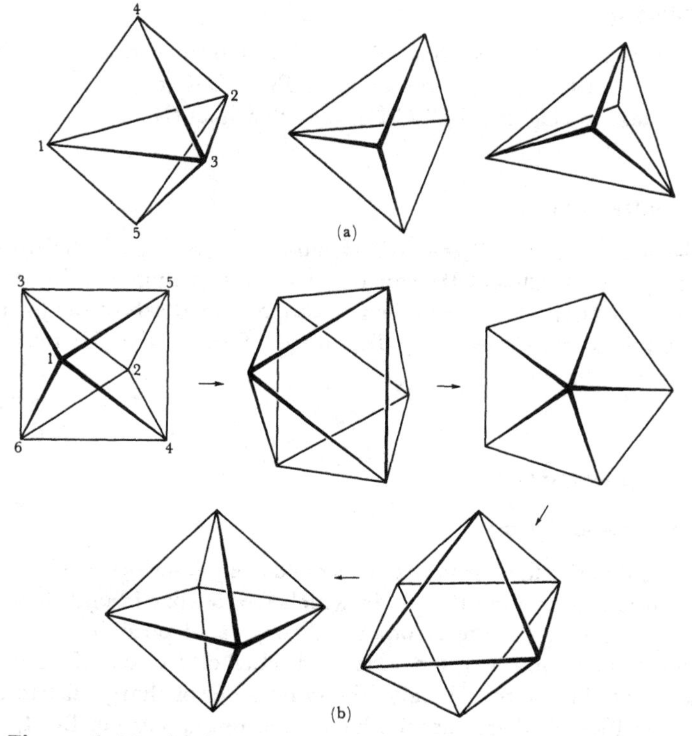


**Figure 5-28** *(a) Rearrangement of trigonal bipyramidal structure so that atoms 2 and 3 become nonadjacent. (b) Rearrangement of octahedral structure so that atoms 4 and 5 become nonadjacent.*

C—C proximity is preserved. The path of transformation of *cis*-$C_2B_4H_6$ to *trans*-$C_2B_4H_6$ may occur through the pentagonal pyramidal intermediate (Fig. 5-28b), but it is not yet possible to rule out the trigonal prism of $D_{3h}$ symmetry as the intermediate. Here, however, the 5-6 isomer is predicted on theoretical grounds[121] to be only somewhat more stable than the 4-5 isomer. As in the $C_2B_3H_5^=$ example, we find here orbital degeneracy, and implied instability for $C_2B_4H_6^=$ in $D_{4h}$ symmetry.

### Rearrangement Chemistry of $B_{10}C_2H_{12}$

So far only the theoretical suggestion of rearrangement of the boron carbon skeleton has been suggested, and it is hoped that experimental studies will be forthcoming.

### Boron Chemistry of $B_{10}C_2H_{12}$

The ortho $B_{10}C_2H_{12}$ (two carbons adjacent) has been shown[277] to react with $Cl_2$ to form $B_{10}H_8Cl_2C_2H_2$ (two isomers), $B_{10}H_7Cl_3C_2H_2$, $B_{10}H_6Cl_4C_2H_2$ (two isomers), $B_{10}H_4Cl_6C_2H_2$, $B_{10}H_2Cl_8C_2H_2$, and $B_{10}Cl_{10}C_2H_2$.

### Carbon Chemistry of $B_{10}C_2H_{12}$

In ethanol-water solutions, $B_{10}H_2Cl_8C_2H_2$ and $B_{10}Cl_{10}C_2H_2$ can be titrated as a diprotic acid, and $B_{10}Cl_{10}C_2ClH$ can be titrated as a monoprotic acid.[277] The corresponding $HNEt_3^+$ salts are known. Also, a second salt of the formula $B_{10}Cl_{10}C_2HNEt_3Cl$ is known. A large literature of derivative chemistry is promised.

## 5–16 Concluding Remarks

### Structure and Valence Theory

Three valence theories for boron hydrides have been described here, (1) the two-center and three-center bond theory, in which the localized bonds have the symmetry of the molecule; (2) the resonance theory of all possible two-center and three-center bonds; and (3) the extended Hückel LCAO-MO theory. The second and third of these theories give the same order of charge distribution among B atoms, and where these theories have been tested, e.g., in $B_{10}H_{14}$, the results are in agreement with electrophilic and nucleophilic substitution processes. No doubt the third theory is the best of the three, but it tends to make each molecule a special case and it is difficult to apply to molecules as large as $B_{18}H_{22}$ or iso-$B_{18}H_{22}$, where the resonance theory has proved applicable. The first theory is most successful in the simpler open hydrides, $B_2H_6$, $B_4H_{10}$, $B_5H_{11}$, $B_3H_8^-$, etc., especially if one requires only a closed-shell description and not the details of a charge distribution. Probably the only real value of the first theory in the more complex hydrides is that it provides a simple language which can be used for writing formulas, tautomers, supposed mechanisms, and reactions, but its value here should not be regarded as greater than, for example, one of the Kekulé structures for a large aromatic hydrocarbon.

A most important aspect of boron hydrides, ions, and related derivatives that has not yet been taken into account is the detailed stereochemistry in conjunction with one or another of the valence theories. After one forms an external B—H, B—X (halogen), or B—L (ligand) bond, the remaining atoms of the boron compounds based upon a single polyhedron or polyhedral fragment fall onto a spherical surface. The electron deficiency of bonding reduces exchange repulsions to the point that within this surface there are essentially only direct attractive valence forces, with negligible bending force constants. For example, the use of a Urey-Bradley force field in the bridge region of $B_2H_6$

(E. Wu and B. L. Crawford, Jr., private communication, 1962) yields no bending-force constants, and all stretching-force constants are positive along all lines between the two B atoms and the two bridge H atoms. The tendency for these atoms to flow over the surface of the sphere leads to rearrangements of the polyhedral types of molecules as described earlier in the text. But there are equilibrium interatomic distances, and sharply rising repulsive forces as these distances become small. Hence one must impose upon any valence theory such requirements as no H · · · H distances within the molecule which are less than about 2.0 A, and no nonbonded B · · · H distances smaller than about 2.4 A. Neither must there be open regions where electron pair donors can easily attack B atoms. These requirements have yet to be introduced into the detailed valence theory. Indeed, the structural information which indicates these principles and which makes clear the preservation of geometrical features has only recently become available. A recent unpublished advance is the development by R. Hoffmann of a computer program for the three-center-bond logic, but the geometrical restrictions have yet to be introduced, and hence the predictions of such a theory are, at present, multitudinous as one goes to the $B_7$ and $B_8$ hydrides.

A final remark on the theory is that a reexamination of the idealized hydrides is probably worth the effort if the theory is modified to place the bridge H atoms on the surface of the sphere determined by the B atoms, rather than out from this sphere above the B · · · B line. Thus further developments of the extended Hückel LCAO-MO theory may be possible in predictions of stable species when large energy gaps exist between filled and lowest unfilled energy levels, and in the examination of intermediate geometries of molecules undergoing transformation or reaction. Again, these results will best be obtained by use of large computers.

### Recent Experimental Results

A brief summary now follows in which recent results are summarized in an order which refers to the number of boron atoms in that molecule which is the major point of the study.

Valence states of atomic B, C, N, and O have recently been studied [G. Pilcher and H. A. Skinner, *Inorg. Chem.*, **24**, 937 (1963)].

The charge distribution in the B—N bond in many compounds has been studied in the extended Hückel LCAO-MO approximation (R. Hoffmann, International Symposium, Boron-Nitrogen Chemistry, Durham, N.C., April 23–25, 1963). The remarkable conclusion, that B is slightly positive and N slightly negative, is contrary to the usual conventional description. However, the reverse polarity $B^{\delta-}$—$N^{\delta+}$ applies to the $\pi$ system when B and N are $sp^2$-hybridized. A most interesting analogue of a cyclobutadiene derivative has been prepared [M. F. Lappert and M. K. Majumdar, *Proc. Chem. Soc.*, **88** (1963)] in which there are alternating B and N atoms in a four-membered ring. The compound is 1,3 diaza-2,4 boretane. Of course the relatively large Coulomb integral on N removes from the $B_2N_2$ ring the degeneracy present in the one-electron approximation to the valence theory of planar $C_4H_4$.

The behavior of $B_3H_8^-$ in various solvents (R. Schaeffer, International Symposium, Boron-Nitrogen Chemistry, Durham, N.C., April 23–25, 1963) is remarkable in that various electron-pair donors appear to open the triangle of B atoms by addition to B adjacent to a bridge H atom. These results were obtained from the study of the $B^{11}$ NMR spectra.

An addition compound between $B_4Cl_4$ and $4N(C_2H_5)_3$ has been prepared, and suggested to be a four-membered $B_4$ ring (T. Wartik, private communication, 1963).

The preparation of $B_5H_9$ (W. V. Hough, private communication, 1963) by storage at 500 psi at room temperature of $B_2H_6$ with a trace of $N(CH_3)_3$ for 5 hours produces little $B_4H_{10}$, and is a significant advance over pyrolytic or discharge methods. In fact, the use of Lewis bases for transformations and preparations of boron hydrides and their derivatives merits much more extensive study. A recent example of the probable function of a Lewis base in a rearrangement is that $B_5H_8R$ rearranges as $B_5H_7R^-$ after removal of a proton by the Lewis base L [W. V. Hough, L. J. Edwards, and A. F. Stang, *J. Am. Chem. Soc.*, **85,** 831 (1963)].

The preparation of $B_9H_{14}^-$ has been reported (W. V. Hough, International Symposium, Boron-Nitrogen Chemistry, Durham, N.C., April 23–25, 1963) according to the equation

$$N_2R_4 + B_{10}H_{14} \xrightarrow{ROH} N_2R_4H^+ \cdot B_9H_{14}^- + H_2$$

This ion, expected by analogy to the more stable $B_9H_{13}L$ compounds, loses hydrogen slowly at room temperature.

The possibility that $B_5H_9$ is an intermediate in the pyrolytic preparation of $B_{10}H_{14}$ has received further study. Specifically the $B_5H_9$—$B_2H_6$ reaction was subjected to further tracer studies [M. Hillman, D. J. Mangold, and J. H. Norman, *J. Inorg. Nucl. Chem.*, **24,** 1565 (1962)]. There is a possibility that a $B_{10}H_{16}$-like intermediate is formed. An extended Hückel LCAO-MO study of the conformation of $B_{10}H_{16}$ (R. Hoffmann, private communication, 1963) suggests a small (1.0 kcal/mole) barrier in favor of the staggered conformation, but in the solid the conformation is certainly eclipsed because the molecule is centrosymmetric.

A proposed revision of the numbering system for polyhedral species (A.C.S. Nomenclature Committee, 1963) is exemplified for $B_{10}H_{10}^{=}$ (Fig. 1–15). Apical B atoms on the $S_8$ axis are 1 and 10. The four B's bonded to 1 are 2, 3, 4, and 5, and the four B's bonded to 10 are 6, 7, 8, and 9, both in clockwise order when viewed along the 1 · · · 10 axis. Atom 6 is bonded to atoms 2, 3, 7, 9, and 10. The C atoms in *o*-$B_{10}C_2H_{12}$ are to be numbered 1 and 2!

The long paper on $B_{12}Cl_{11}$ has appeared [E. P. Schram and G. Urry, *Inorg. Chem.*, **2,** 405 (1963)]. It may be suggested that the $B_{10}H_8$ unit has two extra B's attached on adjacent B atoms of the $B_{10}$ polyhedron and that each B has one Cl attached by a single bond, and that the remaining Cl is a bridge between these two BCl units.

The preparation of $B_{12}H_{16}L_2$ according to the equation

$$B_{10}H_{12}L_2 + 2B_2H_6 + 2SMe_2 \rightarrow B_{12}H_{16}L_2 + H_2 + 2H_3BSMe_2$$

is a most interesting development (W. V. Hough, private communication, 1963). A structure determination would be of importance here.

Several outstanding problems among the complex borides [e.g., $MB_{70}$, R. W. Johnson and A. H. Daane, *J. Chem. Phys.*, **38,** 425 (1963), also J. S. Kasper, private communication, 1960] may receive an interpretation as a result of the model-building solution of the structure of $\beta$-rhombohedral boron [R. E. Hughes, C. H. L. Kennard, D. B. Sullenger, H. A. Weakleim, D. E. Sands, and J. L. Hoard, *J. Am. Chem. Soc.*, **85,** 361 (1963)]. There are 91 atoms in coordination number CN6, 12 in CN8, and two in CN9 in this remarkable structure, which is, for the most part, based upon icosahedra. In addition, we note the possible occurrence of icosahedra in the crystal structure of $AlB_{10}$ [G. Will, Abstract 4.31, Sixth International Union of Crystallography, Rome Meeting, Sept. 9–19, 1963].

Private communications in mid-1963 from authors of various carborane papers have indicated the discovery in 1957, by M. M. Fein, J. Bobinski, N. Mayes, N. Schwartz, and M. S. Cohen, of the reaction of isopropenyl acetylene with $(CH_3CN)_2B_{10}H_{12}$ to form $CH_2C(CH_3)CCHB_{10}H_{10}$. The various papers so far received as preprints are listed below:

1. Synthesis of carboranes from dihydrocarboranes. T. Onak, F. J. Gerhart, and R. E. Williams. *J. Am. Chem. Soc.*

2. A new series of organoboranes: I. The chlorination of 11,12-dicarbadodecaborane, H. Schroeder, T. L. Heying, and J. R. Reiner. *Inorg. Chem.*

3. A new series of organoboranes: II. Carboranes from the reaction of decaborane with acetylenic compounds. T. L. Heying, J. W. Ager, Jr., S. L. Clark, D. J. Mangold, H. L. Goldstein, M. Hillman, R. J. Polak, and J. W. Szmanski. *Inorg. Chem.*

4. A new series of organoboranes: III. Some reactions of 11,12-dicarbadodecaborane(12) (carborane) and its derivatives. T. L. Heying, J. W. Ager, Jr., S. L. Clark, R. P. Alexander, S. Papetti, J. A. Reid, and S. I. Trotz. *Inorg. Chem.*

5. A new series of organoboranes: IV. The participation of the 11,12-dicarbadodecaborane nucleus in some novel heteratomic ring systems. S. Papetti and T. L. Heying. *Inorg. Chem.*

6. Chemistry of decaborane-phosphorous compounds: IV. Monomeric, oligomeric, and cyclic phosphino-carboranes. R. P. Alexander and H. Schroeder. *Inorg. Chem.*

7. Structures for the $C_2H_{12}B_{10}$ isomers. D. Grafstein. *Inorg. Chem.*

8. Neodecacarboranes, a new family of stable cyclocarboranes isomeric with the decacarboranes. D. Grafstein and J. Dvorak. *Inorg. Chem.*

Further studies of the reaction of $B_{10}H_{10}^{=}$ ion have been made by M. F. Hawthorne (private communication). These reactions are summarized in the following diagrams:

$$B_{10}H_{10}^{=} \xrightarrow{Fe^{3+}} B_{20}H_{18}^{=} \xrightarrow{2OH^-} H_2O + B_{20}H_{17}OH^{4-}\ (2\ \text{isomers}) \underset{-H^+}{\overset{H^+}{\rightleftharpoons}} B_{20}H_{18}OH^{3-}\ (2\ \text{isomers}) \xrightarrow{[O]} B_{20}H_{18}OH^{-}$$

$$B_{10}H_{10}^{=} \xrightarrow{Ce^{4+}} B_{20}H_{18}^{=}\cdot B_{20}H_{19}^{3-}\cdot 5Et_3NH^+ \xrightarrow{Me_4N^+} B_{20}H_{19}^{3-} \underset{H^+}{\overset{OH^-}{\rightleftharpoons}} B_{20}H_{18}^{4-} \xrightarrow{[O]} B_{20}H_{18}^{=}$$

Thus the compound labeled in the text $B_{10}H_9OH^{=}$(?) is most probably $B_{20}H_{17}OH^{4-}$.

A new boron hydride of the formula $B_6H_{12}$ has been prepared from $B_3H_8^-$ in acidic solution by R. Schaeffer (private communication) and his students. The NMR spectra suggests that it may have the structure shown in Fig. 2-20b, page 57.

Finally, R. Schaeffer and co-workers have communicated a novel preparation of $B_{12}H_{12}^{=}$ from the reaction of $B_2H_6$ with $BH_4^-$ in diglyme under reflux conditions for periods up to 1 day.

# References


1. V. D. Aftandilian, H. C. Miller, and E. L. Muetterties, *J. Am. Chem. Soc.*, **83,** 2471 (1961). Chemistry of boranes. I. Reactions of boron hydrides with metal and amine salts.
2. V. D. Aftandilian, H. C. Miller, G. W. Parshall, and E. L. Muetterties, *Inorg. Chem.*, **1,** 734 (1962). Chemistry of boranes. V. First example of a $B_{11}$ hydride, the $B_{10}H_{14}^-$ anion.
3. M. Atoji and W. N. Lipscomb, *Acta Cryst.*, **6,** 547 (1953). The crystal and molecular structure of $B_4Cl_4$.
4. M. Atoji and W. N. Lipscomb, *J. Chem. Phys.*, **21,** 172 (1953). The molecular structure of $B_4Cl_4$.
5. M. Atoji, P. J. Wheatley, and W. N. Lipscomb, *J. Chem. Phys.*, **23,** 1176 (1955). Molecular structure of diboron tetrachloride.
6. M. Atoji, P. J. Wheatley, and W. N. Lipscomb, *J. Chem. Phys.*, **27,** 196 (1957). The crystal and molecular structure of diboron tetrachloride.
7. E. B. Baker, R. B. Ellis, and W. S. Wilcox, *J. Inorg. Nucl. Chem.*, **23,** 41 (1961). Sodium borohydride-borane complex.
8. S. H. Bauer, *J. Am. Chem. Soc.*, **59,** 1096 (1937). The structure of diborane.
9. S. H. Bauer, *J. Am. Chem. Soc.*, **60,** 805 (1938). The structure of the hydrides of boron. V. Tetraborane $B_4H_{10}$ and the pentaborane $B_5H_{11}$.
10. S. H. Bauer, *Chem. Rev.*, **31,** 46 (1942). Structures and physical properties of the hydrides of boron and of their derivatives.
11. S. H. Bauer, *J. Am. Chem. Soc.*, **72,** 622 (1950). Reanalysis of the electron diffraction data on $Be(BH_4)_2$ and $Al(BH_4)_3$.
12. S. H. Bauer and L. Pauling, *J. Am. Chem. Soc.*, **58,** 2403 (1936). The structure of pentaborane $B_5H_9$.

13. W. H. Bauer and S. E. Wiberley, *Abstracts, 133rd Meeting Am. Chem. Soc.*, San Francisco, April 1958, p. 13L. Explosive oxidation of the boranes.
14. S. H. Bauer, A. Shepp, and R. E. McCoy, *J. Am. Chem. Soc.*, **75,** 1003 (1953). Thermodynamic and kinetic constants for the diborane-borine equilibrium.
15. J. Y. Beach and S. H. Bauer, *J. Am. Chem. Soc.*, **62,** 3440 (1940). The structure of the hydrides of boron. V. $AB_3H_{12}$.
16. H. C. Beachell and D. E. Hoffmann, *J. Am. Chem. Soc.*, **84,** 180 (1962). The reaction of decaborane with amines and related compounds.
17. R. P. Bell and H. C. Longuet-Higgins, *Proc. Roy. Soc. (London)*, **A183,** 357 (1945). The normal vibrations of bridged $X_2Y_6$ molecules.
18. F. Bertaut and P. Blum, *Compt. Rend.*, **229,** 666 (1949). La Structure des borures d'unanium.
19. N. J. Blay, I. Dunstan, and R. L. Williams, *J. Chem. Soc.*, **1960,** 430. Boron hydride derivatives. III. Electrophilic substitution in pentaborane and decaborane.
20. N. J. Blay, J. Williams, and R. L. Williams, *J. Chem. Soc.*, **1960,** 424. Boron hydride derivatives. II. The separation and identification of some ethylated pentaboranes and decaboranes.
21. A. L. Bloom and J. N. Schoolery, *Phys. Rev.*, **97,** 1261 (1955). Effects of perturbing radiofrequency fields on nuclear spin coupling.
22. J. L. Boone and A. B. Burg, *J. Am. Chem. Soc.*, **80,** 1519 (1958). Interconversion of volatile boranes by basic reagents.
23. J. L. Boone and A. B. Burg, *J. Am. Chem. Soc.*, **81,** 1766 (1959). Synthesis of hexaborane.
24. *Borax to Boranes*, Advances in Chemistry Series, American Chemical Society, Washington, D. C., 1961.
    a. R. M. Adams, p. 60. Preparation of diborane.
    b. W. S. Koski, p. 78. Mechanisms of isotopic exchanges in the boron hydrides.
    c. I. Shapiro, C. O. Wilson, J. F. Ditter, and W. J. Lehman, p. 127. Mass spectrometry in boron chemistry.
    d. M. Hillman, D. J. Mangold, and J. H. Norman, p. 151. Interaction of boranes. A tracer study of the diborane-pentaborane(9) reaction.
    e. W. V. Hough and L. J. Edwards, p. 184. Metal boron hydrides.
    f. S. H. Bauer, p. 88. Kinetics and mechanism for acid-base reactions involving boranes.
    g. J. P. Faust, p. 69. Tetraborane, a review.
25. J. K. Bragg, L. V. McCarty, and F. J. Norton, *J. Am. Chem. Soc.*, **73,** 2134 (1951). Kinetics of pyrolysis of diborane.
26. G. L. Brennan and R. Schaeffer, *J. Inorg. Nucl. Chem.*, **20,** 205 (1961). Interconversion of boranes. IV. A kinetic study of the decomposition of tetraborane-10.
27. H. C. Brown, E. J. Mead, and P. A. Tierney, *J. Am. Chem. Soc.*, **79,** 5400 (1957). The reaction of sodium hydride with methyl borate in solvents. Convenient new procedures for the synthesis of sodium borohydride.
28. H. C. Brown, P. F. Stehle, and P. A. Tierney, *J. Am. Chem. Soc.*, **79,** 2020 (1957). Singly-bridged compounds of the boron halides and boron hydrides.
29. A. B. Burg, *J. Am. Chem. Soc.*, **79,** 2129 (1957). Amine chemistry of the pentaborane $B_5H_9$.
30. A. B. Burg, *Discussion, 133rd Meeting Am. Chem. Soc.*, San Francisco, April 1958, p. 31L. Introductory remarks.

31. A. B. Burg and R. Kratzer, *Inorg. Chem.*, **1,** 725 (1962). The synthesis of nonaborane, $B_9H_{15}$.
32. A. B. Burg and C. L. Randolph, Jr., *J. Am. Chem. Soc.*, **71,** 3451 (1949). The N-methyl derivatives of $B_2H_7N$.
33. A. B. Burg and H. I. Schlesinger, *J. Am. Chem. Soc.*, **55,** 4009 (1933). Hydrides of boron. II. The preparation of $B_5H_{11}$: Its thermal decomposition and reaction with hydrogen.
34. A. B. Burg and H. I. Schlesinger, *J. Am. Chem. Soc.*, **62,** 3425 (1940). Metallo borohydrides. II. Beryllium borohydride.
35. A. B. Burg and F. G. A. Stone, *J. Am. Chem. Soc.*, **75,** 228 (1953). The behavior of tetraborane toward trimethylamine and ethanol.
36. L. A. Burkhardt and N. R. Fetter, *Chem. Ind. (London)*, **1959,** 1191. Complexes of decaborane with pyridine and 2-bromo-pyridine.
37. G. W. Campbell, Jr., *Abstracts, 133rd Meeting Am. Chem. Soc.*, San Francisco, April 1958, p. 28L. Unusual salts derived from the boron hydrides.
38. *Chem. Eng. News*, **32,** 1441 (1954). Nomenclature.
39. *Chem. Eng. News*, **34,** 560 (1956). Nomenclature.
40. B. H. Chirgwin and C. A. Coulson, *Proc. Roy. Soc. (London)*, **A201,** 196 (1950). The electronic structure of conjugated systems. VI.
41. C. C. Clark, U.S. Patent 3,062,756, Nov. 6, 1962.
42. H. K. Clark and J. L. Hoard, *J. Am. Chem. Soc.*, **65,** 2115 (1943). The crystal structure of boron carbide.
43. R. P. Clarke and R. N. Pease, *J. Am. Chem. Soc.*, **73,** 2132 (1951). A preliminary study of the kinetics of pyrolysis of diborane.
44. W. L. Clinton and B. Rice, *J. Chem. Phys.*, **29,** 445 (1958). Electronic structure of $BH_3$.
45. C. D. Cornwell, *J. Chem. Phys.*, **18,** 1118 (1950). Microwave spectra of bromodiborane and vinyl bromide.
46. C. A. Coulson, in the V. Henri Memorial Volume, *Contribution a l'etude de la structure moleculaire*, Desoer, Liege, 1948. The atomic radius of carbon.
47. B. F. Decker and J. S. Kasper, *Acta Cryst.*, **12,** 503 (1959). The crystal structure of a simple rhombohedral form of boron.
48. P. G. Dickens and J. W. Linnett, *Quart. Rev. (London)*, **11,** 291 (1957). Electron correlation and chemical consequences.
49. R. E. Dickerson and W. N. Lipscomb, *J. Chem. Phys.*, **27,** 212 (1957). A semitopological approach to boron hydride structures.
50. R. E. Dickerson, P. J. Wheatley, P. A. Howell, and W. N. Lipscomb, *J. Chem. Phys.*, **27,** 200 (1957). The crystal and molecular structure of $B_9H_{15}$.
51. R. E. Dickerson, P. J. Wheatley, P. A. Howell, W. N. Lipscomb, and R. Schaeffer, *J. Chem. Phys.*, **25,** 606 (1956). The boron arrangement in a $B_9$ hydride.
52. W. Dilthey, *Z. Angew. Chem.*, **34,** 596 (1921). Über die Konstitution des Wassers.
53. J. F. Ditter and I. Shapiro, *Abstracts, 134th Meeting Am. Chem. Soc.*, Chicago, Sept. 1958, p. 22N. Identification of an intermediate compound in the partial oxidation of pentaborane-9.
54. J. F. Ditter and I. Shapiro, *J. Am. Chem. Soc.*, **81,** 1022 (1959). Identification of an intermediate compound in the partial oxidation of pentaborane-9.
55. R. D. Dobrott and W. N. Lipscomb, *J. Chem. Phys.*, **37,** 1779 (1962). Structure of $Cu_2B_{10}H_{10}$.

56. G. H. Duffey, *J. Chem. Phys.*, **19,** 963 (1951), ftnt. p. 964. A seemingly forbidden octacovalent structure.
57. G. H. Duffey, *J. Chem. Phys.*, **21,** 761 (1953). The structure of $B_4Cl_4$.
58. W. J. Dulmage and W. N. Lipscomb, *J. Am. Chem. Soc.*, **73,** 3539 (1951). The molecular structure of pentaborane.
59. W. J. Dulmage and W. N. Lipscomb, *Acta Cryst.*, **5,** 260 (1952). The crystal and molecular structure of pentaborane.
60. I. Dunstan and J. V. Griffiths, *J. Chem. Soc.*, **1962,** 1344. The relative rates of reaction of decaborane with some nitriles and sulphides.
61. I. Dunstan, N. J. Blay, and R. L. Williams, *J. Chem. Soc.*, **1960,** 5016. Boron hydride derivatives. VI. Decaborane Grignard reagents.
62. I. Dunstan, R. L. Williams, and N. J. Blay, *J. Chem. Soc.*, **1960,** 5012. Boron hydride derivatives. V. Nucleophilic substitution in decaborane.
63. J. A. Dupont and M. F. Hawthorne, *J. Am. Chem. Soc.*, **81,** 4998 (1959). Deuterium exchange of decaborane with deuterium chloride under electrophilic conditions.
64. J. A. Dupont and M. F. Hawthorne, *Abstracts, 138th Meeting Am. Chem. Soc.*, New York, Sept. 1960. A study of electrophilic and nucleophilic deuterium exchange between DC1 and $B_{10}H_{14}$ under kinetic conditions.
65. J. A. Dupont and M. F. Hawthorne, *Chem. Ind.* (*London*), **1962,** p. 405. The $B_{10}H_{15}^-$ anion.
66. J. A. Dupont and M. F. Hawthorne, *J. Am. Chem. Soc.*, **84,** 1804 (1962). The nature of the electrophilic deuterium exchange reaction of decaborane with deuterium chloride.
67. J. A. Dupont, M. F. Hawthorne, A. R. Pitochelli, and R. Ettinger, *J. Am. Chem. Soc.*, **84,** 1057 (1962). Observations on the mechanism of $B_{10}H_{10}^=$ formation.
68. G. Eaton, A. Kaczmarczyk, and W. N. Lipscomb, unpublished results, 1962.
69. W. H. Eberhardt, B. L. Crawford, Jr., and W. N. Lipscomb, *J. Chem. Phys.*, **22,** 989 (1954). The valence structure of the boron hydrides.
70. R. E. Enrione and R. Schaeffer, *J. Inorg. Nucl. Chem.*, **18,** 103 (1961). Interconversion of boranes. II. Dueterium isotope effect in the decomposition of diborane.
71. K. Eriks, W. N. Lipscomb, and R. Schaeffer, *J. Chem. Phys.*, **22,** 754 (1954). The boron arrangement in a $B_6$ hydride.
72. N. R. Fetter and L. A. Burkardt, *Abstracts, 135th Meeting Am. Chem. Soc.*, Boston, April 1959, p. 45M. Reaction of decaborane and pyridine.
73. B. Figgis and R. L. Williams, *Spectrochim. Acta*, **13,** 331 (1959). The structures of bromopentaborane and of ethylpentaborane.
74. S. J. Fitch and A. W. Laubengayer, *J. Am. Chem. Soc.*, **80,** 5911 (1958). The reaction of decaborane with dimethylamine.
75. P. T. Ford and R. E. Richards, *Discussions Faraday Soc.*, **19,** 230 (1955). Proton magnetic resonance spectra of crystalline borohydrides of sodium, potassium, and rubidium.
76. M. W. Forsyth, W. V. Hough, M. D. Ford, G. T. Hefferan, and L. J. Edwards, *Abstracts, 135th Meeting Am. Chem. Soc.*, Boston, April 1959, p. 40M. Reactions of Lewis bases with pentaborane-11.
77. J. Gallaghan and B. Siegel, *J. Am. Chem. Soc.*, **81,** 504 (1959). Grignard synthesis of alkyl decaboranes.
78. S. G. Gibbins, private communication, 1960.
79. S. G. Gibbins and I. Shapiro, *J. Am. Chem Soc.*, **82,** 2968 (1960). A new hexaborane.

80. B. M. Graybill and M. F. Hawthorne, *J. Am. Chem. Soc.*, **83,** 2673 (1961). The nature of the colored 6,9-bis-pyridine decaborane molecule, $B_{10}H_{12}Py_2$.
81. B. M. Graybill and J. K. Ruff, *J. Am. Chem. Soc.*, **84,** 1062 (1962). Cleavage of trimethylamine triborane.
82. B. M. Graybill, A. R. Pitochelli, and M. F. Hawthorne, *Inorg. Chem.*, **1,** 622 (1962). The preparation and reactions of $B_{10}H_{13}$ (ligand) anions.
83. B. M. Graybill, A. R. Pitochelli, and M. F. Hawthorne, *Inorg. Chem.*, **1,** 626 (1962). The preparation and reactions of $B_9H_{13}$ (ligand) compounds.
84. B. M. Graybill, J. K. Ruff, and M. F. Hawthorne, *J. Am. Chem. Soc.*, **83,** 2669 (1961). A novel synthesis of the triborohydride anion, $B_3H_8^-$.
85. R. N. Grimes and W. N. Lipscomb, *Proc. Natl. Acad. Sci. U.S.*, **48,** 496 (1962). Decaborane (16): its rearrangement to decaborane (14) and cleavage.
86. R. N. Grimes and W. N. Lipscomb, unpublished work. Also see R. N. Grimes, Ph. D. thesis, University of Minnesota, 1962.
87. R. N. Grimes, F. E. Wang, R. Lewin, and W. N. Lipscomb, *Proc. Natl. Acad. Sci. U.S.*, **47,** 996 (1961). A new type of boron hydride, $B_{10}H_{16}$.
88. S. R. Gunn and L. G. Green, *J. Phys. Chem.*, **65,** 2173 (1961). The heats of decomposition of some higher boron hydrides.
89. G. A. Guter and G. W. Schaeffer, *J. Am. Chem. Soc.*, **78,** 3546 (1956). The strong acid behavior of decaborane.
90. G. A. Guter and G. W. Schaeffer, *Abstracts, 131st Meeting Am. Chem. Soc.*, Miami, 1957, p. 3R. Ionization of decaborane in mixed solvents.
91. A. Haaland, Ph.D. thesis, Georgia Institute of Technology, August, 1961.
92. A. Haaland and W. H. Eberhardt, *J. Chem. Phys.*, **36,** 2386 (1962). Electronic spectrum of decaborane.
93. L. H. Hall and W. S. Koski, *J. Am. Chem. Soc.*, **84,** 4205 (1962). On the nature of some higher boron hydrides produced in radiation-induced reactions in pentaborane and decaborane.
94. F. Halla and R. Weil, *Z. Krist.*, **101,** 435 (1939). Röntgenographische Untersuchung von "kristallisiertem Bor."
95. W. C. Hamilton, *Proc. Roy. Soc. (London)*, **A235,** 395 (1956). A molecular orbital treatment of diborane as a four-center, four-electron problem.
96. B. C. Harrison, I. J. Solomon, R. D. Hites, and M. J. Klein, *Abstracts, 135th Meeting Am. Chem. Soc.*, Boston, April 1959, p. 38M. Reaction of ethylene with tetraborane. Preparation, structure, and properties of dimethylenetetraborane.
97. B. C. Harrison, I. J. Solomon, R. D. Hites, and M. J. Klein, *J. Inorg. Nucl. Chem.*, **14,** 195 (1960). The reaction of ethylene with tetraborane: The preparation, structure, and properties of dimethylenetetraborane.
98. M. F. Hawthorne, *J. Am. Chem. Soc.*, **80,** 3480 (1958). The reaction of phosphine methylenes with boron hydrides.
99. M. F. Hawthorne and J. J. Miller, *J. Am. Chem. Soc.*, **80,** 754 (1958). Deuterium exchange of decaborane with deuterium oxide and dueterium chloride.
100. M. F. Hawthorne and A. R. Pitochelli, *J. Am. Chem. Soc.*, **80,** 6685 (1958). Displacement reactions on the $B_{10}H_{12}$ unit.
101. M. F. Hawthorne and A. R. Pitochelli, *J. Am. Chem. Soc.*, **81,** 5519 (1959). The reactions of bis-acetonitrile decaborane with amines.
102. M. F. Hawthorne, B. M. Graybill, and A. R. Pitochelli, *138th Meeting Am. Chem. Soc.*, New York, Sept. 1960, p. 46N. Preparation and reactions of $B_9H_{15}$ derivatives.

103. M. F. Hawthorne, A. R. Pitochelli, R. D. Strahm, and J. J. Miller, *J. Am. Chem. Soc.*, **82,** 1825 (1960). The preparation and characterization of salts which contain the $B_{10}H_{13}$ anion.
104. K. Hedberg as quoted by Linevsky, Shull, Mann, and Wartik, *J. Am. Chem. Soc.*, **75,** 3287 (1953). The vibrational spectrum of tetrachlorodiboron.
105. K. Hedberg and V. Schomaker, *J. Am. Chem. Soc.*, **73,** 1482 (1951). A reinvestigation of the structures of diborane and ethane by electron diffraction.
106. K. Hedberg and A. J. Stosick, *J. Am. Chem. Soc.*, **74,** 954 (1952). An electron diffraction investigation of the structures of the aminodiboranes $(CH_3)_2NB_2H_5$ and $H_2NB_2H_5$.
107. K. Hedberg, M. E. Jones, and V. Schomaker, *J. Am. Chem. Soc.*, **73,** 3538 (1951). On the structure of stable pentaborane.
108. K. Hedberg, M. E. Jones, and V. Schomaker, *Proc. Natl. Acad. Sci. U.S.*, **38,** 679 (1952). The structure of stable pentaborane.
109. K. Hedberg, V. Schomaker, and M. E. Jones, *Abstracts, 118th Meeting Am. Chem. Soc.*, Chicago, September 1950. Electron diffraction investigations of the molecular structure of the pentaborane $B_5H_9$ and some other electron-deficient compounds of boron.
110. W. H. Hill and M. S. Johnson, *Anal. Chem.*, **27,** 1300 (1955). Determination of decaborane.
111. M. Hillman, *Abstracts, 135th Meeting Am. Chem. Soc.*, Boston, April 1959, p. 44M. The chemistry of decaborane-iodination studies.
112. M. Hillman, *J. Am. Chem. Soc.*, **82,** 1096 (1960). The chemistry of decaborane. Iodination studies.
113. M. Hillman, D. J. Mangold, and J. H. Norman, *Abstracts, 133rd Meeting Am. Chem. Soc.*, San Francisco, April 1958, p. 18L. Interaction of the boranes. Diborane-pentaborane-9 reaction. A tracer study.
114. F. L. Hirshfeld, K. Eriks, R. E. Dickerson, E. L. Lippert, Jr., and W. N. Lipscomb, *J. Chem. Phys.*, **28,** 56 (1958). Molecular and crystal structure of $B_6H_{10}$.
115. J. L. Hoard and A. E. Newkirk, *J. Am. Chem. Soc.*, **82,** 70 (1960). An analysis of polymorphism in boron based upon X-ray diffraction results.
116. J. L. Hoard, S. Geller, and R. E. Hughes, *J. Am. Chem. Soc.*, **73,** 1892 (1951). On the structure of elementary boron.
117. J. L. Hoard, R. E. Hughes, and D. E. Sands, *J. Am. Chem. Soc.*, **80,** 4507 (1958). The structure of tetragonal boron.
118. R. Hoffmann, unpublished results, 1962.
119. R. Hoffmann and M. Gouterman, *J. Chem. Phys.*, **36,** 2189 (1962). Theory of polyhedral molecules. II. A crystal field model.
120. R. Hoffmann and W. N. Lipscomb, *J. Chem. Phys.*, **36,** 2179 (1962). Theory of polyhedral molecules. I. Physical factorization of the secular equation.
121. R. Hoffmann and W. N. Lipscomb, *J. Chem. Phys.*, **36,** 3489 (1962). Theory of polyhedral molecules. III. Population analyses and reactivities for the carboranes.
122. R. Hoffmann and W. N. Lipscomb, *J. Chem. Phys.*, **37,** 520 (1962). Sequential substitution reactions on $B_{10}H_{10}^{=}$ and $B_{12}H_{12}^{=}$.
123. R. Hoffmann and W. N. Lipscomb, *J. Chem. Phys.*, **37,** 2872 (1962). The boron hydrides: LCAO-MO and resonance studies.
124. R. Hoffmann and W. N. Lipscomb, *Inorg. Chem.*, **2,** 231 (1963). Intramolecular

isomerization and transformation in carboranes and substituted polyhedral molecules.

125. W. V. Hough and L. J. Edwards, *Abstracts, 133rd Meeting Am. Chem. Soc.*, San Francisco, April 1958, p. 28L. Metal boron hydrides.
126. W. V. Hough, L. J. Edwards, and A. D. McElroy, *J. Am. Chem. Soc.*, **78,** 689 (1956). The sodium-diborane reaction.
127. W. V. Hough, L. J. Edwards, and A. D. McElroy, *J. Am. Chem. Soc.*, **80,** 1828 (1958). The sodium-diborane reaction.
128. W. V. Hough, M. D. Ford, and L. J. Edwards, Paper presented before the Division of Inorganic Chemistry, *132nd Meeting Am. Chem. Soc.*, New York, Sept. 1957, p. 15N. Reactions of tetraborane with Lewis bases.
129. W. V. Hough, M. D. Ford, and L. J. Edwards, *Abstracts, 135th Meeting Am. Chem. Soc.*, Boston, April 1959, p. 38M. Reactions of pentaborane-9 with Lewis bases.
130. W. V. Hough, M. D. Ford, and L. J. Edwards, *Abstracts, 135th Meeting Am. Chem. Soc.*, Boston, April 1959, p. 38M. Reaction of pentaborane-9 with Lewis bases.
131. H. J. Hrostowski, R. J. Myers, and G. C. Pimentel, *J. Chem. Phys.*, **20,** 518 (1952). The microwave spectra and dipole moment of stable pentaborane.
132. E. W. Hughes, *J. Am. Chem. Soc.*, **78,** 502 (1956). The crystal structure of ammonia-triborane, $H_3NBH_3$.
133. E. W. Hughes and C. Fritchie, Jr., Abstract of Paper F-9, American Crystallographic Association, June 18–22, 1962, Villanova, Pennsylvania. The structure of di-2-bromopyridine decaborane.
134. R. A. Jacobson and W. N. Lipscomb, *J. Am. Chem. Soc.*, **80,** 5571 (1958). The $B_8Cl_8$ structure: A new boron polyhedron in small molecules.
135. R. A. Jacobson and W. N. Lipscomb, *J. Chem. Phys.*, **31,** 605 (1959). The molecular and crystal structure of $B_8Cl_8$. II. Solution of the three-dimensional structure.
136. R. A. Jacobson and W. N. Lipscomb, unpublished results.
137. W. L. Jolly and R. E. Mesmer, *Abstracts, 141st Meeting Am. Chem. Soc.*, Washington, March 1962, p. 4M. Evidence for the intermediate $BH_5$.
138. M. E. Jones and R. E. Marsh, *J. Am. Chem. Soc.*, **76,** 1434 (1954). The preparation and structure of magnesium boride, $MgB_2$.
139. M. E. Jones, K. Hedberg, and V. Schomaker, *J. Am. Chem. Soc.*, **75,** 4116 (1953). On the structure of tetraborane.
140. M. E. Jones, K. Hedberg, and V. Schomaker, private communication, 1954.
141. A. Kaczmarczyk, R. Dobrott, and W. N. Lipscomb, *Proc. Natl. Acad. Sci. U.S.*, **48,** 729 (1962). Reactions of $B_{10}H_{10}^{=}$ ion.
142. J. S. Kasper, C. M. Lucht, and D. Harker, *Acta Cryst.*, **3,** 436 (1950). The crystal structure of decaborane, $B_{10}H_{14}$. A preliminary report of the boron positions by these authors is published in the *J. Am. Chem. Soc.*, **70,** 881 (1948). Interatomic distances in the present paper have been recalculated from the final parameters of the 1950 paper.
143. J. J. Kaufman, private communication, 1959.
144. J. J. Kaufman and W. S. Koski, *Abstracts, 130th Meeting Am. Chem. Soc.*, Atlantic City, Sept. 1956, p. 33R. Infrared study of the exchange of deuterium between decaborane and diborane.
145. J. Kelly, Jr., J. Ray, and R. A. Ogg, Jr., *Phys. Rev.*, **94,** 767A (1954). Nuclear magnetic resonance spectra of boranes and derivatives.

146. C. W. Kern and W. N. Lipscomb, *J. Chem. Phys.*, **37**, 275 (1962). Proton shielding in diborane.
147. R. Kiessling, *Acta Chem. Scand.*, **4**, 209 (1950). The borides of some transition elements.
148. W. H. Knoth and E. L. Muetterties, *J. Inorg. Nucl. Chem.*, **20**, 66 (1961). Chemistry of boranes. II. Decaborane derivatives based on the $B_{10}H_{12}$ structural unit.
149. W. H. Knoth, H. C. Miller, D. C. England, G. W. Parshall, and E. L. Muetterties, *J. Am. Chem. Soc.*, **84**, 1056 (1962). Derivative chemistry of $B_{10}H_{10}^{=}$ and $B_{12}H_{12}^{=}$.
150. G. Kodama, Ph. D. dissertation, University of Michigan, 1957.
151. G. Kodama and W. N. Lipscomb, unpublished results.
152. G. Kodama and R. W. Parry, *16th International Congress of Pure and Applied Chemistry*, Paris, 1957, Butterworth Scientific Publications, London, 1958, p. 483. $BH_2(NH_3)_2{}^+B_3H_8{}^-$.
153. G. Kodama and R. W. Parry, *J. Am. Chem. Soc.*, **82**, 6250 (1960). The preparation and structure of the diammoniate of tetraborane.
154. G. Kodama, R. W. Parry, and J. C. Carter, *J. Am. Chem. Soc.*, **81**, 3534 (1959). The preparation and properties of ammonia-triborane.
155. J. A. Kohn, G. Katz, and A. A. Giardini, *Z. Krist.*, **111**, 53 (1958–59). $AlB_{10}$, a new phase, and a critique on the aluminum borides.
156. W. S. Koski, *Abstracts, 133rd Meeting, Am. Chem. Soc.*, San Francisco, April 1958, p. 18L. Mechanisms of isotopic exchanges in the boron hydrides.
157. W. S. Koski and J. E. Todd, *Abstracts, 134th Meeting Am. Chem. Soc.*, Chicago, Sept. 1958, p. 22N. Deuterium exchange between tetraborane and diborane.
158. W. S. Koski, J. J. Kaufman, and P. C. Lauterbur, *J. Am. Chem. Soc.*, **79**, 2382 (1957). Nuclear magnetic resonance study of the $B_2D_6$—$B_5H_9$ exchange reaction.
159. W. V. Kotlensky and R. Schaeffer, *Abstracts, 132nd Meeting Am. Chem. Soc.*, New York, Sept. 1957, p. 8S. Decomposition of diborane in a silent discharge.
160. W. V. Kotlensky and R. Schaeffer, *J. Am. Chem. Soc.*, **80**, 4517 (1958). Decomposition of diborane in a silent discharge. Isolation of $B_6H_{10}$ and $B_9H_{15}$.
161. W. C. Kreye and R. A. Marcus, *Abstracts, 134th Meeting Am. Chem. Soc.*, Chicago, Sept. 1958, p. 65. The photolysis of diborane.
162. W. C. Kreye and R. A. Marcus, *J. Chem. Phys.*, **37**, 419 (1962). Photolysis of diborane at 1849 A.
163. A. W. Laubengayer and R. Bottei, *J. Am. Chem. Soc.*, **74**, 1618 (1952). The dipole moment of decaborane.
164. L. Lavine and W. N. Lipscomb, *J. Chem. Phys.*, **22**, 614 (1954). The crystal and molecular structure of $B_5H_{11}$.
165. J. E. Lennard-Jones, *Proc. Roy. Soc. (London)*, **A198**, 1, 14 (1949). The molecular orbital theory of chemical valency. I. The determination of molecular orbitals. II. Equivalent orbitals in molecules of known symmetry.
166. A. Levy, J. E. Williamson, and L. W. Steiger, *J. Inorg. Nucl. Chem.*, **17**, 26 (1961). Gamma-induced reactions between boron hydrides and hydrogen.
167. R. Lewin, P. G. Simpson, and W. N. Lipscomb, *J. Am. Chem. Soc.*, **85**, 478 (1963). The structure of $C_2H_5NH_2B_8H_{11}NHC_2H_5$. Also *J. Chem. Phys.*, in press.
168. E. L. Lippert, Jr., and W. N. Lipscomb, *J. Am. Chem. Soc.*, **78**, 503 (1956). The structure of $H_3NBH_3$.
169. W. N. Lipscomb, *J. Chem. Phys.*, **22**, 985 (1954). Structures of the boron hydrides.

170. W. N. Lipscomb, *J. Chem. Phys.*, **25,** 38 (1956). Charge distribution in the pentaboranes.
171. W. N. Lipscomb, *J. Chim. Phys.*, **53,** 515 (1956). La Valence dans les hydrures de bore.
172. W. N. Lipscomb, *J. Phys. Chem.*, **61,** 23 (1957). Valence in the boron hydrides.
173. W. N. Lipscomb, *J. Chem. Phys.*, **28,** 170 (1958). Molecular orbitals in the $B_4H_7^-$ and $B_6H_{11}^+$ ions.
174. W. N. Lipscomb, *J. Phys. Chem.*, **62,** 381 (1958). Possible boron hydride ions.
175. W. N. Lipscomb, *Advances in Inorganic and Radiochemistry*, Vol. 1, Academic Press, New York, 1959, pp. 132–133.
176. W. N. Lipscomb, *J. Inorg. Nucl. Chem.*, **11,** 1 (1959). Structure and reactions of the boron hydrides.
177. W. N. Lipscomb, *Tetrahedron Letters*, No. 18, 20 (1959). Tetrahedral cyclobutadiene?
178. W. N. Lipscomb, *J. Phys. Chem.*, **65,** 1064 (1961). Topologies of $B_6$ and $B_7$ hydrides.
179. W. N. Lipscomb, *Proc. Natl. Acad. Sci. U.S.*, **47,** 1791 (1961). Extensions of the valence theory of boron compounds.
180. W. N. Lipscomb and D. Britton, *J. Chem. Phys.*, **33,** 275 (1960). Valence structure of the higher borides.
181. W. N. Lipscomb and A. Kaczmarczyk, *Proc. Natl. Acad. Sci. U.S.*, **47,** 1796 (1961). Interactions of boranes and their ions with paramagnetic S-state ions.
182. W. N. Lipscomb, A. R. Pitochelli, and M. F. Hawthorne, *J. Am. Chem. Soc.*, **81,** 5833 (1959). Probable structure of the $B_{10}H_{10}^{=}$ ion.
183. J. C. Lockhart, *J. Chem. Soc.*, **1962,** 1197. Mechanism of substitution at a boron atom.
184. H. C. Longuet-Higgins, *J. Chim. Phys.*, **46,** 275 (1949), discussion. Substance hydrogenees avec defaut d'electrons.
185. H. C. Longuet-Higgins, *Quart. Rev. (London)*, **11,** 121 (1957). The structures of electron-deficient molecules.
186. H. C. Longuet-Higgins and R. P. Bell, *J. Chem. Soc.*, **1943,** 250. The structure of the boron hydrides.
187. H. C. Longuet-Higgins and M. de V. Roberts, *Proc. Roy. Soc. (London)*, **A224,** 336 (1954). The electronic structure of the borides $MB_6$.
188. H. C. Longuet-Higgins and M. de V. Roberts, *Proc. Roy. Soc. (London)*, **A230,** 110 (1955). The electronic structure of an icosahedron of boron atoms.
189. C. M. Lucht, *J. Am. Chem. Soc.*, **73,** 2373 (1951). An analysis of the electron diffraction data for decaborane.
190. P. C. Maybury and W. S. Koski, *J. Chem. Phys.*, **21,** 742 (1953). Kinetics of two exchange reactions involving diborane.
191. L. V. McCarty, J. S. Kasper, F. H. Horn, B. F. Decker, and A. E. Newkirk, *J. Am. Chem. Soc.*, **80,** 2592 (1958). A new crystalline modification of boron.
192. J. J. Miller and M. F. Hawthorne, *J. Am. Chem. Soc.*, **81,** 4501 (1959). The course of base-catalyzed hydrogen exchange in decaborane.
193. J. J. Miller and M. F. Hawthorne, *J. Am. Chem. Soc.*, **82,** 500 (1960). The alkoxylation of decaborane.
194. W. Moffitt, *Proc. Roy. Soc. (London)*, **A196,** 510 (1949). Molecular orbitals and the Hartree field.
195. E. B. Moore, Jr., *J. Chem. Phys.*, **37,** 675 (1962). Molecular orbitals in $B_{10}H_{14}$.

196. E. B. Moore, Jr., and W. N. Lipscomb, *Acta Cryst.*, **9,** 668 (1956). The crystal and molecular structure of $Cl_2BCH_2CH_2BCl_2$.
197. E. B. Moore, Jr., R. E. Dickerson, and W. N. Lipscomb, *J. Chem. Phys.*, **27,** 209 (1957). Least-squares refinements of $B_{10}H_{14}$, $B_4H_{10}$, $B_5H_{11}$.
198. E. B. Moore, Jr., L. L. Lohr, Jr., and W. N. Lipscomb, *J. Chem. Phys.*, **35,** 1329 (1961). Molecular orbitals in some boron compounds.
199. E. L. Muetterties, private communication, 1961.
200. E. L. Muetterties, private communication, 1962.
201. E. L. Muetterties and V. D. Aftandilian, *Inorg. Chem.*, **1,** 731 (1962). Chemistry of boranes. IV. Phosphine derivatives of $B_{10}H_{14}$ and $B_9H_{15}$
202. R. S. Mulliken, *J. Chem. Phys.*, **2,** 782 (1934). A new electroaffinity scale; together with data on valence states and on valence ionization potentials and electron affinities.
203. R. S. Mulliken, *Chem. Rev.*, **41,** 207 (1947). The structure of diborane and related molecules.
204. R. S. Mulliken, *J. Chem. Phys.*, **23,** 1833, 1841, 2338, 2343 (1955). Electronic population analysis on LCAO-MO molecular wave functions. I–IV.
205. C. Naar-Colin and T. L. Heying, private communication, 1962.
206. St. v. Naray-Szabo, *Z. Krist.*, **94,** 367 (1936). Röntgenographische Untersuchung des Aluminum-12-borids $AlB_{12}$.
207. C. E. Nordman and W. N. Lipscomb, *J. Am. Chem. Soc.*, **75,** 4116 (1953). The molecular structure of $B_4H_{10}$.
208. C. E. Nordman and W. N. Lipscomb, *J. Chem. Phys.*, **21,** 1856 (1953). The crystal and molecular structure of tetraborane.
209. C. E. Nordman and C. Riemann, *J. Am. Chem. Soc.*, **81,** 3538 (1959). The structure of $B_3H_7NH_3$.
210. C. E. Nordman, C. Riemann, and C. R. Peters, *Abstracts, 133rd Meeting Am. Chem. Soc.*, San Francisco, April 1958, p. 46L. Structures of $NH_3B_3H_7$ and $(NH_3)_2BH_2Cl$.
211. H. G. Norment, *Abstracts, 4th Intern. Congr. Crystallography*, Montreal, 1957, Paper 6.4. The structure of $B_3H_7NMe_3$.
212. R. A. Ogg, Jr., *J. Chem. Phys.*, **22,** 1933 (1954). Nuclear magnetic resonance spectra and structure of borohydride ion and diborane.
213. R. A. Ogg, Jr., and J. D. Ray, *Disc. Faraday Soc.*, **19,** 239 (1955). Nuclear magnetic resonance spectrum and molecular structure of aluminum borohydride.
214. T. P. Onak, *J. Am. Chem. Soc.*, **83,** 2584 (1961). Rearrangement of 1-methyl and 1-ethylpentaborane-9 to 2-methyl and 2-ethylpentaborane-9.
215. T. P. Onak, private communication, 1962. $B_4H_7^-$.
216. T. P. Onak and F. J. Gerhart, *Inorg. Chem.*, **1,** 742 (1962). Thermal rearrangement of 1-alkylpentaboranes: A one-step synthesis of 2-alkylpentaboranes from pentaborane.
217. T. P. Onak and R. E. Williams, *Inorg. Chem.*, **1,** 106 (1962). $B_5H_9$—DCl exchange catalyzed by $AlCl_3$.
218. T. P. Onak, R. E. Williams, and H. G. Weiss, *J. Am. Chem. Soc.*, **84,** 2830 (1962). $B_4C_nH_{2n+4}$. The synthesis of $B_4C_nH_{2n+4}$ compounds from pentaborane-9 and alkynes catalyzed by 2,6-dimethylpyridine.
219. T. P. Onak, H. Landesman, R. E. Williams, and I. Shapiro, *J. Phys. Chem.*, **63,** 1533 (1959). The $B^{11}$ nuclear magnetic resonance chemical shifts and spin coupling values for various compounds.

220. A. J. Owen, *J. Chem. Soc.*, **1961,** 5438. The pyrolysis of decaborane.
221. R. J. Pace, J. Williams, and R. E. Williams, *J. Chem. Soc.*, **1961,** 2196. Boron hydride derivatives. The characterization of some decaborane derivatives of the type $B_{10}H_{12}\cdot 2M$.
222. R. J. F. Palchak, J. H. Norman, and R. E. Williams, *J. Am. Chem. Soc.*, **83,** 3380 (1961). Decaborane, "6-benzyl" $B_{10}H_{13}$ chemistry.
223. R. W. Parry and L. J. Edwards, *Abstracts, 133rd Meeting Am. Chem. Soc.*, San Francisco, April 1958, p. 8L. Systematics in the chemistry of the boron hydrides.
224. R. W. Parry and L. J. Edwards, *J. Am. Chem. Soc.*, **81,** 3554 (1959). Systematics in the chemistry of the boron hydrides.
225. R. W. Parry and S. G. Shore, *J. Am. Chem. Soc.*, **80,** 15, (1958). Chemical evidence for the structure of the "diammoniate of diborane." IV. The reaction of sodium with Lewis acids in liquid ammonia.
226. E. Parthe and J. T. Norton, *Z. Krist.*, **110,** 167 (1958). Note on the cell and symmetry of so-called monoclinic $AlB_{12}$.
227. L. Pauling, *The Nature of the Chemical Bond*, Cornell University Press, Ithaca, N.Y., 1960, p. 378.
228. L. Pauling and B. Kamb, *Z. Krist.*, **112,** 472 (1959). The discussion of tetragonal boron by the resonating-valence-bond theory of electron-deficient substances.
229. L. Pauling and S. Weinbaum, *Z. Krist.*, **87,** 181 (1934). The structure of calcium boride.
230. C. R. Peters and C. E. Nordman, *J. Am. Chem. Soc.*, **82,** 5758 (1960). Structure of the $B_3H_8^-$ ion.
231. W. D. Phillips, H. C. Miller, and E. L. Muetterties, *J. Am. Chem. Soc.*, **81,** 4496 (1959). $B^{11}$ magnetic resonance study of boron compounds.
232. G. C. Pimentel and K. S. Pitzer, *J. Chem. Phys.*, **17,** 882 (1949). The ultraviolet absorption and luminescence of decaborane.
233. A. R. Pitochelli and M. F. Hawthorne, *J. Am. Chem. Soc.*, **82,** 3228 (1960). The isolation of the icosahedral $B_{12}H_{12}^{=}$ ion.
234. A. R. Pitochelli and M. F. Hawthorne, *J. Am. Chem. Soc.*, **84,** 3218 (1962). The preparation of a new boron hydride $B_{18}H_{22}$.
235. A. R. Pitochelli and M. F. Hawthorne, private communication, 1962.
236. A. R. Pitochelli, W. N. Lipscomb, and M. F. Hawthorne, *J. Am. Chem. Soc.*, **84,** 3026 (1962). Isomers of $B_{20}H_{18}^{=}$.
237. K. S. Pitzer, *J. Am. Chem. Soc.*, **67,** 1126 (1946). Electron deficient molecules. I. The principles of hydroboron structures.
238. J. R. Platt, H. B. Klevens, and G. W. Schaeffer, *J. Chem. Phys.*, **15,** 598 (1947). Absorption spectrum of borazole in the vacuum ultraviolet.
239. R. J. Polak and T. L. Heying, *J. Org. Chem.*, **27,** 1483 (1962). The preparation of phosphite and phosphinite decaboranes.
240. B. Post and F. W. Glaser, *J. Metals*, **1952,** 631. Crystal structure of $ZrB_{12}$.
241. B. Post, D. Moscowitz, and F. W. Glaser, *J. Am. Chem. Soc.*, **78,** 1800 (1956). Borides of rare earth metals.
242. W. C. Price, *J. Chem. Phys.*, **15,** 614 (1947). The structure of diborane.
243. W. C. Price, *J. Chem. Phys.*, **16,** 894 (1948). The absorption spectrum of diborane.
244. W. C. Price, *J. Chem. Phys.*, **17,** 1044 (1949). The infrared absorptional spectra of some metal borohydrides.
245. A Quayle, *J. Appl. Chem.*, **9,** 395 (1959). The mass spectra of some boron hydrides.

246. J. Reddy and W. N. Lipscomb, *Abstracts, 135th Meeting Am. Chem. Soc.*, Boston, April 1959, p. 45M. Structure of $B_{10}H_{12}(NCCH_3)_2$.

247. J. Reddy and W. N. Lipscomb, *J. Am. Chem. Soc.*, **81,** 754 (1959). Molecular structure of $B_{10}H_{12}(CH_3CN)_2$.

248. J. Reddy and W. N. Lipscomb, *J. Chem. Phys.*, **31,** 610 (1959). The molecular structure of $B_{10}H_{12}(CH_3CN)_2$.

249. J. S. Rigden, R. C. Hopkins, and J. D. Baldeschwieler, *J. Chem. Phys.*, **35,** 1532 (1961). Absence of $B^{10}$—$B^{11}$ coupling in $B_4H_{10}$.

250. M. G. Rossmann, R. A. Jacobson, F. L. Hirshfeld, and W. N. Lipscomb, *Acta Cryst.*, **12,** 530 (1959). An account of some computing experiences.

251. R. E. Rundle, *J. Chem. Phys.*, **17,** 671 (1949). Electron-deficient compounds. II. Relative energies of "half bonds."

252. R. E. Rundle and P. H. Lewis, *J. Chem. Phys.*, **20,** 132 (1952). Electron deficient compounds. VI. The structure of beryllium chloride.

253. J. Russell, R. Hirst, F. A. Kanda, and A. J. King, *Acta Cryst.*, **6,** 870 (1953). An X-ray study of the magnesium borides.

254. G. V. Sampsonov and L. Y. Markovskii, *Usp. Khim.*, **25,** 190 (1956). Chemistry of borides.

255. D. E. Sands and J. L. Hoard, *J. Am. Chem. Soc.*, **79,** 5582 (1957). Rhombohedral elemental boron.

256. D. E. Sands and A. Zalkin, *Acta Cryst.*, **15,** 410 (1962). The crystal structure of $B_{10}H_{12}[S(CH_3)_2]_2$.

257. A. W. Schaeffer, K. H. Ludlum, and S. E. Wiberley, *J. Am. Chem. Soc.*, **81,** 3157 (1958). Mass spectrometric evidence for heptaborane.

258. G. W. Schaeffer, J. J. Burns, T. J. Klinger, L. A. Martincheck, and R. W. Rozett, *Abstracts, 135th Meeting Am. Chem. Soc.*, Boston, April 1959, p. 44M. Ultraviolet absorption spectra of decaborane and its ionization products.

259. R. Schaeffer, private communication, 1953.

260. R. Schaeffer, *J. Am. Chem. Soc.*, **79,** 1006 (1957). A new type of substituted borane.

261. R. Schaeffer, *J. Am. Chem. Soc.*, **79,** 2726 (1957). The molecular structure of $B_{10}H_{12}I_2$.

262. R. Schaeffer, *J. Chem. Phys.*, **26,** 1349 (1957). Mechanism of formation of pentaborane-9 from pentaborane-11.

263. R. Schaeffer, private communication, 1961. $B_4H_8CO$, NMR spectrum.

264. R. Schaeffer, private communication, 1962. $B_{10}H_{13}^- + BH_3$ scrambles B atoms in $B_{10}H_{13}^-$. See R. Schaeffer and F. Tebbe, *J. Am. Chem. Soc.*, **85,** 2020 (1963).

265. R. Schaeffer and F. Tebbe, *J. Am. Chem. Soc.*, **84,** 3974 (1962). Synthesis of a boron labeled tetraborane.

266. R. Schaeffer, J. N. Shoolery, and R. Jones, *J. Am. Chem. Soc.*, **79,** 4606 (1957). Nuclear magnetic resonance spectra of boranes.

267. R. Schaeffer, J. N. Shoolery, and R. Jones, *J. Am. Chem. Soc.*, **80,** 2670 (1958). Structures of halogen substituted boranes.

268. J. R. Scherer and B. L. Crawford, Jr., private communication, 1959.

269. H. I. Schlesinger, University of Chicago, ONR Contract N173S-9820, Final Rept. 1945–6.

270. H. I. Schlesinger and H. C. Brown, *J. Am. Chem. Soc.*, **62,** 3421 (1940). Metallo borohydrides. I. Aluminum borohydride.

271. H. I. Schlesinger and H. C. Brown, *J. Am. Chem. Soc.*, **62,** 3429 (1940). Metallo borohydrides. III. Lithium borohydride.

272. H. I. Schlesinger and A. O. Walker, *J. Am. Chem. Soc.*, **57,** 621 (1935). Hydrides of boron. IV. The methyl derivatives of diborane.

273. H. I. Schlesinger, R. T. Sanderson, and A. B. Burg, *J. Am. Chem. Soc.*, **61,** 536 (1939). A volatile compound of aluminum, boron, and hydrogen.

274. H. Schmied and W. S. Koski, *Abstracts, 135th Meeting Am. Chem. Soc.*, Boston, April 1959, p. 38M. Radiation-induced chemical reactions in pentaborane.

275. V. Schomaker, *J. Chim. Phys.*, **46,** 262 (1949). Remarque sur les molécules avec défaut d'electrons. Indications sur la structure moléculaire de $B_5H_9$.

276. J. N. Shoolery, *Discussions Faraday Soc.*, **19,** 215 (1955). The relation of high resolution nuclear magnetic resonance spectra to molecular structures.

277. H. Schroeder, T. L. Heying, and J. R. Reiner, *Inorg. Chem.*, **2** (1963), in press. A new series of organoboranes. I. The chlorination of 11,12 dicarbadodecaborane.

278. D. R. Schultz, S. G. Shore, R. W. Parry, G. Kodama, and P. R. Girardot, *J. Am. Chem. Soc.*, **80,** 4 (1958). Chemical evidence for the structure of the "diammoniate of diborane."

279. W. Shand, Jr., private communication, 1945. $B_2H_2Me_4$.

280. I. Shapiro and B. Keilin, *J. Am. Chem. Soc.*, **76,** 3864 (1954). Mass spectrum of octaborane.

281. I. Shapiro and B. Keilin, *J. Am. Chem. Soc.*, **77,** 2663 (1955). Self-exchange of boron in boron hydrides.

282. I. Shapiro, C. D. Good, and R. E. Williams, *J. Am. Chem. Soc.*, **84,** 3837 (1962). The carborane series: $B_nC_2H_{n+2}$. I. $B_3C_2H_5$.

283. I. Shapiro, M. Lustig, and R. E. Williams, *J. Am. Chem. Soc.*, **81,** 838 (1959). Exchange sites in the deuteration of decaborane.

284. S. G. Shore and R. W. Parry, *J. Am. Chem. Soc.*, **77,** 6084 (1955). The crystalline compound, ammonia borane, $H_3NBH_3$.

285. B. Siegel, J. L. Mack, J. V. Lowe, and J. Gallaghan, *J. Am. Chem. Soc.*, **80,** 4523 (1958). Decaborane Grignard reagents.

286. G. Silbiger and S. H. Bauer, *J. Am. Chem. Soc.*, **68,** 312 (1946). The structure of the hydrides of boron. VII. Beryllium borohydride, $BeB_2H_8$.

287. G. Silbiger and S. H. Bauer, *J. Am. Chem. Soc.*, **70,** 115 (1948). The structures of the hydrides of boron. VII. Decaborane.

288. A. H. Silver and P. J. Bray, *Bull. Am. Phys. Soc.*, [II] **2,** 387 (1957). Nuclear quadrupole coupling of boron-11 and aluminum-27 in ionic crystals.

289. A. H. Silver and P. J. Bray, *J. Chem. Phys.*, **32,** 288 (1960). NMR study of bonding in some solid boron compounds.

290. P. G. Simpson and W. N. Lipscomb, *J. Chem. Phys.*, **35,** 1340 (1961). Refinement of the $B_9H_{15}$ structure and test for solid solutions.

291. P. G. Simpson and W. N. Lipscomb, *Proc. Natl. Acad. Sci. U.S.*, **48,** 1490 (1962). Molecular structure of $B_{18}H_{22}$. Also *J. Chem. Phys.*, **39,** 26 (1963).

292. A. M. Soldate, *J. Am. Chem. Soc.*, **69,** 987 (1947). Crystal structure of sodium borohydride.

293. I. J. Solomon, M. J. Klein, and Kiyo Hattori, *Abstracts, 133rd Meeting, Am. Chem. Soc.*, San Francisco, April 1958, p. 38L. Preparation and reactions of mono- and diethylidiboranes.

294. C. Spencer and W. N. Lipscomb, *J. Chem. Phys.*, **28,** 355 (1958). Errata: B—Cl distance in boron trichloride.
295. W. von Stackelberg and F. Neumann, *Z. Physik. Chem. (Frankfurt)*, **B19,** 314 (1932). Die Kristallstructure der Boride der Zusammensetzung $MeB_6$.
296. R. D. Stewart and R. G. Alder, *Abstracts, 134th Meeting Am. Chem. Soc.*, Chicago, Sept. 1958, p. 6S. Pyrolysis of diborane.
297. F. Stitt, *J. Chem. Phys.*, **8,** 981 (1940). The gaseous heat capacity and restricted internal rotation of diborane.
298. F. Stitt, *J. Chem. Phys.*, **9,** 780 (1941). Infrared and Raman spectra of polyatomic molecules. XV. Diborane.
299. A. Stock, *Hydrides of Boron and Silicon*, Cornell University Press, Ithaca, N.Y., 1933.
300. A. Stock and H. Laudenklos, *Z. Anorg. Allgem. Chem.*, **228,** 178 (1936). Borwasserstoffe, zur Kenntnis der Boronsalze.
301. A. Stock and W. Siecke, *Ber.*, **57,** 566 (1924). Borwasserstoffe. VII[1] Pentabor-Hydride.
302. A. Stock, W. Sutterlin, and F. Kurzen, *Z. Anorg. Allgem. Chem.*, **255,** 225, 243 (1935). Borwasserstoffe: Diborankolium, zur Kenntnis der Kaliumverbindungen des $B_4H_{10}$ und des $B_5H_9$.
303. A. Streitweiser, *Molecular Orbital Theory for Organic Chemists*, Wiley, New York, 1961.
304. C. P. Talley, B. Post, and S. LaPlaca, *Conference on Boron*, U.S. Army Signal Corps Research and Development Laboratory, Fort Monmouth, N.J., Sept. 1959.
305. L. H. Thomas and K. Umeda, *J. Chem. Phys.*, **24,** 1113 (1956). Dependence on atomic number of the diamagnetic susceptibility calculated from the Thomas-Fermi-Dirac model.
306. R. H. Toniskoetter, Ph. D. thesis, St. Louis University, 1956.
307. R. H. Toniskoetter, G. W. Schaeffer, E. C. Evers, R. E. Hughes, and G. E. Bagley, *Abstracts, 134th Meeting Am. Chem. Soc.*, Chicago, Sept. 1958, pp. 23–24. Preparation and characterization of a sodium salt derived from decaborane.
308. P. Torkington, *J. Chem. Phys.*, **19,** 528 (1951). The general valence orbitals derivable by $sp_3$ hybridization.
309. L. Trefonas and W. N. Lipscomb, *J. Chem. Phys.*, **28,** 54 (1958). Crystal and molecular structure of diboron tetrafluoride.
310. G. Urry, E. P. Schram, and S. I. Weissman, *Abstracts, 139th Meeting Am. Chem. Soc.*, St. Louis, March 1961, p. 14R. A stable free radical boron subchloride ($B_1Cl_{0.9}$).
311. G. Urry, E. P. Schram, and S. I. Weissman, *J. Am. Chem. Soc.*, **84,** 2654 (1962). A new system of stable free radicals.
312. G. Urry, T. Wartik, and H. I. Schlesinger, *J. Am. Chem. Soc.*, **74,** 5809 (1952). A new sub-chloride of boron, $B_4Cl_4$.
313. G. Urry, J. Kerrigan, T. D. Parsons, and H. I. Schlesinger, *J. Am. Chem. Soc.*, **76,** 5299 (1954). Diboron tetrachloride, $B_2Cl_4$, as a reagent for the synthesis of organoboron compounds. I. The reaction of diboron tetrachloride with ethylene.
314. G. Urry, T. Wartik, R. Moore, and H. I. Schlesinger, *J. Am. Chem. Soc.*, **76,** 5293 (1954). The preparation and some of the properties of diboron tetrachloride, $B_2Cl_4$.
315. U.S. Patent 3,030,423, April 3, 1962.
316. U.S. Patent 3,028,432, April 17, 1962.
317. A. D. Walsh, *J. Chem. Soc.*, **1947,** 89. Coordinate links formed by bonding electrons; a suggestion regarding the structure of diborane.

318. F. E. Wang, L. L. Lohr, Jr., and W. N. Lipscomb, unpublished studies, 1961.
319. F. E. Wang, P. G. Simpson, and W. N. Lipscomb, *J. Am. Chem. Soc.*, **83,** 491 (1961). The molecular structure of $B_9H_{13}NCCH_3$.
320. F. E. Wang, P. G. Simpson, and W. N. Lipscomb, *J. Chem. Phys.*, **35,** 1335 (1961). Molecular structure of $B_9H_{13}(CH_3CN)$.
321. T. Wartik, private communication, 1961.
322. T. Wartik, private communication, 1962. $(B_{10}I_8)_x$.
323. J. R. Weaver, C. W. Heitsch, and R. W. Parry, *J. Chem. Phys.*, **30,** 1075 (1959). Dipole moment of tetraborane.
324. J. R. Weaver, S. G. Shore, and R. W. Parry, *J. Chem. Phys.*, **29,** 1 (1958). The dipole moment of $H_3NBH_3$.
325. A. T. Whatley and R. N. Pease, *J. Am. Chem. Soc.*, **76,** 835 (1954). A kinetic study of the diborane-ethylene reaction.
326. R. A. Wiesboeck, A. R. Pitochelli, and M. F. Hawthorne, *J. Am. Chem. Soc.*, **83,** 4108 (1961). Oxazahydroborate ions.
327. R. L. Williams, private communication, 1960. Propagation of coupling through a single B—B bond.
328. R. E. Williams, *J. Inorg. Nucl. Chem.*, **20,** 198 (1961). Tautomerism and exchange in the boron hydrides; $^{11}B$ and $^{1}H$ NMR spectra.
329. R. E. Williams, private communication, 1961.
330. R. E. Williams and I. Shapiro, *J. Chem. Phys.*, **29,** 677 (1958). Reinterpretation of nuclear magnetic resonance spectra of decaborane.
331. R. L. Williams, I. Dunstan, and N. J. Blay, *J. Chem. Soc.*, **1960,** 5006. Boron hydride derivatives. IV. Friedel-Crafts methylation of decaborane.
332. R. E. Williams, S. G. Gibbins, and I. Shapiro, *J. Am. Chem. Soc.*, **81,** 6164 (1959). N.m.r. spectra of $B_4H_{10}$: Correlation of boron hydride spectra.
333. R. E. Williams, S. G. Gibbins, and I. Shapiro, *J. Chem. Phys.*, **30,** 320 (1959). Reinterpretation of the nuclear magnetic resonance spectra of unstable pentaborane—$B_5H_{11}$.
334. R. E. Williams, S. G. Gibbins, and I. Shapiro, *J. Chem. Phys.*, **30,** 333 (1959). Nuclear magnetic resonance spectra of hexaborane.
335. R. E. Williams. C. D. Good, and I. Shapiro, *Abstracts, 140th Meeting Am. Chem. Soc.*, Chicago, Sept. 1961, p. 14N. Carboranes. I. $B_3C_2H_5$.
336. J. Wunderlich and W. N. Lipscomb, *J. Am. Chem. Soc.*, **82,** 4427 (1960). Structure of $B_{12}H_{12}^{=}$ ion.
337. M. Yamazaki, *J. Chem. Phys.*, **27,** 1401 (1957). Electronic structure of diborane.
338. A. Zalkin and D. H. Templeton, *Acta Cryst.*, **6,** 269 (1953). The crystal structures of $CeB_4$, $ThB_4$, and $UB_4$.
339. G. S. Zhdanov and N. G. Sevast'yanov, *Compt. Rend. Acad. Sci. U.S.S.R.*, **32,** 432 (1941). Crystal structure of boron carbide.

# Recent Reviews of Boron Chemistry

D. C. Bradley, The stereochemistry of some elements of Group III, *Progress in Stereochemistry*, Butterworths, London, 1962.

W. Gerrard, *The Organic Chemistry of Boron*, Academic Press, New York, 1961.

M. F. Hawthorne, Decaborane-14 and its derivatives, *Advances in Inorganic and Radiochemistry*, Vol. 5, Academic Press, New York, in press.

H. G. Heal, Recent studies in boron chemistry, *Roy. Inst. Chem., Lectures, Monographs, and Reports, No. 1*, 1960.

A. K. Holliday and A. G. Massey, Boron subhalides and related compounds with boron-boron bonds, *Chem. Rev.*, **62,** 303 (1962).

M. F. Lappert, Organic compounds of boron, *Chem. Rev.*, **56,** 959 (1956).

W. N. Lipscomb, Recent studies of the boron hydrides, *Advances in Inorganic and Radiochemistry*, Vol. 1, Academic Press, New York, 1959, p. 117. Ref. 175.

W. N. Lipscomb, Structure and reactions of the boron hydrides, *J. Inorg. Nucl. Chem.*, **11,** 1 (1959). Ref. 176

H. C. Longuet-Higgins, Structures of electron deficient molecules, *Quart. Rev. (London)*, **11,** 121 (1957). Ref. 185.

F. G. A. Stone, Chemical reactivity of the boron hydrides and related compounds, *Advances in Inorganic and Radiochemistry*, Vol. 2, Academic Press, New York, 1960, p. 279.

F. G. A. Stone, Chemistry of the boron hydrides, *Quart. Rev. (London)*, **9,** 174 (1955).

# Appendix A

# Some Three-Center Bond Structures for Ions of —1 and —2 Charges

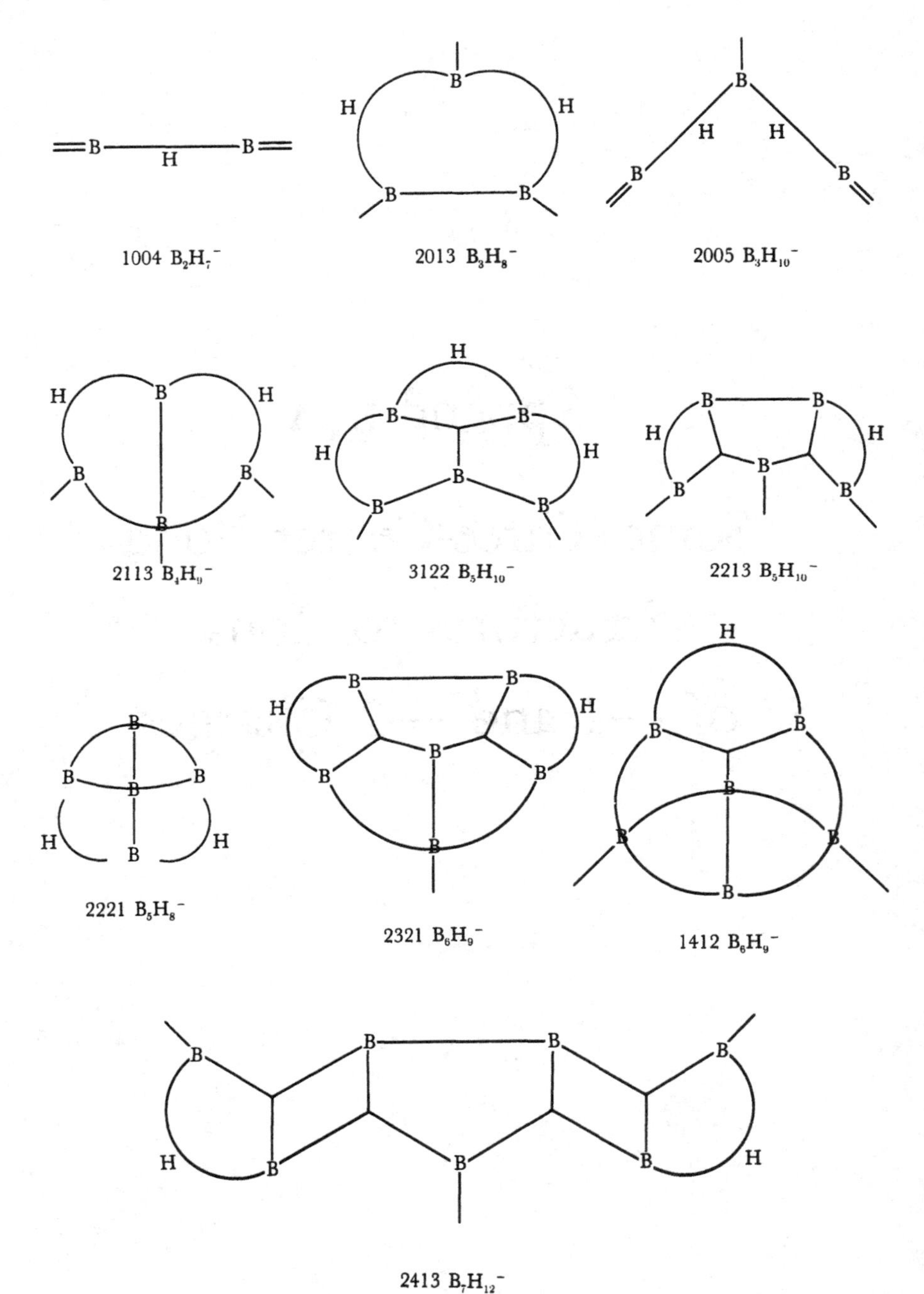


**Figure A–1** *Some of the interesting possibilities for −1 ions and intermediates based upon icosahedral fragments. The* $H_3$ *atoms of a* $BH_3$ *group may be counted as* $x = 2$ *in the equations of balance.*

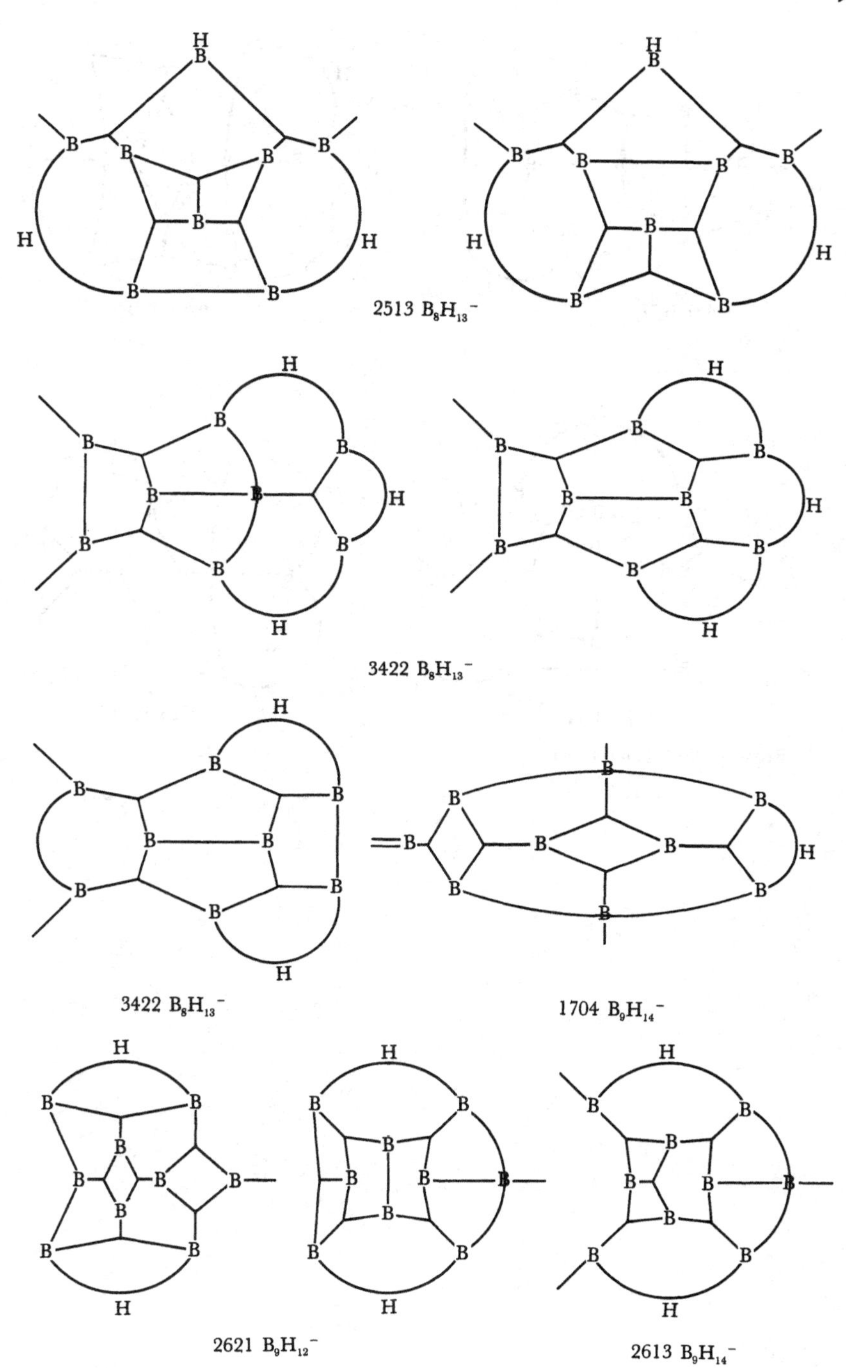


**Figure A-1** *(continued)*

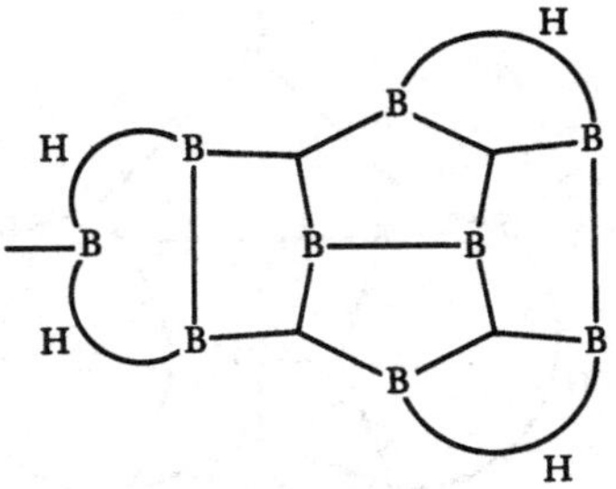

4431 $B_9H_{14}^-$

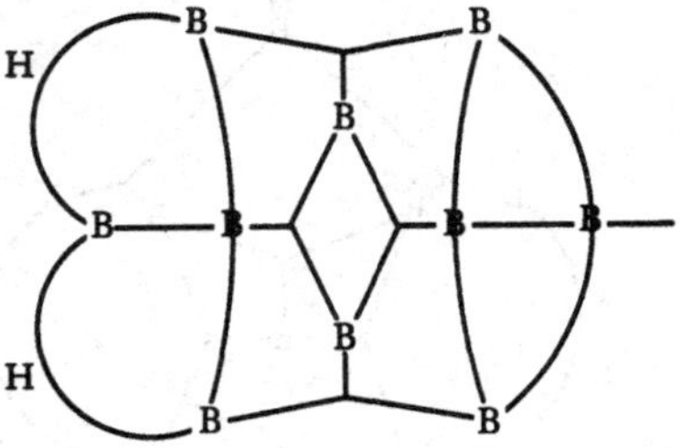

2721 $B_{10}H_{13}^-$

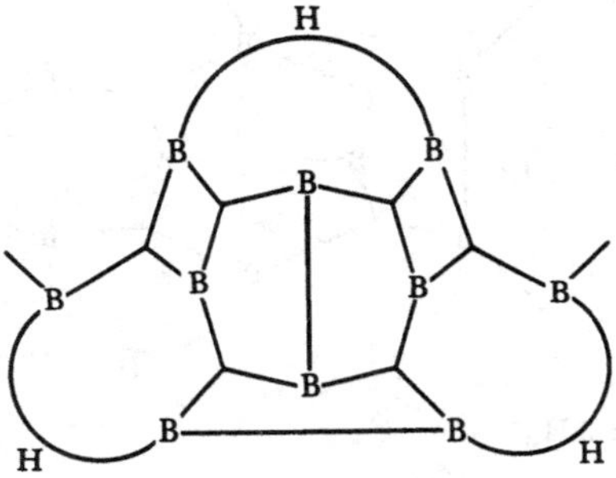

3622 $B_{10}H_{15}^-$

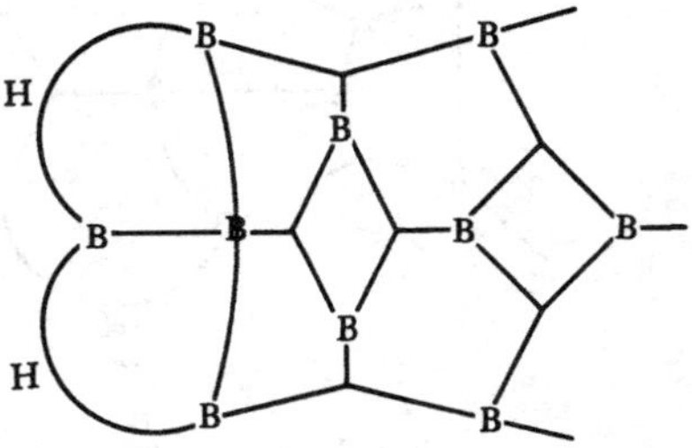

2713 $B_{10}H_{15}^-$

**Figure A-1** (*continued*)

0030 $B_2H_2^{=}$

0022 $B_2H_4^{=}$

0014 $B_2H_6^{=}$

0006 $2BH_4^{-}$

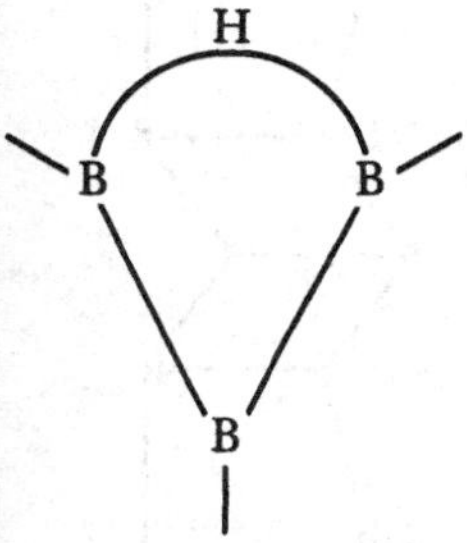

1023 $B_3H_7^{=}$

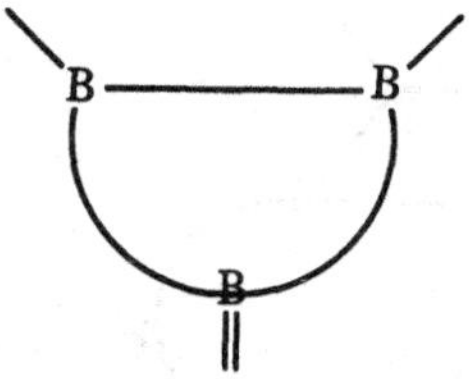

0114 $B_3H_7^{=}$

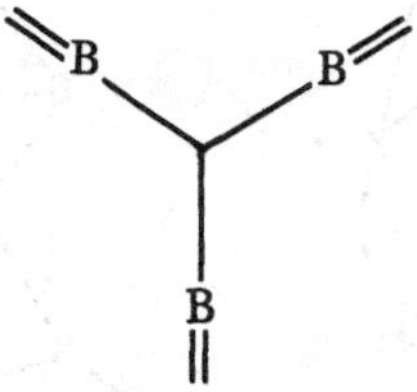

0106 $B_3H_9^{=}$

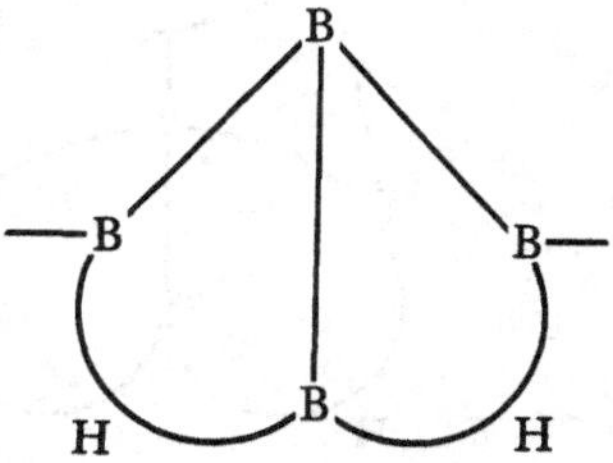

2032 $B_4H_8^{=}$

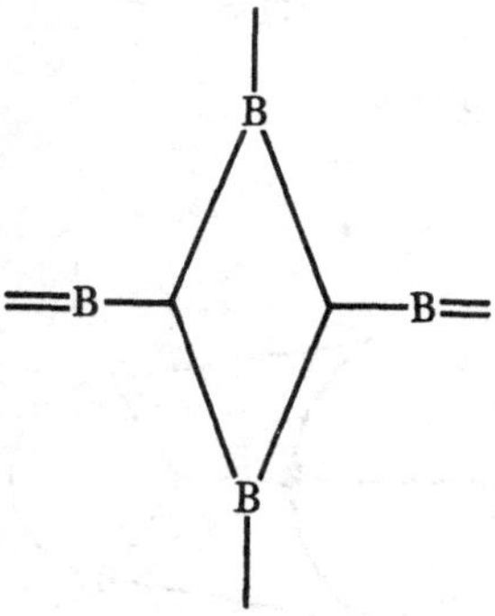

0206 $B_4H_{10}^{=}$

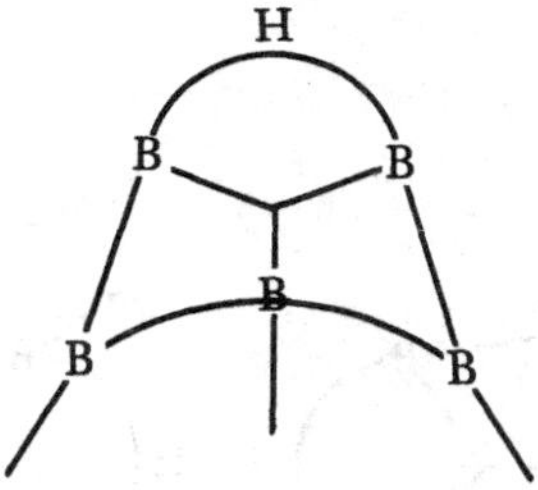

1223 $B_5H_9^{=}$

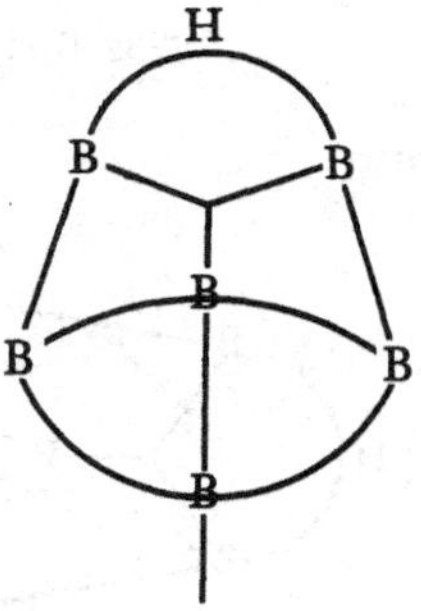

1331 $B_6H_8^{=}$

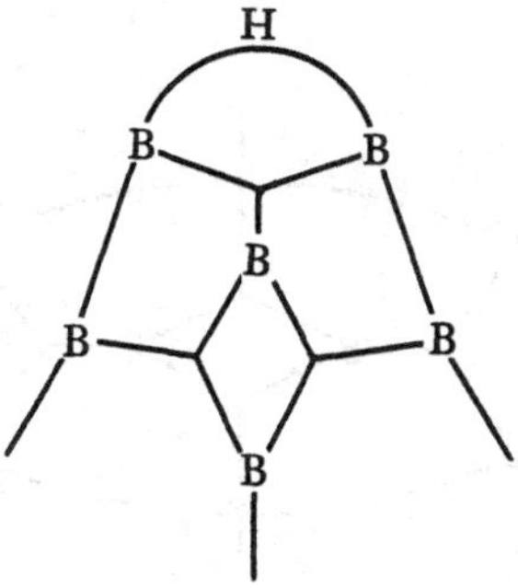

1323 $B_6H_{10}^{=}$

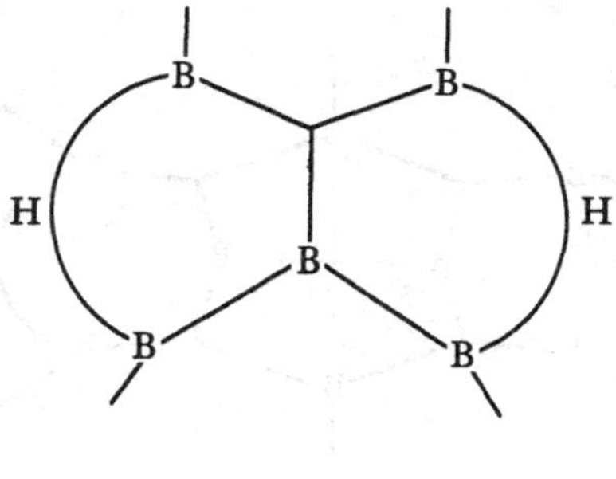

2124 $B_5H_{11}^{=}$

**Figure A–2** *Some plausible −2 ions and intermediates, all assumed to have at least a plane of symmetry, and to be based on icosahedral fragments.*

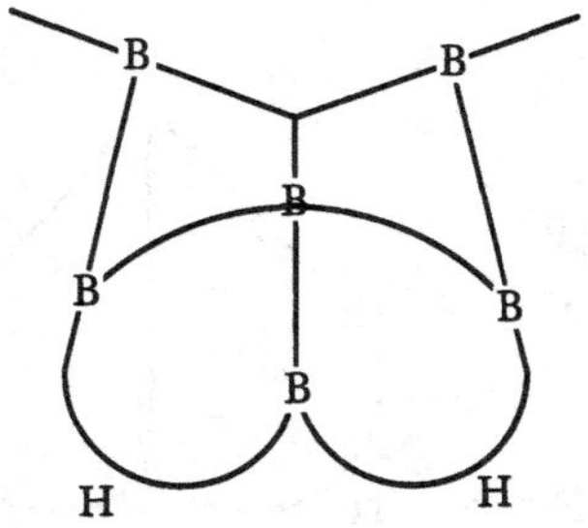

2232 $B_6H_{10}^-$

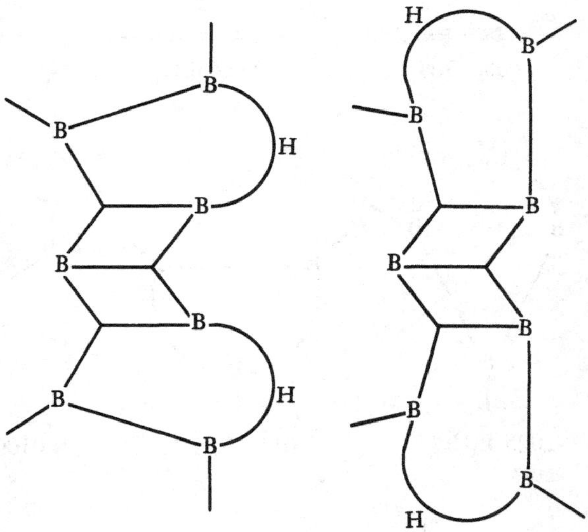

2324 $B_7H_{13}^-$

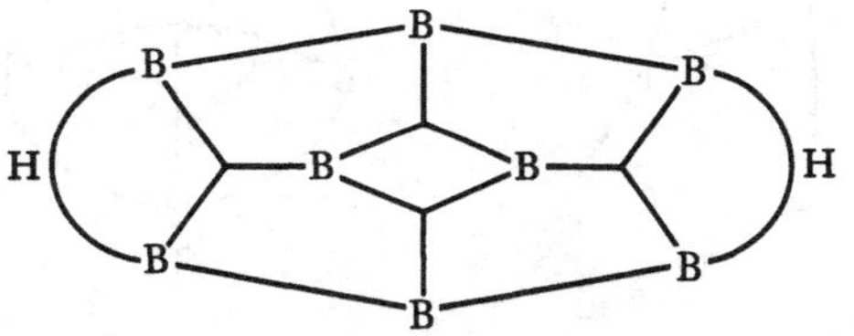

2440 $B_8H_{10}^-$

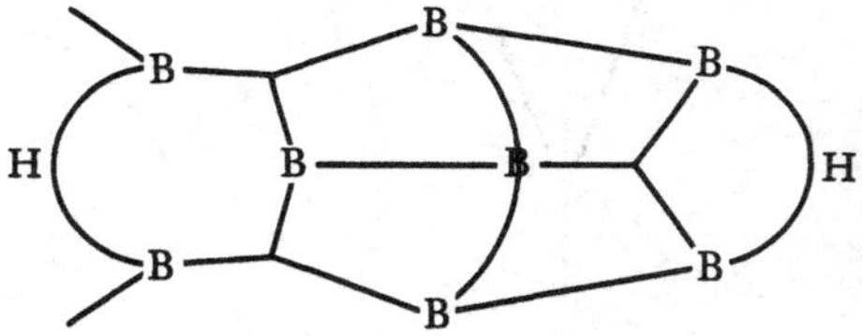

2432 $B_8H_{12}^-$

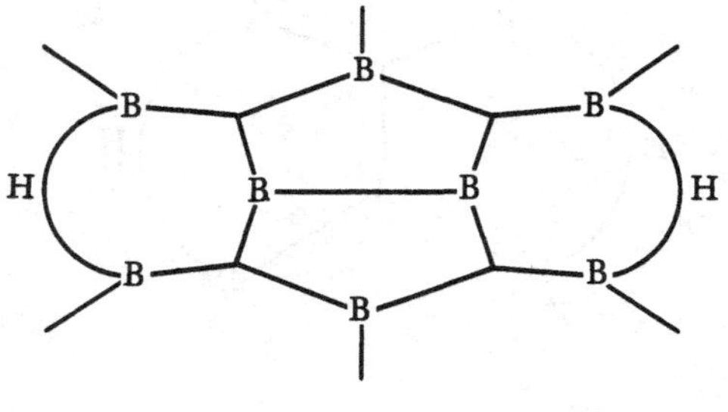

2416 $B_8H_{16}^-$

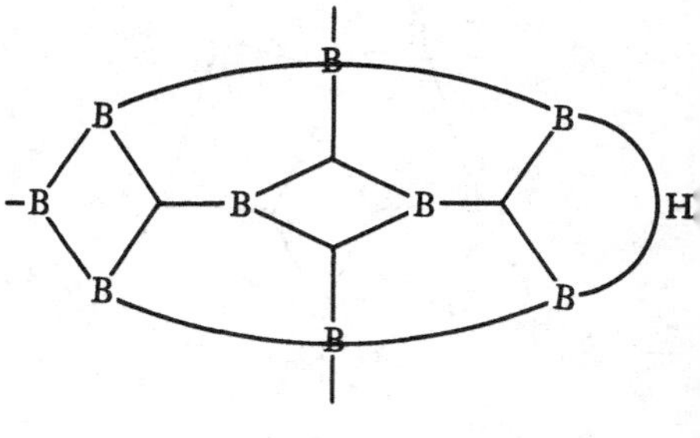

1623 $B_9H_{13}^-$

**Figure A-2** (*continued*)

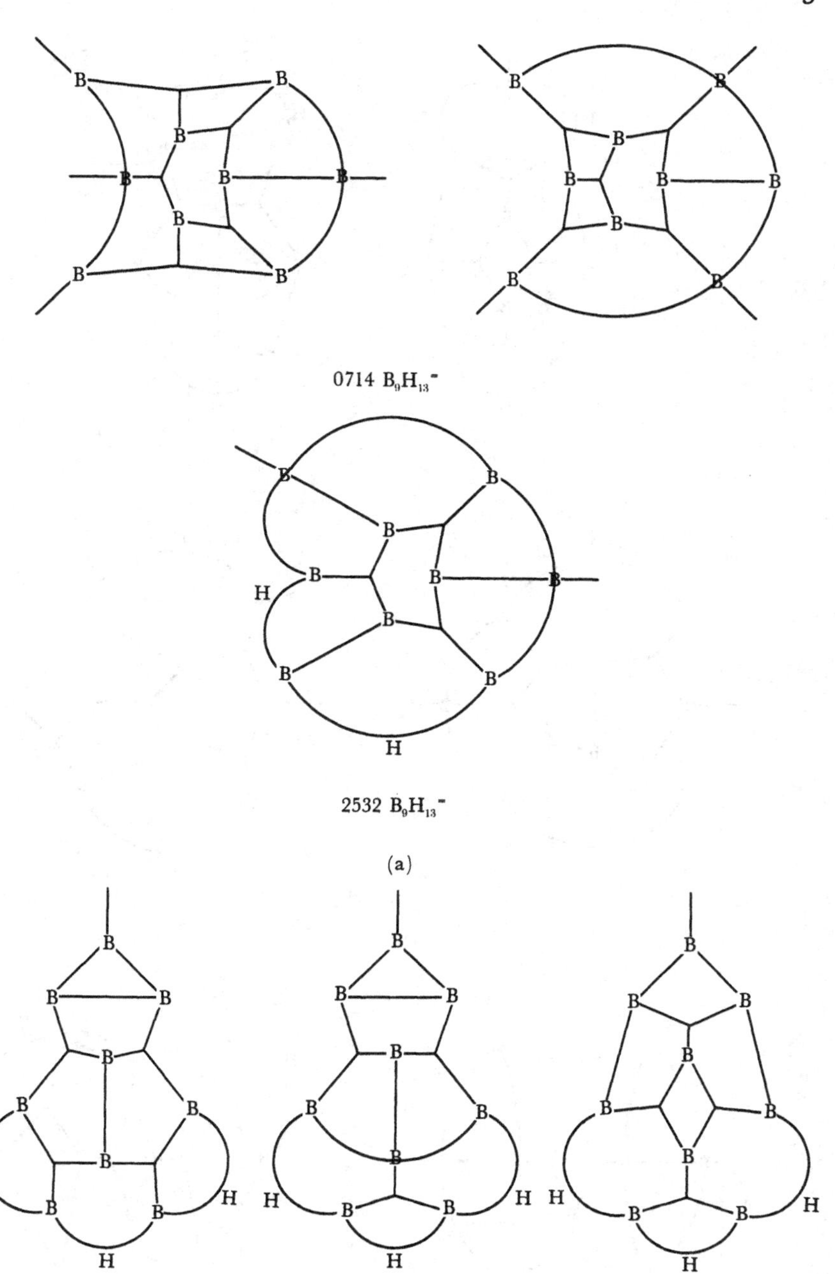


**Figure A–2** *(continued)*

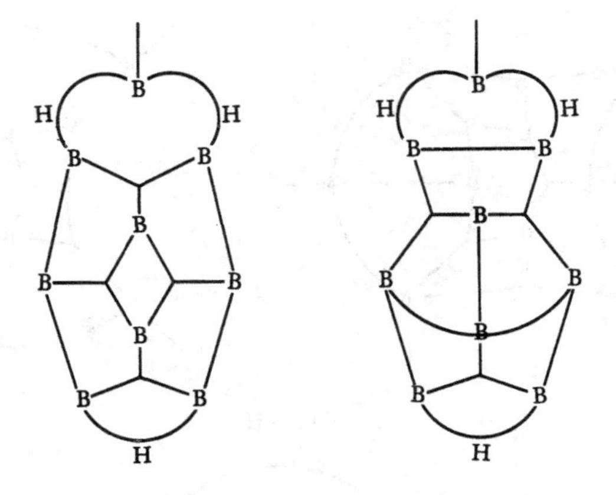

3441 $B_9H_{13}^-$

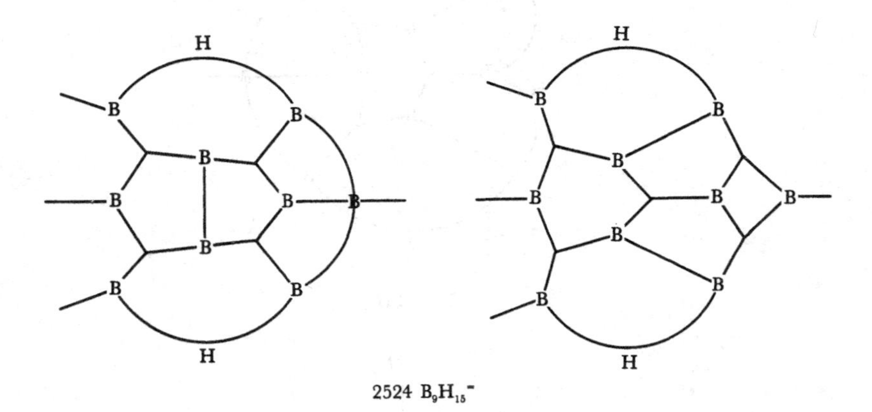

2524 $B_9H_{15}^-$

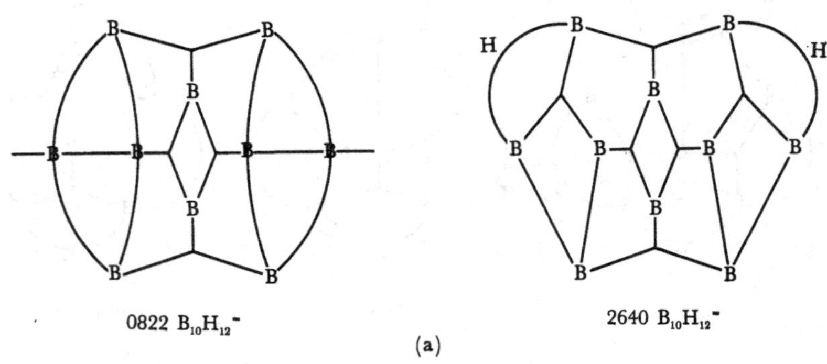

0822 $B_{10}H_{12}^-$

2640 $B_{10}H_{12}^-$

(a)

**Figure A-2** (*continued*)

4450 $B_{10}H_{14}^{-}$

(b)

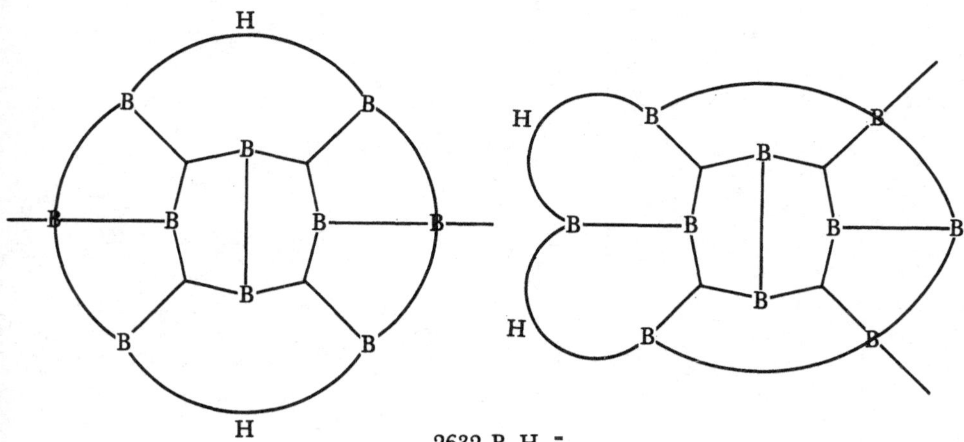

2632 $B_{10}H_{14}^{-}$

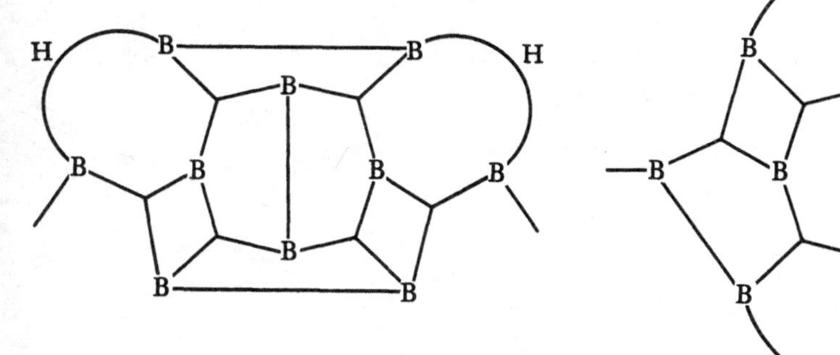

2632 $B_{10}H_{14}^{-}$

**Figure A–2** (*continued*)

# Appendix B

# Spherical Coordinates for Boron Hydride Models

Precise models of the boron hydrides may be constructed from the spherical coordinates and related descriptions given in Figs. B–1 and B–2.

A convenient demonstration scale is 10 cm per A. It is reasonable to drill holes in green balls about $1\frac{3}{4}$ inches in diameter for B atoms, and in blue (terminal) and red (bridge) balls about $\frac{3}{4}$ inch in diameter for H atoms. Brass rods about $\frac{1}{8}$ inch in diameter may be used for connections. In Tables B–1 to B–13 those H atoms are not listed separately when only one hole needs to be drilled in each, but they may be found listed under the B atom to which each is bonded.

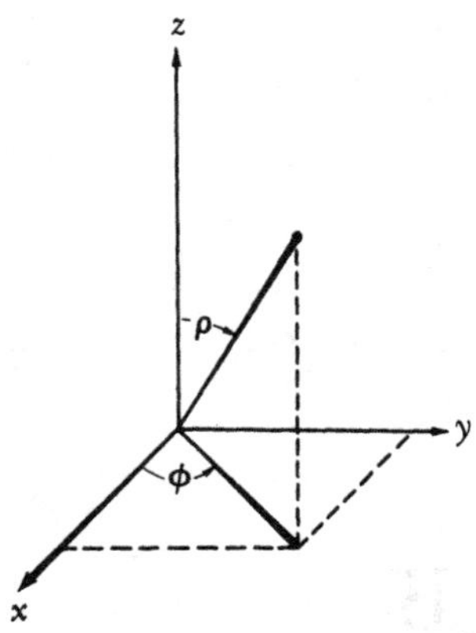


**Figure B–1** *Spherical coordinates.*

**Table B–1** *Diborane $B_2H_6$*

| Center atom | Coordinating atom | Phi* | Rho* | Distance, A |
|---|---|---|---|---|
| 1 B1 | | | | |
| | 2 B2 | 180.0 | 127.1 | 1.770 |
| | 5 H3 | 90.0 | 0 | 1.190 |
| | 6 H4 | 0 | 105.8 | 1.190 |
| | 7 H5 | 234.6 | 113.7 | 1.330 |
| | 8 H6 | 125.4 | 113.7 | 1.330 |
| 2 B2 | | | | |
| | 1 B1 | 180.0 | 127.1 | 1.770 |
| | 3 H1 | 0 | 105.8 | 1.190 |
| | 4 H2 | 90.0 | 0 | 1.190 |
| | 7 H5 | 125.4 | 113.7 | 1.330 |
| | 8 H6 | 234.6 | 113.7 | 1.330 |
| 7 H5 | | | | |
| | 1 B1 | 90.0 | 0 | 1.330 |
| | 2 B2 | 0 | 83.4 | 1.330 |
| 8 H6 | | | | |
| | 1 B1 | 90.0 | 0 | 1.330 |
| | 2 B2 | 0 | 83.4 | 1.330 |

* See Fig. B–1.

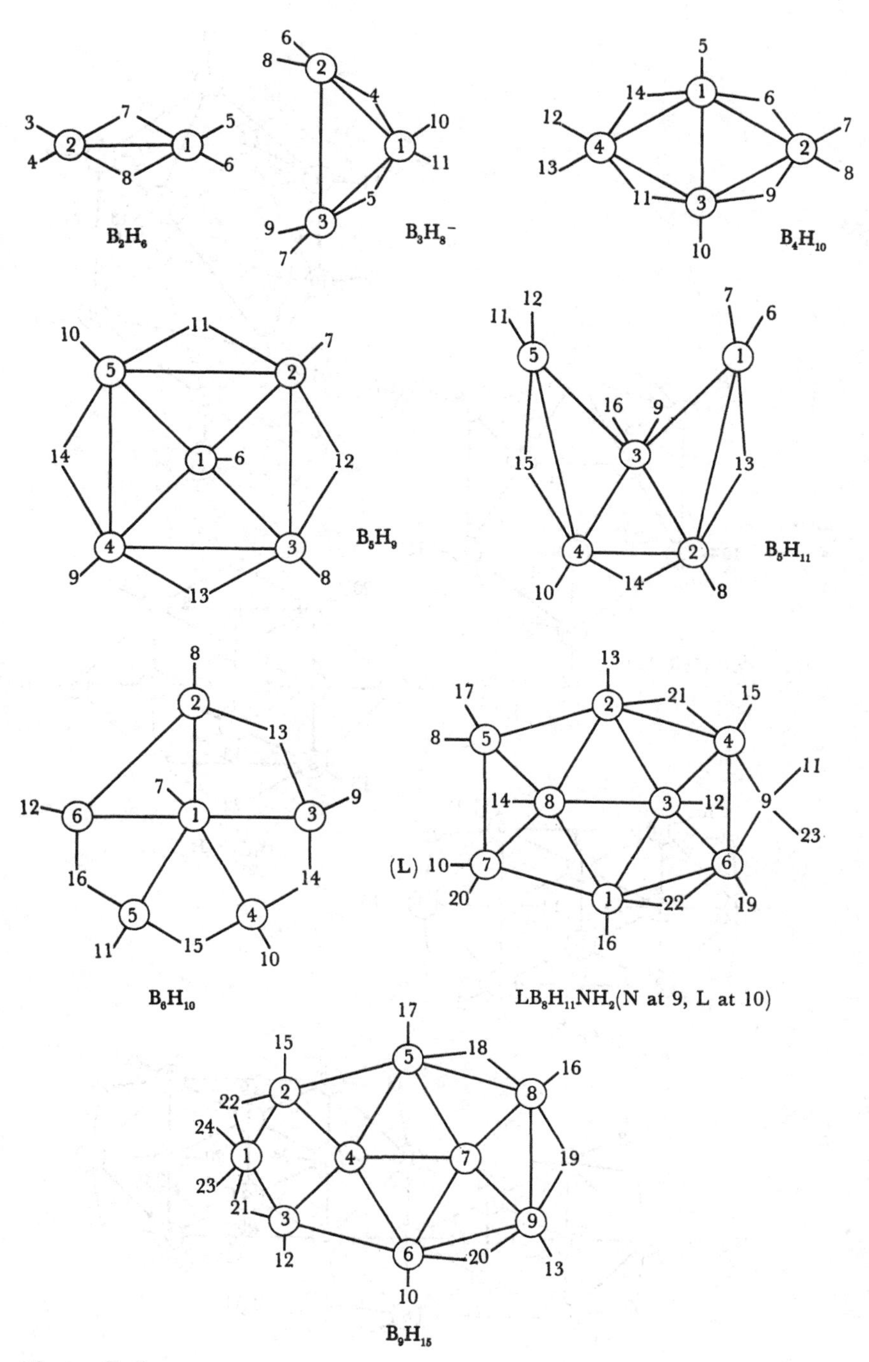

$B_2H_6$
$B_3H_8^-$
$B_4H_{10}$
$B_5H_9$
$B_5H_{11}$
$B_6H_{10}$
$LB_8H_{11}NH_2$(N at 9, L at 10)
(L)
$B_9H_{15}$


**Figure B–2**

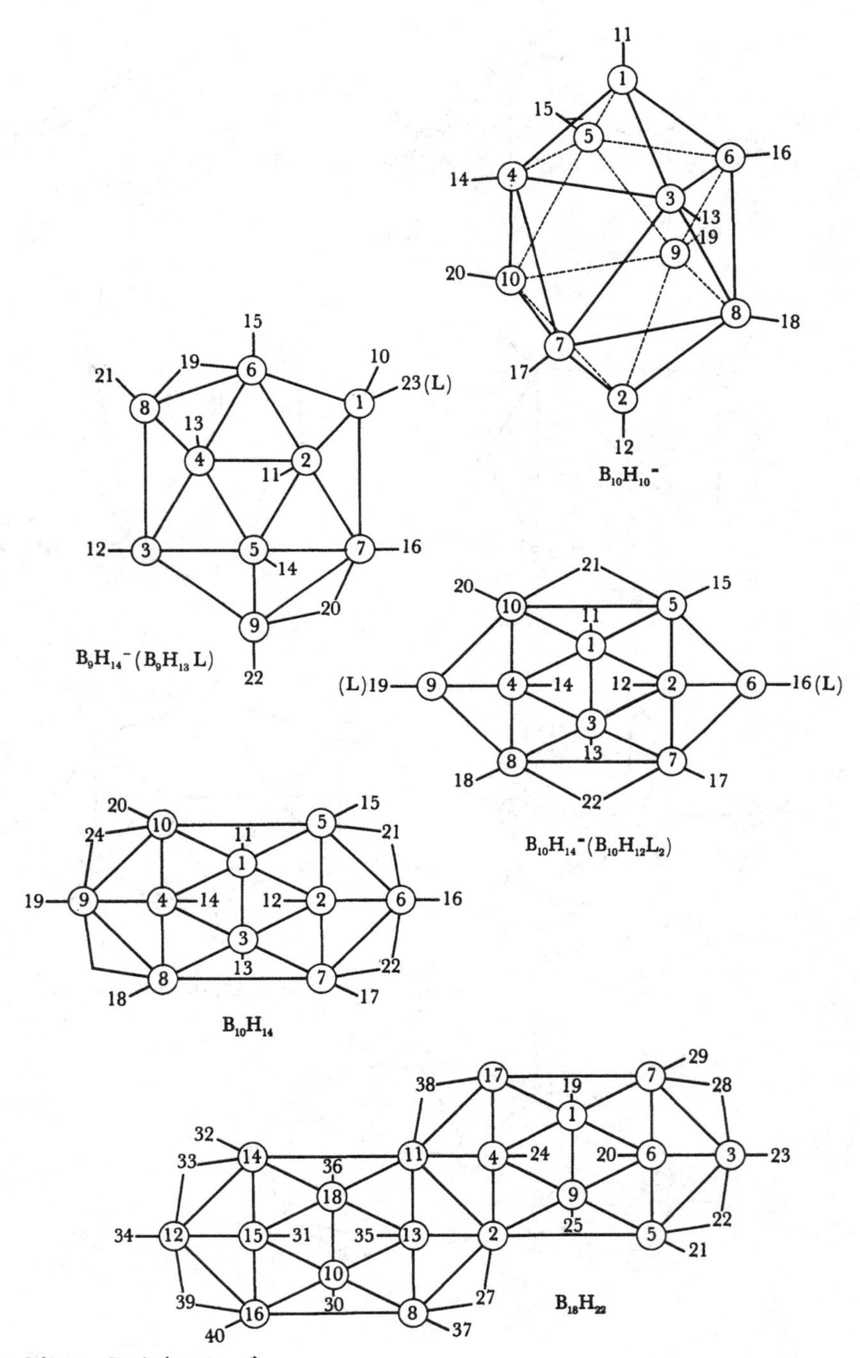


**Figure B–2** (*continued*)

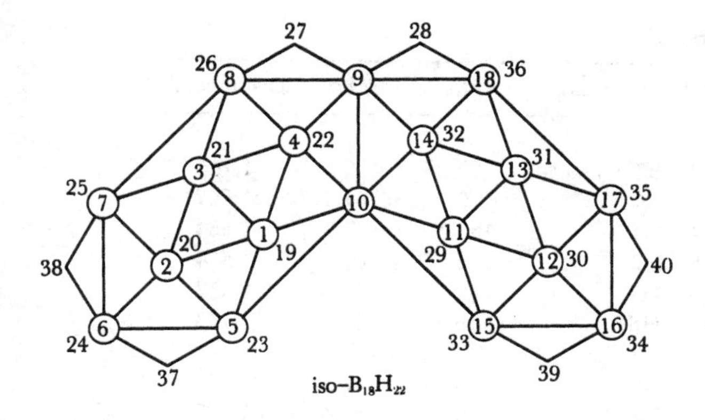


iso-$B_{18}H_{22}$

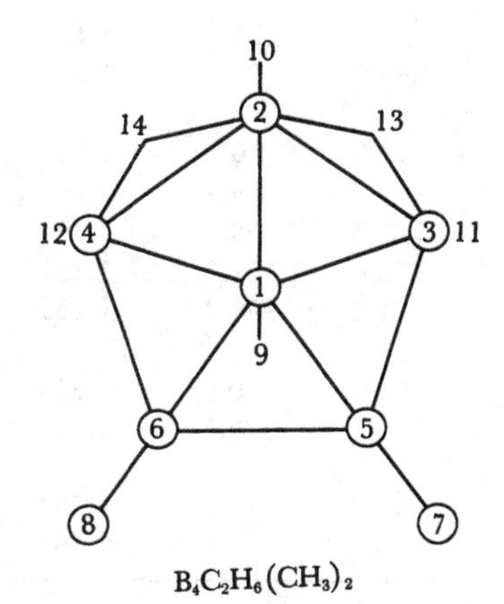


$B_4C_2H_6(CH_3)_2$

**Figure B–2** *(continued)*

**Table B–2** $B_3H_8^-$

| Center atom | Coordinating atom | Phi | Rho | Distance, A |
|---|---|---|---|---|
| 1 B1 | | | | |
| | 2 B2 | 90.0 | 0 | 1.766 |
| | 3 B3 | 0 | 61.1 | 1.767 |
| | 4 H1 | 182.3 | 41.2 | 1.554 |
| | 5 H2 | 358.5 | 102.3 | 1.554 |
| | 10 H7 | 249.7 | 120.5 | 1.037 |
| | 11 H8 | 105.4 | 114.3 | 1.143 |
| 2 B2 | | | | |
| | 1 B1 | 90.0 | 0 | 1.766 |
| | 3 B3 | 0 | 59.4 | 1.797 |
| | 4 H1 | 177.7 | 59.7 | 1.185 |
| | 6 H3 | 99.4 | 121.3 | 1.157 |
| | 8 H5 | 259.0 | 121.4 | 1.156 |
| 3 B3 | | | | |
| | 1 B1 | 90.0 | 0 | 1.767 |
| | 2 B2 | 0 | 59.4 | 1.797 |
| | 5 H2 | 182.3 | 59.7 | 1.184 |
| | 7 H4 | 260.6 | 121.3 | 1.157 |
| | 9 H6 | 101.0 | 121.4 | 1.156 |
| 4 H1 | | | | |
| | 1 B1 | 90.0 | 0 | 1.554 |
| | 2 B2 | 0 | 79.1 | 1.185 |
| 5 H2 | | | | |
| | 1 B1 | 90.0 | 0 | 1.554 |
| | 3 B3 | 0 | 79.1 | 1.184 |

**Table B–3** *Tetraborane* $B_4H_{10}$

| Center atom | Coordinating atom | Phi | Rho | Distance, A |
|---|---|---|---|---|
| 1 B1 | | | | |
| | 2 B2 | 0 | 98.9 | 1.845 |
| | 3 B3 | 51.7 | 62.2 | 1.709 |
| | 4 B4 | 90.0 | 0 | 1.843 |
| | 5 H1 | 155.5 | 128.8 | 1.157 |
| | 6 H2 | 310.0 | 117.1 | 1.129 |
| | 14 H10 | 239.2 | 44.7 | 1.173 |
| 2 B2 | | | | |
| | 1 B1 | 252.9 | 37.7 | 1.845 |
| | 3 B3 | 244.4 | 92.6 | 1.833 |
| | 6 H2 | 90.0 | 0 | 1.424 |
| | 7 H3 | 0 | 104.0 | 1.137 |
| | 8 H4 | 130.4 | 96.3 | 1.085 |
| | 9 H5 | 229.6 | 130.6 | 1.413 |
| 3 B3 | | | | |
| | 1 B1 | 90.0 | 0 | 1.709 |
| | 2 B2 | 115.7 | 62.6 | 1.833 |
| | 4 B4 | 234.3 | 62.5 | 1.838 |
| | 9 H5 | 124.9 | 112.3 | 1.185 |
| | 10 H6 | 0 | 115.6 | 1.063 |
| | 11 H7 | 225.5 | 108.4 | 1.141 |
| 4 B4 | | | | |
| | 1 B1 | 0 | 55.3 | 1.843 |
| | 3 B3 | 90.0 | 0 | 1.838 |
| | 11 H7 | 191.5 | 38.2 | 1.344 |
| | 12 H8 | 101.0 | 112.6 | 1.118 |
| | 13 H9 | 255.0 | 115.9 | 1.023 |
| | 14 H10 | 355.2 | 94.4 | 1.304 |
| 6 H2 | | | | |
| | 1 B1 | 90.0 | 0 | 1.129 |
| | 2 B2 | 0 | 91.8 | 1.424 |
| 9 H5 | | | | |
| | 2 B2 | 90.0 | 0 | 1.413 |
| | 3 B3 | 0 | 89.3 | 1.185 |
| 11 H7 | | | | |
| | 3 B3 | 90.0 | 0 | 1.141 |
| | 4 B4 | 0 | 95.1 | 1.344 |
| 14 H10 | | | | |
| | 1 B1 | 90.0 | 0 | 1.173 |
| | 4 B4 | 0 | 96.0 | 1.304 |

**Table B–4** *Pentaborane* $B_5H_9$

| Center atom | Coordinating atom | Phi | Rho | Distance, A |
|---|---|---|---|---|
| 1 B1 | | | | |
| | 2 B2 | 90.0 | 0 | 1.687 |
| | 3 B3 | 0 | 64.5 | 1.687 |
| | 4 B4 | 56.7 | 98.0 | 1.687 |
| | 5 B5 | 113.4 | 64.5 | 1.687 |
| | 6 H1 | 236.7 | 131.0 | 1.220 |
| 2 B2 | | | | |
| | 1 B1 | 90.0 | 0 | 1.687 |
| | 3 B3 | 0 | 57.8 | 1.800 |
| | 5 B5 | 246.6 | 57.8 | 1.800 |
| | 7 H2 | 123.3 | 136.8 | 1.220 |
| | 11 H6 | 258.9 | 104.5 | 1.350 |
| | 12 H7 | 347.7 | 104.5 | 1.350 |
| 3 B3 | | | | |
| | 1 B1 | 90.0 | 0 | 1.687 |
| | 2 B2 | 246.6 | 57.8 | 1.800 |
| | 4 B4 | 0 | 57.8 | 1.800 |
| | 8 H3 | 123.3 | 136.8 | 1.220 |
| | 12 H7 | 258.9 | 104.5 | 1.350 |
| | 13 H8 | 347.7 | 104.5 | 1.350 |
| 4 B4 | | | | |
| | 1 B1 | 90.0 | 0 | 1.687 |
| | 3 B3 | 246.6 | 57.8 | 1.800 |
| | 5 B5 | 0 | 57.8 | 1.800 |
| | 9 H4 | 123.3 | 136.8 | 1.220 |
| | 13 H8 | 258.9 | 104.5 | 1.350 |
| | 14 H9 | 347.7 | 104.5 | 1.350 |
| 5 B5 | | | | |
| | 1 B1 | 90.0 | 0 | 1.687 |
| | 2 B2 | 0 | 57.8 | 1.800 |
| | 4 B4 | 246.6 | 57.8 | 1.800 |
| | 10 H5 | 123.3 | 136.8 | 1.220 |
| | 11 H6 | 347.7 | 104.5 | 1.350 |
| | 14 H9 | 258.9 | 104.5 | 1.350 |
| 11 H6 | | | | |
| | 2 B2 | 90.0 | 0 | 1.350 |
| | 5 B5 | 0 | 83.6 | 1.350 |
| 12 H7 | | | | |
| | 2 B2 | 90.0 | 0 | 1.350 |
| | 3 B3 | 0 | 83.6 | 1.350 |
| 13 H8 | | | | |
| | 3 B3 | 90.0 | 0 | 1.350 |
| | 4 B4 | 0 | 83.6 | 1.350 |
| 14 H9 | | | | |
| | 4 B4 | 90.0 | 0 | 1.350 |
| | 5 B5 | 0 | 83.6 | 1.350 |

Table B–5 $B_5H_{11}$

| Center atom | Coordinating atom | Phi | Rho | Distance, A |
|---|---|---|---|---|
| 1 B1 | | | | |
| | 2 B2 | 0 | 56.9 | 1.751 |
| | 3 B3 | 90.0 | 0 | 1.867 |
| | 6 H1 | 259.5 | 114.9 | 1.136 |
| | 7 H2 | 114.9 | 113.6 | 1.139 |
| | 13 H8 | 20.1 | 88.9 | 1.168 |
| 2 B2 | | | | |
| | 1 B1 | 90.0 | 0 | 1.751 |
| | 3 B3 | 0 | 64.9 | 1.728 |
| | 4 B4 | 36.7 | 111.1 | 1.771 |
| | 8 H3 | 244.8 | 123.1 | 1.125 |
| | 13 H8 | 145.3 | 40.7 | 1.081 |
| | 14 H9 | 83.6 | 124.1 | 1.015 |
| 3 B3 | | | | |
| | 1 B1 | 90.0 | 0 | 1.867 |
| | 2 B2 | 0 | 58.2 | 1.728 |
| | 4 B4 | 322.4 | 108.7 | 1.706 |
| | 5 B5 | 259.9 | 108.3 | 1.868 |
| | 9 H4 | 108.5 | 125.4 | 1.020 |
| | 16 H11 | 210.3 | 67.7 | 1.085 |
| 4 B4 | | | | |
| | 2 B2 | 90.0 | 0 | 1.771 |
| | 3 B3 | 0 | 59.6 | 1.706 |
| | 5 B5 | 41.9 | 110.7 | 1.765 |
| | 10 H5 | 258.4 | 120.7 | 1.029 |
| | 14 H9 | 157.2 | 34.0 | 1.242 |
| | 15 H10 | 100.2 | 124.7 | 1.367 |
| 5 B5 | | | | |
| | 3 B3 | 90.0 | 0 | 1.868 |
| | 4 B4 | 0 | 55.9 | 1.765 |
| | 11 H6 | 240.3 | 113.6 | 1.048 |
| | 12 H7 | 96.8 | 108.6 | 1.195 |
| | 15 H10 | 343.1 | 102.5 | 1.436 |
| 13 H8 | | | | |
| | 1 B1 | 90.0 | 0 | 1.168 |
| | 2 B2 | 0 | 102.3 | 1.081 |
| 14 H9 | | | | |
| | 2 B2 | 90.0 | 0 | 1.015 |
| | 4 B4 | 0 | 102.9 | 1.242 |
| 15 H10 | | | | |
| | 4 B4 | 90.0 | 0 | 1.367 |
| | 5 B5 | 0 | 78.0 | 1.436 |

**Table B–6** $B_6H_{10}$

| Center atom | Coordinating atom | Phi | Rho | Distance, A |
|---|---|---|---|---|
| 1 B1 | | | | |
| | 2 B2 | 90.0 | 0 | 1.795 |
| | 3 B3 | 0 | 58.6 | 1.752 |
| | 4 B4 | 320.5 | 107.4 | 1.740 |
| | 5 B5 | 256.8 | 101.2 | 1.752 |
| | 6 B6 | 221.5 | 52.8 | 1.795 |
| | 7 H1 | 100.9 | 135.8 | 1.245 |
| 2 B2 | | | | |
| | 1 B1 | 131.0 | 63.6 | 1.795 |
| | 3 B3 | 168.6 | 110.6 | 1.737 |
| | 6 B6 | 90.0 | 0 | 1.596 |
| | 8 H2 | 0 | 137.8 | 1.283 |
| | 13 H7 | 220.7 | 113.1 | 1.355 |
| 3 B3 | | | | |
| | 1 B1 | 90.0 | 0 | 1.752 |
| | 2 B2 | 0 | 61.9 | 1.737 |
| | 4 B4 | 223.6 | 58.8 | 1.794 |
| | 9 H3 | 106.4 | 119.4 | 1.178 |
| | 13 H7 | 331.0 | 103.9 | 1.310 |
| | 14 H8 | 246.1 | 99.9 | 1.483 |
| 4 B4 | | | | |
| | 1 B1 | 318.6 | 59.4 | 1.740 |
| | 3 B3 | 90.0 | 0 | 1.794 |
| | 5 B5 | 0 | 103.5 | 1.794 |
| | 10 H4 | 203.6 | 125.8 | 1.138 |
| | 14 H8 | 107.1 | 54.3 | 1.324 |
| | 15 H9 | 56.6 | 111.7 | 1.324 |
| 5 B5 | | | | |
| | 1 B1 | 320.4 | 58.8 | 1.752 |
| | 4 B4 | 90.0 | 0 | 1.794 |
| | 6 B6 | 0 | 107.6 | 1.737 |
| | 11 H5 | 214.9 | 126.5 | 1.178 |
| | 15 H9 | 108.9 | 46.5 | 1.483 |
| | 16 H10 | 53.7 | 111.9 | 1.310 |
| 6 B6 | | | | |
| | 1 B1 | 37.6 | 63.6 | 1.795 |
| | 2 B2 | 90.0 | 0 | 1.596 |
| | 5 B5 | 0 | 110.6 | 1.737 |
| | 12 H6 | 168.6 | 137.8 | 1.283 |
| | 16 H10 | 307.9 | 113.1 | 1.355 |
| 13 H7 | | | | |
| | 2 B2 | 90.0 | 0 | 1.355 |
| | 3 B3 | 0 | 81.3 | 1.310 |

**Table B–6** (*continued*)

| Center atom | Coordinating atom | Phi | Rho | Distance, A |
|---|---|---|---|---|
| 14 H8 | | | | |
| | 3 B3 | 90.0 | 0 | 1.483 |
| | 4 B4 | 0 | 79.2 | 1.324 |
| 15 H9 | | | | |
| | 4 B4 | 90.0 | 0 | 1.324 |
| | 5 B5 | 0 | 79.2 | 1.483 |
| 16 H10 | | | | |
| | 5 B5 | 90.0 | 0 | 1.310 |
| | 6 B6 | 0 | 81.3 | 1.355 |

**Table B-7** *($B_8H_{12}^-$)NHEt, i.e., Et $NH_2B_8H_{11}$NHEt*

| Center atom | Coordinating atom | Phi | Rho | Distance, A |
|---|---|---|---|---|
| 1 B91 | | | | |
| | 3 B4 | 90.0 | 0 | 1.798 |
| | 6 B98 | 0 | 59.4 | 1.860 |
| | 7 B101 | 110.1 | 110.5 | 1.933 |
| | 8 B82 | 135.5 | 60.5 | 1.753 |
| | 16 H91 | 245.3 | 118.6 | 1.096 |
| | 22 X–91–98 | 8.0 | 107.5 | 1.308 |
| 2 B5 | | | | |
| | 3 B4 | 90.0 | 0 | 1.774 |
| | 4 B90 | 0 | 60.4 | 1.847 |
| | 5 B92 | 251.7 | 109.8 | 1.907 |
| | 8 B82 | 221.0 | 61.6 | 1.720 |
| | 13 H5 | 103.7 | 108.7 | 1.107 |
| | 21 X–5–90 | 353.7 | 106.9 | 1.296 |
| 3 B4 | | | | |
| | 1 B91 | 90.0 | 0 | 1.798 |
| | 2 B5 | 0 | 109.4 | 1.774 |
| | 4 B90 | 67.4 | 116.6 | 1.821 |
| | 6 B98 | 107.3 | 62.0 | 1.814 |
| | 8 B82 | 331.7 | 58.5 | 1.790 |
| | 12 H4 | 221.0 | 121.7 | 1.235 |
| 4 B90 | | | | |
| | 2 B5 | 90.0 | 0 | 1.847 |
| | 3 B4 | 0 | 57.8 | 1.821 |
| | 6 B98 | 324.2 | 102.8 | 1.998 |
| | 9 N103 | 272.9 | 114.1 | 1.581 |
| | 15 H90 | 98.4 | 123.8 | 1.078 |
| | 21 X–5–90 | 188.3 | 44.6 | 1.349 |

Table B–7 (*continued*)

| Center atom | Coordinating atom | Phi | Rho | Distance, A |
|---|---|---|---|---|
| 5 B92 | | | | |
| | 2 B5 | 90.0 | 0 | 1.907 |
| | 7 B101 | 0 | 106.5 | 1.924 |
| | 8 B82 | 333.7 | 56.2 | 1.728 |
| | 17 H92(T) | 233.1 | 114.5 | 1.128 |
| | 18 H92(2) | 101.3 | 98.1 | 1.103 |
| 6 B98 | | | | |
| | 1 B91 | 90.0 | 0 | 1.860 |
| | 3 B4 | 0 | 58.6 | 1.814 |
| | 4 B90 | 32.9 | 105.9 | 1.998 |
| | 9 N103 | 86.7 | 118.3 | 1.552 |
| | 19 H98 | 250.1 | 120.4 | 1.173 |
| | 22 X–91–98 | 169.8 | 44.6 | 1.398 |
| 7 B101 | | | | |
| | 1 B91 | 90.0 | 0 | 1.933 |
| | 5 B92 | 0 | 103.2 | 1.924 |
| | 8 B82 | 33.4 | 56.9 | 1.728 |
| | 10 H87 | 131.6 | 117.0 | 1.562 |
| | 20 H101 | 252.0 | 100.5 | 1.021 |
| 8 B82 | | | | |
| | 1 B91 | 90.0 | 0 | 1.753 |
| | 2 B5 | 0 | 114.2 | 1.720 |
| | 3 B4 | 30.3 | 61.0 | 1.790 |
| | 5 B92 | 283.2 | 120.5 | 1.728 |
| | 7 B101 | 239.5 | 67.5 | 1.728 |
| | 14 H82 | 140.9 | 116.7 | 1.215 |
| 9 N103 | | | | |
| | 4 B90 | 90.0 | 0 | 1.581 |
| | 6 B98 | 0 | 79.3 | 1.552 |
| | 11 H103 | 245.4 | 116.3 | 0.938 |
| | 23 ethyl | 117.3 | 115.8 | 1.490 |
| 21 X–5–90 | | | | |
| | 2 B5 | 90.0 | 0 | 1.296 |
| | 4 B90 | 0 | 88.5 | 1.349 |
| 22 X–91–98 | | | | |
| | 1 B91 | 90.0 | 0 | 1.308 |
| | 6 B98 | 0 | 86.7 | 1.398 |

**Table B–8** $B_9H_{15}$

| Center atom | Coordinating atom | Phi | Rho | Distance, A |
|---|---|---|---|---|
| 1 B1 | | | | |
| | 2 B2 | 90.0 | 0 | 1.832 |
| | 3 B3 | 0 | 58.1 | 1.877 |
| | 21 H12 | 343.7 | 88.8 | 1.485 |
| | 22 H13 | 187.2 | 49.7 | 1.564 |
| | 23 H14 | 85.0 | 106.4 | 1.088 |
| | 24 H15 | 255.8 | 122.2 | 0.912 |
| 2 B2 | | | | |
| | 1 B1 | 142.9 | 62.2 | 1.832 |
| | 3 B3 | 90.0 | 0 | 1.801 |
| | 4 B4 | 0 | 58.0 | 1.792 |
| | 5 B5 | 31.3 | 104.8 | 1.963 |
| | 15 H6 | 262.1 | 118.8 | 1.221 |
| | 22 H13 | 136.2 | 117.3 | 1.448 |
| 3 B3 | | | | |
| | 1 B1 | 90.0 | 0 | 1.877 |
| | 2 B2 | 0 | 59.7 | 1.801 |
| | 4 B4 | 325.7 | 110.7 | 1.741 |
| | 6 B6 | 263.3 | 115.5 | 1.954 |
| | 12 H3 | 117.7 | 119.0 | 1.103 |
| | 21 H12 | 209.9 | 52.2 | 1.062 |
| 4 B4 | | | | |
| | 2 B2 | 90.0 | 0 | 1.792 |
| | 3 B3 | 0 | 61.3 | 1.741 |
| | 5 B5 | 142.5 | 67.1 | 1.758 |
| | 6 B6 | 40.1 | 116.8 | 1.775 |
| | 7 B7 | 113.0 | 122.4 | 1.764 |
| | 11 H2 | 250.4 | 126.0 | 1.177 |
| 5 B5 | | | | |
| | 2 B2 | 90.0 | 0 | 1.963 |
| | 4 B4 | 0 | 57.3 | 1.758 |
| | 7 B7 | 25.6 | 111.1 | 1.817 |
| | 8 B8 | 89.8 | 122.4 | 1.799 |
| | 17 H8 | 237.9 | 107.9 | 1.035 |
| | 18 H9 | 124.3 | 89.6 | 1.351 |
| 6 B6 | | | | |
| | 3 B3 | 90.0 | 0 | 1.954 |
| | 4 B4 | 0 | 55.4 | 1.775 |
| | 7 B7 | 335.1 | 108.9 | 1.834 |
| | 9 B9 | 271.5 | 122.7 | 1.777 |
| | 10 H1 | 126.5 | 116.9 | 1.333 |
| | 20 H11 | 243.0 | 105.0 | 1.270 |

**Table B–8** (*continued*)

| Center atom | Coordinating atom | Phi | Rho | Distance, A |
|---|---|---|---|---|
| 7 B7 | | | | |
| | 4 B4 | 90.0 | 0 | 1.764 |
| | 5 B5 | 0 | 58.8 | 1.817 |
| | 6 B6 | 233.4 | 59.1 | 1.834 |
| | 8 B8 | 324.5 | 109.0 | 1.742 |
| | 9 B9 | 257.8 | 108.3 | 1.754 |
| | 14 H5 | 113.5 | 106.9 | 1.195 |
| 8 B8 | | | | |
| | 5 B5 | 90.0 | 0 | 1.799 |
| | 7 B7 | 0 | 61.7 | 1.742 |
| | 9 B9 | 38.5 | 107.4 | 1.820 |
| | 16 H7 | 225.5 | 121.1 | 1.009 |
| | 18 H9 | 147.5 | 48.4 | 1.302 |
| | 19 H10 | 90.3 | 105.0 | 1.313 |
| 9 B9 | | | | |
| | 6 B6 | 90.0 | 0 | 1.777 |
| | 7 B7 | 37.8 | 62.6 | 1.754 |
| | 8 B8 | 0 | 108.0 | 1.820 |
| | 13 H4 | 164.0 | 132.4 | 1.149 |
| | 19 H10 | 311.9 | 106.1 | 1.394 |
| | 20 H11 | 242.3 | 43.7 | 0.955 |
| 18 H9 | | | | |
| | 5 B5 | 90.0 | 0 | 1.351 |
| | 8 B8 | 0 | 85.4 | 1.302 |
| 19 H10 | | | | |
| | 8 B8 | 90.0 | 0 | 1.313 |
| | 9 B9 | 0 | 84.5 | 1.394 |
| 20 H11 | | | | |
| | 6 B6 | 90.0 | 0 | 1.270 |
| | 9 B9 | 0 | 105.1 | 0.955 |
| 21 H12 | | | | |
| | 1 B1 | 90.0 | 0 | 1.485 |
| | 3 B3 | 0 | 93.4 | 1.062 |
| 22 H13 | | | | |
| | 1 B1 | 90.0 | 0 | 1.564 |
| | 2 B2 | 0 | 74.8 | 1.448 |

**Table B–9** $B_9H_{14}^-$ ($B_9H_{13}NCCH_3$)

| Center atom | Coordinating atom | Phi | Rho | Distance, A |
|---|---|---|---|---|
| 1 B1 | | | | |
| | 2 B2 | 327.4 | 57.9 | 1.736 |
| | 6 B6 | 90.0 | 0 | 1.873 |
| | 7 B7 | 0 | 106.7 | 1.873 |
| | 10 H1 | 101.4 | 98.4 | 1.134 |
| | 23 H+ | 229.1 | 116.0 | 1.197 |
| 2 B2 | | | | |
| | 1 B1 | 90.0 | 0 | 1.736 |
| | 4 B4 | 0 | 114.6 | 1.763 |
| | 5 B5 | 69.4 | 114.6 | 1.763 |
| | 6 B6 | 323.5 | 65.0 | 1.752 |
| | 7 B7 | 106.0 | 65.0 | 1.752 |
| | 11 H2 | 214.7 | 118.4 | 0.957 |
| 3 B3 | | | | |
| | 4 B4 | 90.0 | 0 | 1.761 |
| | 5 B5 | 0 | 62.4 | 1.761 |
| | 8 B8 | 212.3 | 56.5 | 1.867 |
| | 9 B9 | 331.3 | 111.6 | 1.867 |
| | 12 H3 | 113.4 | 123.3 | 1.074 |
| 4 B4 | | | | |
| | 2 B2 | 90.0 | 0 | 1.763 |
| | 3 B3 | 0 | 103.9 | 1.761 |
| | 5 B5 | 39.3 | 58.8 | 1.826 |
| | 6 B6 | 255.1 | 59.2 | 1.783 |
| | 8 B8 | 291.8 | 112.2 | 1.719 |
| | 13 H4 | 141.9 | 126.3 | 1.193 |
| 5 B5 | | | | |
| | 2 B2 | 90.0 | 0 | 1.763 |
| | 3 B3 | 0 | 103.9 | 1.761 |
| | 4 B4 | 320.7 | 58.8 | 1.826 |
| | 7 B7 | 104.9 | 59.2 | 1.783 |
| | 9 B9 | 68.2 | 112.2 | 1.719 |
| | 14 H5 | 218.1 | 126.3 | 1.193 |
| 6 B6 | | | | |
| | 1 B1 | 90.0 | 0 | 1.873 |
| | 2 B2 | 0 | 57.1 | 1.752 |
| | 4 B4 | 325.9 | 107.3 | 1.783 |
| | 8 B8 | 265.4 | 115.0 | 1.846 |
| | 15 H6 | 109.4 | 120.2 | 1.172 |
| | 19 H10 | 226.1 | 87.1 | 1.206 |

**Table B–9** (*continued*)

| Center atom | Coordinating atom | Phi | Rho | Distance, A |
|---|---|---|---|---|
| 7 B7 | | | | |
| | 1 B1 | 90.0 | 0 | 1.873 |
| | 2 B2 | 0 | 57.1 | 1.752 |
| | 5 B5 | 34.1 | 107.3 | 1.783 |
| | 9 B9 | 94.6 | 115.0 | 1.846 |
| | 16 H7 | 250.6 | 120.2 | 1.172 |
| | 20 H11 | 133.9 | 87.1 | 1.206 |
| 8 B8 | | | | |
| | 3 B3 | 90.0 | 0 | 1.867 |
| | 4 B4 | 0 | 58.6 | 1.719 |
| | 6 B6 | 42.5 | 102.5 | 1.846 |
| | 17 H8 | 252.3 | 121.8 | 1.044 |
| | 19 H10 | 83.8 | 114.2 | 1.355 |
| | 21 H12 | 169.5 | 81.9 | 1.080 |
| 9 B9 | | | | |
| | 3 B3 | 90.0 | 0 | 1.867 |
| | 5 B5 | 0 | 58.6 | 1.719 |
| | 7 B7 | 317.5 | 102.5 | 1.846 |
| | 18 H9 | 107.7 | 121.8 | 1.044 |
| | 20 H11 | 276.2 | 114.2 | 1.355 |
| | 22 H13 | 190.5 | 81.9 | 1.080 |
| 19 H10 | | | | |
| | 6 B6 | 90.0 | 0 | 1.206 |
| | 8 B8 | 0 | 92.1 | 1.355 |
| 20 H11 | | | | |
| | 7 B7 | 90.0 | 0 | 1.206 |
| | 9 B9 | 0 | 92.1 | 1.355 |

**Table B–10** $B_{10}H_{10}^{=}$

| Center atom | Coordinating atom | Phi | Rho | Distance, A |
|---|---|---|---|---|
| 1 B1 | | | | |
| | 6 B6 | 90.0 | 0 | 1.733 |
| | 4 B4 | 0 | 98.8 | 1.733 |
| | 5 B5 | 303.1 | 64.9 | 1.733 |
| | 3 B3 | 56.9 | 64.9 | 1.733 |
| | 11 H1 | 180.0 | 130.6 | 1.200 |
| 2 B2 | | | | |
| | 9 B9 | 90.0 | 0 | 1.733 |
| | 8 B8 | 0 | 64.9 | 1.733 |
| | 7 B7 | 56.9 | 98.8 | 1.733 |
| | 10 B10 | 113.9 | 64.9 | 1.733 |
| | 12 H2 | 236.9 | 130.6 | 1.200 |
| 6 B6 | | | | |
| | 1 B1 | 90.0 | 0 | 1.733 |
| | 5 B5 | 0 | 57.5 | 1.860 |
| | 3 B3 | 246.1 | 57.5 | 1.860 |
| | 9 B9 | 336.8 | 112.3 | 1.810 |
| | 8 B8 | 269.3 | 112.3 | 1.810 |
| | 16 H6 | 123.1 | 109.6 | 1.200 |
| 4 B4 | | | | |
| | 1 B1 | 90.0 | 0 | 1.733 |
| | 5 B5 | 0 | 57.5 | 1.860 |
| | 3 B3 | 113.9 | 57.5 | 1.860 |
| | 6 B6 | 90.7 | 112.3 | 1.810 |
| | 10 B10 | 23.2 | 112.3 | 1.810 |
| | 14 H4 | 236.9 | 109.6 | 1.200 |
| 5 B5 | | | | |
| | 1 B1 | 90.0 | 0 | 1.733 |
| | 6 B6 | 0 | 57.5 | 1.860 |
| | 4 B4 | 113.9 | 57.5 | 1.860 |
| | 9 B9 | 23.2 | 112.3 | 1.810 |
| | 10 B10 | 90.7 | 112.3 | 1.810 |
| | 15 H5 | 236.9 | 109.6 | 1.200 |
| 3 B3 | | | | |
| | 1 B1 | 90.0 | 0 | 1.733 |
| | 6 B6 | 0 | 57.5 | 1.860 |
| | 4 B4 | 246.1 | 57.5 | 1.860 |
| | 8 B8 | 336.8 | 112.3 | 1.810 |
| | 7 B7 | 269.3 | 112.3 | 1.810 |
| | 13 H3 | 123.1 | 109.6 | 1.200 |

**Table B–10** (*continued*)

| Center atom | Coordinating atom | Phi | Rho | Distance, A |
|---|---|---|---|---|
| 9 B9 | | | | |
| | 2 B2 | 90.0 | 0 | 1.733 |
| | 6 B6 | 0 | 112.3 | 1.810 |
| | 5 B5 | 292.5 | 112.3 | 1.810 |
| | 8 B8 | 23.2 | 57.5 | 1.860 |
| | 10 B10 | 269.3 | 57.5 | 1.860 |
| | 19 H9 | 146.3 | 109.6 | 1.201 |
| 8 B8 | | | | |
| | 2 B2 | 90.0 | 0 | 1.733 |
| | 6 B6 | 0 | 112.3 | 1.810 |
| | 9 B9 | 336.8 | 57.5 | 1.860 |
| | 7 B7 | 90.7 | 57.5 | 1.860 |
| | 6 B6 | 67.5 | 112.3 | 1.810 |
| | 18 H8 | 213.7 | 109.6 | 1.201 |
| 7 B7 | | | | |
| | 2 B2 | 90.0 | 0 | 1.733 |
| | 10 B10 | 90.7 | 57.5 | 1.860 |
| | 4 B4 | 67.5 | 112.3 | 1.810 |
| | 3 B3 | 0 | 112.3 | 1.810 |
| | 8 B8 | 336.8 | 57.5 | 1.860 |
| | 17 H7 | 213.7 | 109.6 | 1.201 |
| 10 B10 | | | | |
| | 2 B2 | 90.0 | 0 | 1.733 |
| | 4 B4 | 0 | 112.3 | 1.810 |
| | 5 B5 | 67.5 | 112.3 | 1.810 |
| | 9 B9 | 90.7 | 57.5 | 1.860 |
| | 7 B7 | 336.8 | 57.5 | 1.860 |
| | 20 H10 | 213.7 | 109.6 | 1.201 |

**Table B–11** $B_{10}H_{14}^{=}$ $[B_{10}H_{12}(NCCH_3)_2]$

| Center atom | Coordinating atom | Phi | Rho | Distance, A |
|---|---|---|---|---|
| 1 B1 | | | | |
| | 2 B2 | 90.0 | 0 | 1.771 |
| | 3 B3 | 0 | 57.8 | 1.839 |
| | 4 B4 | 321.2 | 103.8 | 1.746 |
| | 5 B5 | 217.5 | 58.6 | 1.795 |
| | 10 B10 | 259.1 | 108.0 | 1.771 |
| | 11 H1 | 105.3 | 123.6 | 1.175 |
| 2 B2 | | | | |
| | 1 B1 | 90.0 | 0 | 1.771 |
| | 3 B3 | 0 | 63.1 | 1.746 |
| | 5 B5 | 142.5 | 61.4 | 1.745 |
| | 6 B6 | 107.8 | 116.7 | 1.742 |
| | 7 B7 | 36.4 | 112.8 | 1.747 |
| | 12 H2 | 246.3 | 126.7 | 1.108 |
| 3 B3 | | | | |
| | 1 B1 | 90.0 | 0 | 1.839 |
| | 2 B2 | 0 | 59.1 | 1.746 |
| | 4 B4 | 225.2 | 57.8 | 1.770 |
| | 7 B7 | 325.3 | 108.5 | 1.771 |
| | 8 B8 | 258.2 | 107.2 | 1.795 |
| | 13 H3 | 111.5 | 118.7 | 1.175 |
| 4 B4 | | | | |
| | 1 B1 | 90.0 | 0 | 1.746 |
| | 3 B3 | 0 | 63.1 | 1.770 |
| | 8 B8 | 324.3 | 113.9 | 1.746 |
| | 9 B9 | 251.9 | 116.6 | 1.742 |
| | 10 B10 | 218.8 | 61.0 | 1.747 |
| | 14 H4 | 117.7 | 123.9 | 1.108 |
| 5 B5 | | | | |
| | 1 B1 | 90.0 | 0 | 1.795 |
| | 2 B2 | 0 | 60.0 | 1.745 |
| | 6 B6 | 30.3 | 109.7 | 1.863 |
| | 10 B10 | 135.1 | 57.6 | 1.877 |
| | 15 H5 | 251.4 | 119.1 | 1.111 |
| | 21 H11 | 128.0 | 96.5 | 1.228 |
| 6 B6 | | | | |
| | 2 B2 | 90.0 | 0 | 1.742 |
| | 5 B5 | 0 | 57.8 | 1.863 |
| | 7 B7 | 222.7 | 58.1 | 1.852 |
| | 16 H6 | 290.4 | 136.1 | 1.061 |
| | 23 H+ | 111.7 | 108.1 | 1.200 |

Table B–11 (*continued*)

| Center atom | Coordinating atom | Phi | Rho | Distance, A |
|---|---|---|---|---|
| 7 B7 | | | | |
| | 2 B2 | 90.0 | 0 | 1.747 |
| | 3 B3 | 0 | 59.5 | 1.771 |
| | 6 B6 | 147.1 | 57.8 | 1.852 |
| | 8 B8 | 38.8 | 105.0 | 1.877 |
| | 17 H7 | 252.5 | 121.9 | 1.057 |
| | 22 H12 | 76.8 | 126.5 | 1.214 |
| 8 B8 | | | | |
| | 3 B3 | 90.0 | 0 | 1.795 |
| | 4 B4 | 0 | 60.0 | 1.746 |
| | 7 B7 | 135.1 | 57.6 | 1.877 |
| | 9 B9 | 30.3 | 109.7 | 1.863 |
| | 18 H8 | 251.4 | 119.1 | 1.111 |
| | 22 H12 | 127.9 | 96.5 | 1.228 |
| 9 B9 | | | | |
| | 4 B4 | 90.0 | 0 | 1.742 |
| | 8 B8 | 0 | 57.8 | 1.863 |
| | 10 B10 | 222.7 | 58.1 | 1.852 |
| | 19 H9 | 290.3 | 136.0 | 1.061 |
| | 24 $H^{++}$ | 111.7 | 108.2 | 1.200 |
| 10 B10 | | | | |
| | 1 B1 | 90.0 | 0 | 1.771 |
| | 4 B4 | 0 | 59.5 | 1.747 |
| | 5 B5 | 225.0 | 58.9 | 1.877 |
| | 9 B9 | 330.7 | 110.0 | 1.852 |
| | 20 H10 | 112.0 | 119.2 | 1.057 |
| | 21 H11 | 232.3 | 98.3 | 1.214 |
| 21 H11 | | | | |
| | 5 B5 | 90.0 | 0 | 1.228 |
| | 10 B10 | 0 | 100.5 | 1.214 |
| 22 H12 | | | | |
| | 7 B7 | 90.0 | 0 | 1.214 |
| | 8 B8 | 0 | 100.5 | 1.228 |

**Table B–12** *Decaborane* $B_{10}H_{14}$

| Center atom | Coordinating atom | Phi | Rho | Distance, A |
|---|---|---|---|---|
| 1 B1 | | | | |
| | 2 B2 | 0 | 116.4 | 1.801 |
| | 3 B3 | 28.9 | 62.0 | 1.710 |
| | 4 B4 | 90.0 | 0 | 1.784 |
| | 5 B5 | 293.1 | 117.3 | 1.784 |
| | 10 B10 | 252.3 | 60.8 | 1.772 |
| | 11 H1 | 155.7 | 116.4 | 1.161 |
| 2 B2 | | | | |
| | 1 B1 | 223.4 | 57.0 | 1.801 |
| | 3 B3 | 90.0 | 0 | 1.784 |
| | 5 B5 | 261.4 | 104.8 | 1.764 |
| | 6 B6 | 325.7 | 109.9 | 1.718 |
| | 7 B7 | 0 | 59.3 | 1.800 |
| | 12 H2 | 127.6 | 122.3 | 1.267 |
| 3 B3 | | | | |
| | 1 B1 | 28.9 | 62.0 | 1.710 |
| | 2 B2 | 90.0 | 0 | 1.784 |
| | 4 B4 | 0 | 116.4 | 1.801 |
| | 7 B7 | 252.3 | 60.8 | 1.772 |
| | 8 B8 | 293.1 | 117.3 | 1.784 |
| | 13 H3 | 155.7 | 116.4 | 1.161 |
| 4 B4 | | | | |
| | 1 B1 | 90.0 | 0 | 1.784 |
| | 3 B3 | 322.0 | 57.0 | 1.801 |
| | 8 B8 | 0 | 104.8 | 1.764 |
| | 9 B9 | 64.3 | 109.9 | 1.718 |
| | 10 B10 | 98.6 | 59.3 | 1.800 |
| | 14 H4 | 226.2 | 122.3 | 1.267 |
| 5 B5 | | | | |
| | 1 B1 | 218.6 | 61.0 | 1.784 |
| | 2 B2 | 90.0 | 0 | 1.764 |
| | 6 B6 | 0 | 58.1 | 1.774 |
| | 10 B10 | 249.5 | 107.7 | 2.007 |
| | 15 H5 | 101.4 | 118.2 | 1.355 |
| | 21 H11 | 342.4 | 112.1 | 1.248 |
| 6 B6 | | | | |
| | 2 B2 | 0 | 60.7 | 1.718 |
| | 5 B5 | 90.0 | 0 | 1.774 |
| | 7 B7 | 315.6 | 105.6 | 1.771 |
| | 16 H6 | 111.5 | 120.5 | 1.282 |
| | 21 H11 | 199.7 | 43.8 | 1.505 |
| | 22 H12 | 264.2 | 127.6 | 1.430 |

Table B–12 (*continued*)

| Center atom | Coordinating atom | Phi | Rho | Distance, A |
|---|---|---|---|---|
| 7 B7 | | | | |
| | 2 B2 | 93.8 | 57.5 | 1.800 |
| | 3 B3 | 60.1 | 108.0 | 1.772 |
| | 6 B6 | 90.0 | 0 | 1.771 |
| | 8 B8 | 0 | 115.3 | 2.007 |
| | 17 H7 | 210.6 | 123.7 | 1.215 |
| | 22 H12 | 283.8 | 52.1 | 1.393 |
| 8 B8 | | | | |
| | 3 B3 | 329.1 | 61.0 | 1.784 |
| | 4 B4 | 90.0 | 0 | 1.764 |
| | 7 B7 | 0 | 107.7 | 2.007 |
| | 9 B9 | 110.5 | 58.1 | 1.774 |
| | 18 H8 | 212.0 | 118.2 | 1.355 |
| | 23 H13 | 92.9 | 112.1 | 1.248 |
| 9 B9 | | | | |
| | 4 B4 | 0 | 60.7 | 1.718 |
| | 8 B8 | 90.0 | 0 | 1.774 |
| | 10 B10 | 315.6 | 105.6 | 1.771 |
| | 19 H9 | 111.5 | 120.5 | 1.282 |
| | 23 H13 | 199.7 | 43.8 | 1.505 |
| | 24 H14 | 264.2 | 127.6 | 1.430 |
| 10 B10 | | | | |
| | 1 B1 | 90.0 | 0 | 1.772 |
| | 4 B4 | 327.3 | 59.9 | 1.800 |
| | 5 B5 | 108.8 | 55.9 | 2.007 |
| | 9 B9 | 0 | 108.0 | 1.771 |
| | 20 H10 | 208.6 | 121.1 | 1.215 |
| | 24 H14 | 53.2 | 137.1 | 1.393 |
| 21 H11 | | | | |
| | 5 B5 | 90.0 | 0 | 1.248 |
| | 6 B6 | 0 | 79.6 | 1.505 |
| 22 H12 | | | | |
| | 6 B6 | 90.0 | 0 | 1.430 |
| | 7 B7 | 0 | 77.7 | 1.393 |
| 23 H13 | | | | |
| | 8 B8 | 90.0 | 0 | 1.248 |
| | 9 B9 | 0 | 79.6 | 1.505 |
| 24 H14 | | | | |
| | 9 B9 | 90.0 | 0 | 1.430 |
| | 10 B10 | 0 | 77.7 | 1.393 |

**Table B-13** $B_{18}H_{22}$

| Center atom | Coordinating atom | Phi | Rho | Distance, A |
|---|---|---|---|---|
| 1 B1 | | | | |
| | 4 B4 | 248.2 | 61.0 | 1.766 |
| | 6 B6 | 143.7 | 118.6 | 1.767 |
| | 7 B7 | 107.2 | 68.1 | 1.763 |
| | 9 B9 | 209.4 | 108.7 | 1.790 |
| | 17 B17 | 90.0 | 0 | 1.768 |
| | 19 H1 | 0 | 118.8 | 1.114 |
| 2 B2 | | | | |
| | 4 B4 | 180.0 | 59.2 | 1.792 |
| | 5 B5 | 273.8 | 119.7 | 1.980 |
| | 8 B8 | 43.5 | 102.3 | 1.825 |
| | 9 B9 | 211.4 | 112.3 | 1.739 |
| | 11 B11 | 90.0 | 0 | 1.816 |
| | 13 B13 | 0 | 59.7 | 1.781 |
| | 27 H9 | 94.4 | 116.6 | 1.260 |
| 3 B3 | | | | |
| | 5 B5 | 238.9 | 38.4 | 1.786 |
| | 6 B6 | 254.8 | 98.2 | 1.729 |
| | 7 B7 | 316.9 | 119.2 | 1.791 |
| | 22 H4 | 90.0 | 0 | 1.221 |
| | 23 H5 | 117.9 | 108.7 | 1.122 |
| | 28 H10 | 0 | 96.1 | 1.406 |
| 4 B4 | | | | |
| | 1 B1 | 33.4 | 111.4 | 1.766 |
| | 2 B2 | 128.8 | 61.1 | 1.792 |
| | 9 B9 | 98.7 | 111.3 | 1.796 |
| | 11 B11 | 90.0 | 0 | 1.781 |
| | 17 B17 | 0 | 61.4 | 1.794 |
| | 24 H6 | 242.8 | 123.8 | 1.062 |
| 5 B5 | | | | |
| | 2 B2 | 120.0 | 95.4 | 1.980 |
| | 3 B3 | 241.1 | 42.4 | 1.786 |
| | 6 B6 | 225.8 | 98.6 | 1.794 |
| | 9 B9 | 167.7 | 130.2 | 1.754 |
| | 21 H3 | 0 | 110.2 | 1.113 |
| | 22 H4 | 90.0 | 0 | 1.124 |
| 6 B6 | | | | |
| | 1 B1 | 102.9 | 60.5 | 1.767 |
| | 3 B3 | 206.2 | 110.5 | 1.729 |
| | 5 B5 | 239.6 | 58.7 | 1.794 |
| | 7 B7 | 142.0 | 106.0 | 1.807 |
| | 9 B9 | 90.0 | 0 | 1.784 |
| | 20 H2 | 0 | 120.1 | 1.149 |

Table B-13 (*continued*)

| Center atom | Coordinating atom | Phi | Rho | Distance, A |
|---|---|---|---|---|
| 7 B7 | | | | |
| | 1 B1 | 198.9 | 125.2 | 1.763 |
| | 3 B3 | 113.0 | 51.1 | 1.791 |
| | 6 B6 | 136.8 | 104.1 | 1.807 |
| | 17 B17 | 243.6 | 87.7 | 1.976 |
| | 28 H10 | 90.0 | 0 | 1.315 |
| | 29 H11 | 0 | 112.1 | 1.159 |
| 8 B8 | | | | |
| | 2 B2 | 108.3 | 109.4 | 1.825 |
| | 10 B10 | 90.0 | 0 | 1.768 |
| | 13 B13 | 140.9 | 59.5 | 1.794 |
| | 16 B16 | 0 | 55.8 | 1.976 |
| | 27 H9 | 62.7 | 128.0 | 1.395 |
| | 37 H19 | 248.6 | 126.0 | 1.185 |
| 9 B9 | | | | |
| | 1 B1 | 114.0 | 59.0 | 1.790 |
| | 2 B2 | 248.5 | 60.9 | 1.739 |
| | 4 B4 | 90.0 | 0 | 1.796 |
| | 5 B5 | 208.2 | 118.2 | 1.754 |
| | 6 B6 | 140.1 | 112.7 | 1.784 |
| | 25 H7 | 0 | 115.7 | 1.053 |
| 10 B10 | | | | |
| | 8 B8 | 110.2 | 61.0 | 1.768 |
| | 13 B13 | 90.0 | 0 | 1.766 |
| | 15 B15 | 0 | 115.1 | 1.767 |
| | 16 B16 | 69.3 | 116.7 | 1.763 |
| | 18 B18 | 333.1 | 60.7 | 1.790 |
| | 30 H12 | 218.2 | 121.2 | 1.114 |
| 11 B11 | | | | |
| | 2 B2 | 245.4 | 59.2 | 1.816 |
| | 4 B4 | 245.4 | 118.9 | 1.781 |
| | 13 B13 | 90.0 | 0 | 1.792 |
| | 14 B14 | 0 | 107.6 | 1.980 |
| | 17 B17 | 170.5 | 135.8 | 1.825 |
| | 18 B18 | 32.0 | 61.1 | 1.739 |
| | 38 H20 | 130.2 | 99.8 | 1.260 |
| 12 B12 | | | | |
| | 14 B14 | 90.0 | 0 | 1.786 |
| | 15 B15 | 44.5 | 61.4 | 1.729 |
| | 16 B16 | 0 | 105.6 | 1.791 |
| | 33 H15 | 242.5 | 38.4 | 1.221 |
| | 34 H16 | 165.3 | 123.6 | 1.122 |
| | 39 H21 | 310.9 | 113.7 | 1.406 |

**Table B–13** (*continued*)

| Center atom | Coordinating atom | Phi | Rho | Distance, A |
|---|---|---|---|---|
| 13 B13 | | | | |
| | 2 B2 | 295.7 | 111.4 | 1.781 |
| | 8 B8 | 261.7 | 59.5 | 1.794 |
| | 10 B10 | 90.0 | 0 | 1.766 |
| | 11 B11 | 0 | 104.7 | 1.792 |
| | 18 B18 | 38.7 | 60.3 | 1.796 |
| | 35 H17 | 143.3 | 118.1 | 1.062 |
| 14 B14 | | | | |
| | 11 B11 | 90.0 | 0 | 1.980 |
| | 12 B12 | 0 | 114.6 | 1.786 |
| | 15 B15 | 60.9 | 104.7 | 1.794 |
| | 18 B18 | 97.0 | 55.1 | 1.754 |
| | 32 H14 | 205.0 | 115.8 | 1.113 |
| | 33 H15 | 320.6 | 95.4 | 1.124 |
| 15 B15 | | | | |
| | 10 B10 | 90.0 | 0 | 1.767 |
| | 12 B12 | 0 | 111.1 | 1.729 |
| | 14 B14 | 295.4 | 106.6 | 1.794 |
| | 16 B16 | 32.8 | 59.1 | 1.807 |
| | 18 B18 | 257.7 | 60.5 | 1.784 |
| | 31 H13 | 160.5 | 116.0 | 1.587 |
| 16 B16 | | | | |
| | 8 B8 | 112.7 | 56.1 | 1.976 |
| | 10 B10 | 90.0 | 0 | 1.763 |
| | 12 B12 | 0 | 108.5 | 1.791 |
| | 15 B15 | 329.1 | 59.3 | 1.807 |
| | 39 H21 | 54.9 | 125.2 | 1.315 |
| | 40 H22 | 214.6 | 119.9 | 1.159 |
| 17 B17 | | | | |
| | 1 B1 | 214.2 | 109.4 | 1.768 |
| | 4 B4 | 247.0 | 59.0 | 1.794 |
| | 7 B7 | 153.6 | 115.5 | 1.976 |
| | 11 B11 | 90.0 | 0 | 1.825 |
| | 26 H8 | 0 | 113.2 | 1.185 |
| | 38 H20 | 88.9 | 43.6 | 1.395 |
| 18 B18 | | | | |
| | 10 B10 | 320.5 | 105.9 | 1.790 |
| | 11 B11 | 90.0 | 0 | 1.739 |
| | 13 B13 | 0 | 60.9 | 1.796 |
| | 14 B14 | 217.6 | 69.1 | 1.754 |
| | 15 B15 | 257.4 | 116.2 | 1.784 |
| | 36 H18 | 104.5 | 120.0 | 1.053 |

**Table B–13** (*continued*)

| Center atom | Coordinating atom | Phi | Rho | Distance, A |
|---|---|---|---|---|
| 22 H4 | | | | |
| | 3 B3 | 90.0 | 0 | 1.221 |
| | 5 B5 | 0 | 99.2 | 1.124 |
| 27 H9 | | | | |
| | 2 B2 | 90.0 | 0 | 1.260 |
| | 8 B8 | 0 | 86.7 | 1.395 |
| 28 H10 | | | | |
| | 3 B3 | 90.0 | 0 | 1.406 |
| | 7 B7 | 0 | 82.2 | 1.315 |
| 33 H15 | | | | |
| | 12 B12 | 90.0 | 0 | 1.221 |
| | 14 B14 | 0 | 99.2 | 1.124 |
| 38 H20 | | | | |
| | 11 B11 | 90.0 | 0 | 1.260 |
| | 17 B17 | 0 | 86.7 | 1.395 |
| 39 H21 | | | | |
| | 12 B12 | 90.0 | 0 | 1.406 |
| | 16 B16 | 0 | 82.2 | 1.315 |

**Table B–14** *iso*-$B_{18}H_{22}$

| Center atom | Coordinating atom | Phi | Rho | Distance, A |
|---|---|---|---|---|
| 1 B1 | | | | |
| | 2 B2 | 90.0 | 0 | 1.795 |
| | 3 B3 | 0 | 59.8 | 1.789 |
| | 4 B4 | 333.8 | 113.1 | 1.786 |
| | 5 B5 | 222.5 | 60.4 | 1.745 |
| | 10 B10 | 264.4 | 116.8 | 1.780 |
| | 19 H1 | 115.3 | 117.4 | 1.022 |
| 2 B2 | | | | |
| | 1 B1 | 90.0 | 0 | 1.795 |
| | 3 B3 | 0 | 59.9 | 1.786 |
| | 5 B5 | 137.5 | 58.4 | 1.782 |
| | 6 B6 | 103.3 | 110.0 | 1.713 |
| | 7 B7 | 39.1 | 104.8 | 1.796 |
| | 20 H2 | 250.0 | 114.7 | 1.144 |
| 3 B3 | | | | |
| | 1 B1 | 90.0 | 0 | 1.789 |
| | 2 B2 | 0 | 60.3 | 1.786 |
| | 4 B4 | 208.3 | 60.4 | 1.760 |
| | 7 B7 | 319.2 | 107.0 | 1.752 |
| | 8 B8 | 247.2 | 109.0 | 1.755 |
| | 21 H3 | 106.4 | 128.1 | 1.064 |
| 4 B4 | | | | |
| | 1 B1 | 90.0 | 0 | 1.786 |
| | 3 B3 | 0 | 60.6 | 1.760 |
| | 8 B8 | 322.5 | 107.4 | 1.794 |
| | 9 B9 | 257.6 | 109.9 | 1.726 |
| | 10 B10 | 224.4 | 58.8 | 1.841 |
| | 22 H4 | 107.5 | 120.4 | 1.093 |
| 5 B5 | | | | |
| | 1 B1 | 90.0 | 0 | 1.745 |
| | 2 B2 | 0 | 61.2 | 1.782 |
| | 6 B6 | 32.6 | 109.4 | 1.778 |
| | 10 B10 | 140.4 | 56.2 | 2.002 |
| | 23 H5 | 243.0 | 119.5 | 1.013 |
| | 37 H19 | 78.5 | 129.1 | 1.250 |
| 6 B6 | | | | |
| | 2 B2 | 90.0 | 0 | 1.713 |
| | 5 B5 | 0 | 61.3 | 1.778 |
| | 7 B7 | 230.6 | 61.7 | 1.786 |
| | 24 H6 | 116.3 | 130.0 | 1.147 |
| | 37 H19 | 346.3 | 104.0 | 1.237 |
| | 38 H20 | 245.1 | 105.2 | 1.194 |

**Table B–14** *(continued)*

| Center atom | Coordinating atom | Phi | Rho | Distance, A |
|---|---|---|---|---|
| 7 B7 | | | | |
| | 2 B2 | 90.0 | 0 | 1.796 |
| | 3 B3 | 0 | 60.4 | 1.752 |
| | 6 B6 | 144.7 | 57.1 | 1.786 |
| | 8 B8 | 30.4 | 108.2 | 1.961 |
| | 25 H7 | 248.7 | 126.5 | 1.075 |
| | 38 H20 | 131.5 | 97.1 | 1.280 |
| 8 B8 | | | | |
| | 3 B3 | 90.0 | 0 | 1.755 |
| | 4 B4 | 0 | 59.4 | 1.794 |
| | 7 B7 | 139.6 | 55.9 | 1.961 |
| | 9 B9 | 33.2 | 107.3 | 1.791 |
| | 26 H8 | 244.5 | 121.0 | 1.078 |
| | 27 H9 | 83.0 | 130.3 | 1.248 |
| 9 B9 | | | | |
| | 4 B4 | 90.0 | 0 | 1.726 |
| | 8 B8 | 0 | 61.3 | 1.791 |
| | 10 B10 | 227.9 | 63.0 | 1.794 |
| | 14 B14 | 228.0 | 125.4 | 1.725 |
| | 18 B18 | 142.5 | 138.4 | 1.793 |
| | 27 H9 | 347.9 | 103.8 | 1.343 |
| | 28 H10 | 109.2 | 102.6 | 1.382 |
| 10 B10 | | | | |
| | 1 B1 | 90.0 | 0 | 1.780 |
| | 4 B4 | 0 | 59.0 | 1.841 |
| | 5 B5 | 218.2 | 54.6 | 2.002 |
| | 9 B9 | 328.8 | 107.1 | 1.794 |
| | 11 B11 | 149.4 | 145.0 | 1.767 |
| | 14 B14 | 273.7 | 147.0 | 1.822 |
| | 15 B15 | 95.7 | 107.6 | 1.978 |
| 11 B11 | | | | |
| | 10 B10 | 90.0 | 0 | 1.767 |
| | 12 B12 | 0 | 116.0 | 1.794 |
| | 13 B13 | 296.1 | 107.0 | 1.773 |
| | 14 B14 | 256.4 | 61.9 | 1.777 |
| | 15 B15 | 38.6 | 68.6 | 1.744 |
| | 29 H11 | 149.2 | 115.2 | 1.069 |

**Table B–14** (*continued*)

| Center atom | Coordinating atom | Phi | Rho | Distance, A |
|---|---|---|---|---|
| 12 B12 | | | | |
| | 11 B11 | 90.0 | 0 | 1.794 |
| | 13 B13 | 0 | 59.7 | 1.771 |
| | 15 B15 | 137.1 | 58.4 | 1.780 |
| | 16 B16 | 103.2 | 110.6 | 1.721 |
| | 17 B17 | 39.0 | 105.1 | 1.795 |
| | 30 H12 | 247.1 | 119.8 | 1.078 |
| 13 B13 | | | | |
| | 11 B11 | 90.0 | 0 | 1.773 |
| | 12 B12 | 0 | 60.8 | 1.771 |
| | 14 B14 | 208.4 | 60.5 | 1.757 |
| | 17 B17 | 319.3 | 107.7 | 1.754 |
| | 18 B18 | 247.2 | 109.7 | 1.754 |
| | 31 H13 | 106.4 | 128.2 | 1.155 |
| 14 B14 | | | | |
| | 9 B9 | 90.0 | 0 | 1.725 |
| | 10 B10 | 0 | 60.7 | 1.822 |
| | 11 B11 | 327.8 | 110.6 | 1.777 |
| | 13 B13 | 263.1 | 109.9 | 1.757 |
| | 18 B18 | 228.9 | 61.0 | 1.803 |
| | 32 H14 | 113.5 | 117.9 | 1.083 |
| 15 B15 | | | | |
| | 10 B10 | 90.0 | 0 | 1.978 |
| | 11 B11 | 0 | 56.3 | 1.744 |
| | 12 B12 | 36.2 | 106.8 | 1.780 |
| | 16 B16 | 97.7 | 115.1 | 1.790 |
| | 33 H15 | 244.3 | 115.3 | 1.078 |
| | 39 H21 | 141.3 | 89.5 | 1.194 |
| 16 B16 | | | | |
| | 12 B12 | 90.0 | 0 | 1.721 |
| | 15 B15 | 0 | 60.9 | 1.790 |
| | 17 B17 | 230.9 | 61.6 | 1.784 |
| | 34 H16 | 111.1 | 129.0 | 1.038 |
| | 39 H21 | 346.7 | 100.7 | 1.357 |
| | 40 H22 | 248.0 | 104.4 | 1.326 |
| 17 B17 | | | | |
| | 12 B12 | 90.0 | 0 | 1.795 |
| | 13 B13 | 0 | 59.9 | 1.754 |
| | 16 B16 | 144.8 | 57.5 | 1.784 |
| | 18 B18 | 30.0 | 108.0 | 1.956 |
| | 35 H17 | 249.3 | 122.6 | 1.037 |
| | 40 H22 | 127.3 | 102.3 | 1.284 |

**Table B–14** (*cantinued*)

| Center atom | Coordinating atom | Phi | Rho | Distance, A |
|---|---|---|---|---|
| 18 B18 | | | | |
| | 9 B9 | 90.0 | 0 | 1.793 |
| | 13 B13 | 0 | 107.0 | 1.754 |
| | 14 B14 | 33.7 | 57.3 | 1.803 |
| | 17 B17 | 300.2 | 113.2 | 1.956 |
| | 28 H10 | 232.2 | 50.1 | 1.276 |
| | 36 H18 | 145.5 | 124.0 | 1.117 |
| 27 H9 | | | | |
| | 8 B8 | 90.0 | 0 | 1.248 |
| | 9 B9 | 0 | 87.4 | 1.343 |
| 28 H10 | | | | |
| | 9 B9 | 90.0 | 0 | 1.382 |
| | 18 B18 | 0 | 84.7 | 1.276 |
| 37 H19 | | | | |
| | 5 B5 | 90.0 | 0 | 1.250 |
| | 6 B6 | 0 | 91.2 | 1.237 |
| 38 H20 | | | | |
| | 6 B6 | 90.0 | 0 | 1.194 |
| | 7 B7 | 0 | 92.4 | 1.280 |
| 39 H21 | | | | |
| | 15 B15 | 90.0 | 0 | 1.194 |
| | 16 B16 | 0 | 88.9 | 1.357 |
| 40 H22 | | | | |
| | 16 B16 | 90.0 | 0 | 1.326 |
| | 17 B17 | 0 | 86.2 | 1.284 |

Table B–15 $B_4C_2H_6(CH_3)_2$

| Center atom | Coordinating atom | Phi | Rho | Distance, A |
|---|---|---|---|---|
| 1 B1 | | | | |
| | 2 B2 | 90.0 | 0 | 1.705 |
| | 3 B3 | 0 | 61.6 | 1.768 |
| | 4 B4 | 237.5 | 61.6 | 1.768 |
| | 5 C1 | 322.9 | 98.0 | 1.763 |
| | 6 C2 | 274.5 | 98.0 | 1.763 |
| | 9 H1 | 118.7 | 121.7 | 1.208 |
| 2 B2 | | | | |
| | 1 B1 | 90.0 | 0 | 1.705 |
| | 3 B3 | 0 | 60.9 | 1.779 |
| | 4 B4 | 122.5 | 60.9 | 1.779 |
| | 10 H2 | 241.3 | 134.8 | 1.224 |
| | 13 H4 | 19.0 | 103.0 | 1.382 |
| | 14 H5 | 103.6 | 103.0 | 1.382 |
| 3 B3 | | | | |
| | 1 B1 | 90.0 | 0 | 1.768 |
| | 2 B2 | 0 | 57.5 | 1.779 |
| | 5 C1 | 230.2 | 64.3 | 1.521 |
| | 11 H3 | 119.8 | 128.9 | 1.428 |
| | 13 H4 | 339.4 | 104.0 | 1.284 |
| 4 B4 | | | | |
| | 1 B1 | 90.0 | 0 | 1.768 |
| | 2 B2 | 0 | 57.5 | 1.779 |
| | 6 C2 | 129.8 | 64.3 | 1.521 |
| | 12 H4 | 240.2 | 128.9 | 1.428 |
| | 14 H5 | 20.6 | 104.0 | 1.284 |
| 5 C1 | | | | |
| | 1 B1 | 90.0 | 0 | 1.763 |
| | 3 B3 | 0 | 64.6 | 1.521 |
| | 6 C2 | 223.4 | 66.0 | 1.432 |
| | 7 ME | 111.5 | 128.3 | 1.507 |
| 6 C2 | | | | |
| | 1 B1 | 90.0 | 0 | 1.763 |
| | 4 B4 | 0 | 64.6 | 1.521 |
| | 5 C1 | 136.6 | 66.0 | 1.432 |
| | 8 ME | 248.5 | 128.3 | 1.507 |
| 13 H4 | | | | |
| | 2 B2 | 90.0 | 0 | 1.382 |
| | 3 B3 | 0 | 83.6 | 1.284 |
| 14 H5 | | | | |
| | 2 B2 | 90.0 | 0 | 1.382 |
| | 4 B4 | 0 | 83.6 | 1.284 |

# Name Index*



* Italic numbers indicate references; this listing begins on page 201. "*Recent review*" refers to the supplementary list of references; this will be found on page 216.

# Compound Index*



* All formulas are listed in the order of number of B atoms, number of H atoms attached to the polyhedral framework, and then number of other elements in alphabetical order.

# Subject Index